The Design Book
Techniques and Solutions for Digital Computer Systems

L. HOWARD POLLARD

University of New Mexico

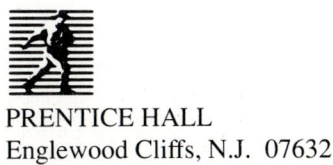

PRENTICE HALL
Englewood Cliffs, N.J. 07632

Library of Congress Cataloging-in-Publication Data

Pollard, L. Howard.
  The design book : techniques and solutions for digital computer
  systems / L. Howard Pollard.
    p.   cm.
  ISBN 0-13-200304-X
  1. Electronic digital computers—Design and construction.
  2. Computer architecture. I. Title.
  TK7888.3.P635 1990   89–23177
  621.39—dc20   CIP

Editorial/production supervision and
    interior design: Fred Dahl
Cover design: 20/20 Services, Inc.
Manufacturing buyer: Mary Noonan

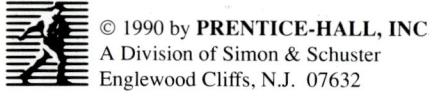 © 1990 by **PRENTICE-HALL, INC.**
A Division of Simon & Schuster
Englewood Cliffs, N.J. 07632

Printed in the United States of America

10 9 8 7 6 5 4 3 2 1

ISBN 0-13-200304-X

Prentice-Hall International (UK) Limited, *London*
Prentice-Hall of Australia Pty. Limited, *Sydney*
Prentice-Hall of Canada Inc., *Toronto*
Prentice-Hall Hispanoamericana, S. A., *Mexico*
Prentice-Hall of India Private Limited, *New Delhi*
Prentice-Hall of Japan, Inc., *Tokyo*
Simon & Schuster Asia Pte. Ltd., *Singapore*
Editora Prentice-Hall do Brasil, Ltda., *Rio de Janeiro*

Contents

Preface

Developing design techniques, instincts, and experience is not an easy process, and many books and materials have been created to present different ideas and methods of design. When I have attempted to discuss design techniques with others, either in a peer situation in industry or in an instructional situation, communication is most effective when an example is available to demonstrate specific ideas. This book presents a number of different techniques which can be utilized in a number of different ways. It is intended to be a supplemental text for an advanced design course, or a design demonstration book which can be used to introduce different ideas to working engineers. In particular, this book was created to be a companion volume for a specific design text (L. H. Pollard, *Computer Design and Architecture*, Englewood Cliffs, NJ: Prentice Hall, 1990.) It does not identify the ''best'' way to do designs; rather, it presents some ideas and techniques which can be utilized in a variety of situations. Also, where appropriate, it compares different design techniques and ideas, one with another, demonstrating how engineering criteria can be used to evaluate one technique with respect to another. Different criteria will apply in different circumstances, but the techniques of evaluation will be similar.

The material in this book covers a number of different topics, and the designs and techniques presented should be evaluated in the arena in which they are presented. Certainly some of the designs can be improved, and where this can be identified, differences between the approach presented in the book and other approaches should be detailed and the reasons for the improvements explained. In this way, students and professionals can evaluate the validity of assumptions, of design techniques, and of implementation methods. Design ideas are valuable only when applied to the solution of real problems, and the solutions can be very helpful when shared with colleagues.

The logic drawings included in this book conform to ideas of polarized mnemonics and logical state indicators demonstrated by Fletcher (William I. Fletcher, *An Engineering Approach to Digital Design*, Englewood Cliffs, NJ: Prentice Hall, 1980) in his book and in practice. These ideas have proved very beneficial in practice; the drawings presented here utilize these ideas to a great degree. Signals are named in a manner which connotes their function. The final two letters of the name indicate the level at which the function is performed. An -H suffix indicates that the function is accomplished when the signal is high; an -L suffix indicates that the function occurs when the signal is low. These names are created to agree with the presence or absence of logical state indicators on the outputs of the gates which create them. A signal which is the output of a gate with a logical state indicator (the little bubble at the output of the gate) will have an -L suffix, while a signal which is the output of a gate without the logical state indicator will have an -H ending. Also, there should be agreement between the creation and use of signals. That is, signals with a logical state indicator at the output of a gate should feed gates (or devices) which also have logical state indicators. Where this does not occur, an incompatibility triangle marks the signal at the input of the gate. This is a communication help, and it allows the creator of a design to tell the reader that the incompatibility was done on purpose.

The material in the book is broken into six different chapters covering different areas of interest. While some of the information may appear trivial to some observers, other readers may gain new insights into old problems by considering the information presented. Chapter 1 contains some information about number systems and information representation. The material presented covers some integer methods for number conversions, basic limitations to information representation, and some floating point exercises. These number systems are basic to the manner in which we deal with computers and should be understood by all users of the machines.

Chapter 2 presents several methods for design of combinational systems. Some will argue that combinational design is not interesting, since much of the design can be turned over to automatic design tools. However, engineers and designers must know how to evaluate the effectiveness of the tools, and hence different design methods are presented. These methods are compared one to another to determine the utilization of system resources, and one design will be selected based on the desired use of a critical resource.

Chapter 3 contains a number of register transfer language (RTL) presentations of different instruction techniques. The techniques begin with a very simple system, and the examples presented progress in their complexity to a fairly complex unit. The intent is to demonstrate the effectiveness of an RTL in describing the work which must be done in a system, as well as to present different techniques for working with information. The techniques presented can be modified or used directly to derive information about an existing system.

Chapter 4, the longest in the book, contains examples in the area of sequential control design. This is an area which has many different possibilities, and questions abound concerning techniques and their implementations. Again, the same design is approached from different aspects, to derive designs with different characteristics. A system designer must be able to make reasonable engineering judgements about an implementation; this chapter progresses through the details of design implementations in order to demonstrate effective sequential control techniques.

Chapter 5 deals with the area of computer interface. Many different techniques have been utilized over the years to communicate with computers, and the information presented here covers only a small part of the total field. Nevertheless, it does tie together some of the ideas presented in earlier chapters to create systems which provide an exchange of digital information. And, as done in other parts of the book, an evaluation method is suggested through the design methodology presented.

Chapter 6 contains designs for memory systems. These start with simple 1-D and 2-D design techniques which can be applicable within memory chips and within larger memory systems. The designs move to progressively more complex implementations, ending with a combination dynamic RAM and cache system.

The designs and techniques presented here provide at least three different possibilities. First, for engineers and students just entering the exciting field of digital design, this book contains a number of methods and ideas which will lead to good working systems. Second, through the comparison techniques presented, different approaches to evaluation of implementations can be explored. It is often as important that an engineer know why a technique is good as it is to identify a good technique. And third, if a designer disagrees with a technique or an idea, a thorough examination of the information presented here and alternative techniques will lead to a better understanding of the underlying issues.

Many of the designs presented here have been implemented in one form or another in the process of solving design problems at a number of locations. My first designs were done at Utah State University, where as a student of Bill Fletcher and Al Despain I attempted to create digital devices for a number of systems. These design ideas were further enhanced by creating stand-alone digital systems for Reticon Corporation. Then I had the opportunity to become involved with a number of computer interfaces while employed by Lockheed Missiles and Space Company. Since that time I have been involved with academic institutions, and have attempted to teach many of these ideas to captive audiences of beleaguered students. However, I have had the privilege of continued involvement in real designs and real computer systems through an association with the Data Systems group (MEE-10) of the Mechanical and Electronics Engineering Division of Los Alamos National Laboratory. I continue to enjoy designing and building digital systems, and to those who have made that possible and fun I extend my thanks.

L. HOWARD POLLARD

The Design Book
Techniques and Solutions for Digital Computer Systems

1

Number Systems

In any representation of information, meaning is attached to patterns of symbols. These patterns are then interpreted by a knowledgeable user according to a decoding scheme. In order for information to be transferred from one individual (or computer, or user, or . . .) to another, it is imperative that both agree on the rules used for creating the patterns. For number systems, we use combinations of k symbols in a radix k system to represent numeric information. If we allow n digits in a radix k system, then k^n different patterns can be created. For example, in a 5 digit base 10 system, 10^5 patterns, or 100,000 patterns, can be represented. Similarly, in a 16 digit base 2 system, more commonly known as a 16-bit base 2 system, 2^{16} patterns, or 65,536 patterns, can be represented.

One of the tasks required by the system is the accurate interpretation of a pattern, and the assignment of an appropriate value to it. In the process of synthesis, where a designer specifies the format for the data and instructions of a system, care must be taken to ensure that the range of information the system is being designed to handle can be dealt with in a reasonable way. In the process of analysis, where patterns of digits are being interpreted, the user of the system must be careful to assign the correct values or other information to the patterns being examined. For number systems, values will generally fall into one of two different systems, integer or floating point. This chapter looks at some methods for dealing with integer or integer-type data, and then looks at some characteristics of floating point systems.

1.1. Some Methods for Integer Systems

Almost all computer systems work with base 2 numbers, while almost all humans use base 10 numbers. One of the tasks required of system users is the conversion of numbers between the two systems. This is a favorite area for instructors to inflict pain and discouragement on their students, since there are a number of methods for converting between numbers represented in any base [Mano88,Flet80]. We will now look at two simple techniques for converting base 10 numbers to binary and octal values, and leave it as an exercise for the reader to reverse the process (binary or octal to base 10), and to extend it to base 16 numbers.

To begin, note that the rudimentary form for (positive) numbers in any base follows a positional representation, where the value is represented as:

$$\text{Value} = \sum_{i=0}^{n-1} d_i \times r^i$$

for a number represented as n digits in base r. For binary numbers, $r = 2$, while for decimal numbers $r = 10$. The least significant digit of the binary number has a value of either 1 or 0, and this alone determines whether the number is odd or even. Using this fact, we can then convert from a base 10 number to a base 2 number according to the following algorithm.

> *Algorithm BI.* Given a value V, create binary digits from lesser to more significant values in the following manner:
>
> > If V is odd, place a ''1'' in the result and subtract one from V;
> >
> > > otherwise, place a ''0'' in the result.
> >
> > Divide V by 2.
> >
> > Repeat the above steps until V is zero.

The above algorithm is very simple and will work for all positive integers. A demonstration of the algorithm is shown in Figure 1.1. The number to be converted to binary is 697_{10}. Since this number is odd, step 1 subtracts one from it and places a ''1'' bit in the result. The second step divides the resulting number (696) by two, with a result of 348. Since this number is even, the third step need not subtract anything, so a ''0'' is placed in the result. And so the process continues, each odd numbered step determining whether the number is odd or even and placing the appropriate digit (bit) in the result, and each even numbered step dividing the resulting value by two. The process terminates when the most significant ''1'' bit is determined.

Another system, which shares many of the characteristics of integer arithmetic, is a fractional system in which the digits of a number are considered part of a fraction, with an assumed radix point to the left of the most significant digit. This also is a positional system, with the value of the number represented by a simple equation:

$$\text{Value} = \sum_{i=-1}^{-n} d_i \times r^i$$

Note that the subscripts and superscripts in the equation are all negative. Putting the integers of the first equation together with the fractions of the second equation results in a number in which digits have positive and negative labels, and associated integer and fractional values: $d_5 d_4 d_3 d_2 d_1 d_0 . d_{-1} d_{-2} d_{-3} d_{-4}$.

The algorithm for converting decimal fractions to binary fractions is very similar to the algorithm for integers. We make the observation that the most significant bit of the fractional part of the number will be a ''1'' if and only if the

Convert 697_{10} to binary, using algorithm BI.

Step	Operation	Resulting Bit
1	697 − 1 = 696	1
2	696 ÷ 2 = 348	
3	348 − 0 = 348	0
4	348 ÷ 2 = 174	
5	174 − 0 = 174	0
6	174 ÷ 2 = 87	
7	87 − 1 = 86	1
8	86 ÷ 2 = 43	
9	43 − 1 = 42	1
10	42 ÷ 2 = 21	
11	21 − 1 = 20	1
12	20 ÷ 2 = 10	
13	10 − 0 = 10	0
14	10 ÷ 2 = 5	
15	5 − 1 = 4	1
16	4 ÷ 2 = 2	
17	2 − 0 = 2	0
18	2 ÷ 2 = 1	
19	1 − 1 = 0	1

Final result is 1010111001.

Figure 1.1. Decimal-to-Binary Conversion with Algorithm BI.

decimal fraction is greater than one-half. Repeated application of this observation results in the following algorithm for converting fractional numbers.

> *Algorithm BF.*   Given a value F, create binary digits from more to lesser significant values in the following manner:
>
>> If F is greater than one-half, place a "1" in the result and subtract one-half from F;
>>
>>> otherwise, place a "0" in the result.
>
> Multiply F by 2.
>
> Repeat the above steps until F is zero, or until the desired significance is reached.

Like the algorithm for generating binary integers (algorithm BI), this algorithm for binary fractions is very simple and can be used to create bit patterns as needed. Two examples of the algorithm are shown in Figure 1.2.

The first fraction used in Figure 1.2 results in a direct equivalent. That is, the decimal value and the binary value represent the same number. When this is the case, the final step of the algorithm will produce a 1, and no further steps are necessary. However, the second fraction used in Figure 1.2 does not end up in the same manner. The decimal value 0.9 does not have a binary equivalent. Note that the operation of step 10 for this conversion is exactly the same as the operation of step 2, both of which result in a value of 0.8. Thus, steps 2 through 9 will be

Generate a binary fraction equivalent of 0.703125_{10} using algorithm BF.

Step	Operation	Resulting Bit
1	$0.703125 - 0.5 = 0.203125$	1
2	$0.203125 \times 2 = 0.40625$	
3	$0.40625 - 0.0 = 0.40625$	0
4	$0.40625 \times 2 = 0.8125$	
5	$0.8125 - 0.5 = 0.3125$	1
6	$0.3125 \times 2 = 0.625$	
7	$0.625 - 0.5 = 0.125$	1
8	$0.125 \times 2 = 0.250$	
9	$0.250 - 0.0 = 0.250$	0
10	$0.250 \times 2 = 0.5$	
11	$0.5 - 0.5 = 0.0$	1

Final result is 0.101101.

Generate a binary fraction equivalent of 0.9.

Step	Operation	Resulting Bit
1	$0.9 - 0.5 = 0.4$	1
2	$0.4 \times 2 = 0.8$	
3	$0.8 - 0.5 = 0.3$	1
4	$0.3 \times 2 = 0.6$	
5	$0.6 - 0.5 = 0.1$	1
6	$0.1 \times 2 = 0.2$	
7	$0.2 - 0.0 = 0.2$	0
8	$0.2 \times 2 = 0.4$	
9	$0.4 - 0.0 = 0.4$	0
10	$0.4 \times 2 = 0.8$	
11	$0.8 - 0.5 = 0.3$	1

Final result is 0.111001 ...

Figure 1.2. Fractional Decimal-to-Binary Conversion with Algorithm BF.

repeated as steps 10 through 17, and they will continue to be repeated until the desired precision has been reached. This will be true of all decimal fractions that do not have an exact binary equivalent.

Conversion from decimal to binary integers can be accomplished with algorithm BI, and algorithm BF converts decimal fractions to binary fractions. However, these algorithms can be time consuming as well as laborious. If a calculator with reasonable precision is available, a slightly different application of the same principles can be used to do the work. The method relies on the common practice of grouping bits together into sets of three and creating octal digits. The same technique can be used to group bits into sets of four, creating hexadecimal digits. That is, the decimal number 697 was converted to the binary representation 1010111001 in Figure 1.1, and grouping into sets of three results in the number 1 010 111 001, or 1271_8. Similarly the fraction 0.703125 was converted to the binary representation 0.101101, which can be grouped into sets of three: 0.101 101. This results in the octal fraction 0.55.

Conversion between decimal and octal fractions can be performed by modifying algorithm BF to work with octal digits rather than binary digits:

> *Algorithm OF.* Given a value F, create octal digits from more to lesser significant values in the following manner:
>
> > Multiply F by 8. Place the whole part of this value (which will be an octal digit) in the result. Then subtract the whole value from F.
> >
> > Repeat the above steps until F is zero, or until the desired significance is reached.

This simple algorithm will convert fractions to any desired significance. Figure 1.3 gives two examples of this process. The first example is converting the decimal fraction 0.4541015625 to octal. Admittedly, this is a contrived example, but it demonstrates the desired method. The second demonstration in the figure is for a decimal value, 0.9_{10}, which gives a repeating pattern in octal and binary. The algorithm can be carried on as long as needed to produce the result to the desired precision.

The calculator method can be extended to work for the integer algorithm above as well. Instead of merely determining if an integer is even or odd, we will need to determine which of the eight octal digits is needed. Thus, the calculator is used to divide the given number by eight, and the fractional part of the result will identify the correct digit. The following algorithm produces the appropriate octal values:

> *Algorithm OI:* Given a value V, create octal digits from lesser to greater significance in the following manner:
>
> > Divide V by eight. Place a digit in the result which is the equivalent of the fractional part times eight. This will be a valid octal digit.
> >
> > Subtract off the fractional part of V.
> >
> > Repeat the above steps until V is a legal octal digit; this will be the most significant result digit.

The application of algorithm OI is given for two integers in Figure 1.4. In either case, each octal digit is developed, then its effect removed from the value V before moving on to develop the next digit. In the first example, 1357049_{10} is converted to octal. Dividing 1357049 by 8 results in a number with a fraction of 0.125, or 1/8. Thus the first (least significant) digit is $1/8 \times 8$, or 1. Then, in step 3, the effect of this digit is removed by subtracting off the fractional part of the number. And the process continues in the remaining steps. The second divide results in a fractional part of 7/8, and this is dealt with in the appropriate manner. The final division, performed in step 16, results in a fractional part of 1/8, which is converted to a "1," and a nonfractional part of 5, which is the most significant digit.

The second example of Figure 1.4 is the conversion to octal of 807872_{10}. This example proceeds as the first one, but notice that the first two divides result in a fractional part of zero. Hence, the two least significant digits are zero. Again, the process completes when the final digit is a legal octal value.

Create an octal representation for the fraction 0.4541015625_{10} using algorithm OF.

Step	Operation	Resulting Octal Digit
1	$0.4541015625 \times 8 = 3.6328125$	3
2	$3.6328125 - 3 = 0.6328125$	
3	$0.6328125 \times 8 = 5.0625$	5
4	$5.0625 - 5 = 0.0625$	
5	$0.0625 \times 8 = 0.5$	0
6	$0.5 - 0 = 0.5$	
7	$0.5 \times 8 = 4.0$	4
8	$4.0 - 4 = 0.0$	

Final result is 0.3504_8, which is 0.011101000100_2.

Generate a fractional equivalent of 0.9, which is correct to 8 octal digits, or 24 binary bits.

Step	Operation	Resulting Octal Digit
1	$0.9 \times 8 = 7.2$	7
2	$7.2 - 7 = 0.2$	
3	$0.2 \times 8 = 1.6$	1
4	$1.6 - 1 = 0.6$	
5	$0.6 \times 8 = 4.8$	4
6	$4.8 - 4 = 0.8$	
7	$0.8 \times 8 = 6.4$	6
8	$6.4 - 6 = 0.4$	
9	$0.4 \times 8 = 3.2$	3
10	$3.2 - 3 = 0.2$	
11	$0.2 \times 8 = 1.6$	1
12	$1.6 - 1 = 0.6$	
13	$0.6 \times 8 = 4.8$	4
14	$4.8 - 4 = 0.8$	
15	$0.8 \times 8 = 6.4$	6

Final result is 0.71463146_8, or $0.111001100110011001100110_2$.

Figure 1.3. Fractional Decimal-to-Octal Conversion with Algorithm OF.

The final integer method to consider is the representation of information with excess codes. The mechanism involved is the representation of some value according to the following formula:

$$S = V + E$$

Excess codes are used mainly in storing exponents in floating point numbers, and S represents the value to be stored or manipulated in the system. The actual value of the system is represented by the quantity V. The final element in the system is the excess, E. Excess codes are not, in general, positional codes like the unsigned

Convert the decimal integer 1357049 to octal using algorithm OI.

Step	Operation			Resulting Octal Digit
1	1357049 ÷ 8	=	169631.125	
2	0.125 × 8	=	1	1
3	169631.125 − 0.125	=	169631	
4	169631 ÷ 8	=	21203.875	
5	0.875 × 8	=	7	7
6	21203.875 − 0.875	=	21203	
7	21203 ÷ 8	=	2650.375	
8	0.375 × 8	=	3	3
9	2650.375 − 0.375	=	2650	
10	2650 ÷ 8	=	331.250	
11	0.250 × 8	=	2	2
12	331.250 − 0.250	=	331	
13	331 ÷ 8	=	41.375	
14	0.375 × 8	=	3	3
15	41.375 − 0.375	=	41	
16	41 ÷ 8	=	5.125	
17	0.125 × 8	=	1	1
18	5.125 − 0.125	=	5	

Final result is 5132371_8.

Convert the decimal integer 807872 to octal using algorithm OI.

Step	Operation			Resulting Octal Digit
1	807872 ÷ 8	=	100984.000	
2	0 × 8	=	0	0
3	100984 − 0.000	=	100984	
4	100984.000 ÷ 8	=	12623.000	
5	0 × 8	=	0	0
6	12623.000 − 0.000	=	12623	
7	12623 ÷ 8	=	1577.875	
8	0.875 × 8	=	7	7
9	1577.875 − 0.875	=	1577	
10	1577 ÷ 8	=	197.125	
11	0.125 × 8	=	1	1
12	197.125 − 0.125	=	197	
13	197 ÷ 8	=	24.625	
14	0.625 × 8	=	5	5
15	24.625 − 0.625	=	24	
16	24 ÷ 8	=	3.000	
17	0 × 8	=	0	0
18	3.000 − 0.000	=	3	

Final result is 3051700_8.

Figure 1.4. Decimal-to-Octal Conversion of Integers with Algorithm OI.

binary system. However, they are very useful in creating a number system with a built-in offset.

Consider, for example, the creation of a number system representing all the temperatures one could expect to find anywhere in the world, in Fahrenheit. This would naturally lead to a discussion of what those temperatures could be, from somewhere in the polar regions to the hottest desert. Since this is a contrived example, we will assume that the range of the system is to be from $-80°$ to $+160°$. . This requires the ability to represent 241 different values. It is possible to represent this many values with 8 bits, which allow 2^8 different values. A system able to store all the desired numbers is an excess 80 code, where each value is contained within 8 bits. From the above equation, the smallest value would be assigned to the pattern of all zeros, and this would be -80. The largest value would be assigned to the pattern with all ones, and this value would be 175. Thus, this number system is not symmetric, because it contains more positive values than negative values.

Excess number systems can be very useful in many different applications, and the system architect can tailor the system to meet specific needs. However, one of the difficulties which arises is doing arithmetic in this system. For example, consider adding two values together, where the information has been stored in an excess format. Let these two numbers be called N_1 and N_2. The sum, $N_1 + N_2$, is then $V_1 + E + V_2 + E$, which is just $V_1 + V_2 + 2 \times E$. Thus, if the result is to be retained in the same excess system, then the excess must be handled appropriately. That is, the excess must be adjusted to remove the extra value which resulted in the addition.

Sum with excess code – method A

01111001	This is 41 in excess 80 code.
+ 10000100	This is 52 in excess 80 code.
11111101	This is the sum, but is not correct.
− 01010000	Subtract out one excess of 80.
10101101	This is the correct result.

Sum with excess code – method B

01111001	This is 41 in excess 80 code.
− 01010000	Subtract out the excess.
00101001	This is 41 in binary.

10000100	This is 52 in excess 80 code.
− 01010000	Subtract out the excess.
00110100	This is 52 in binary.

00101001	To 41 in binary
+ 00110100	add 52 in binary.
01011101	The sum - 93 in binary
+ 01010000	Add in one excess.
10101101	The result, in excess 80.

Figure 1.5. Addition with Excess Codes.

Two examples of addition with excess codes are shown in Figure 1.5. The first demonstrates the method mentioned above. That is, the excess is subtracted out of the sum of the stored values, leaving the proper result. The second method involves more work, but may be beneficial. Method B calls for removing the excess from the individual elements prior to doing the addition. This would be useful if some operation of the system needed the true (not excess represented) value for either input, or for the result.

1.2. Floating Point Operations

Floating point numbers are very wonderful in their realm because they remove from the user the worry of scaling data. Thus, all that a programmer or other user of the machine need do is identify that the number in use is a floating point variable, and the hardware/system takes care of the rest. However, it is important to remember that floating point numbers are not magic, and that, in fact, a floating point system will represent fewer values than an integer system with the same number of bits. However, the range of the numbers will be much larger for the floating point system, since it is able to represent numbers with different magnitudes. This section examines some floating point methods, first looking at representations of numbers in existing systems and converting between those systems. Then it looks at the characteristics of a small floating point system, and at doing arithmetic in the smaller system. These exercises should provide a better understanding of the basics of floating point systems.

Floating point numbers follow the scientific notation method for storing information. The number within the system consists of a sign, an exponent, and a mantissa. The value of the number can then be obtained from the following formula:

$$\text{Value} = (-1)^S \times M \times r_s{}^E$$

where S is the sign of the number, M is the mantissa, and E is the exponent. The radix or base of the system, r_s, is information which is part of the number system itself. For most common floating point systems, r_s is 2. However, other values are possible, such as 16, which is the r_s used in large IBM machines. Other information is needed to fully characterize the information in a floating point number, such as the number of bits used to represent the mantissa and the exponent. Let us now look at three 32-bit floating point number systems, DEC, IBM, and IEEE.

In examining these floating point systems, we will limit ourselves to the representation of normalized numbers. Normalization refers to the fact that there is only one ''normal'' representation of a number. That is, some rule in effect within the system coerces the bit patterns for the system in such a way that there is only one legal representation for a value. For the DEC system this rule is that the left-most digit (bit) must be nonzero. This fact, combined with the fact that the radix point for the system is to the left of the mantissa, establishes the fact that the value of the mantissa will be at least ½, and at most just a little less than 1. For the IEEE system, the normalization rule is very similar, and the left-most digit (bit) of the system must be nonzero. However, in the IEEE system the radix point is not to the left of the entire mantissa, but rather the radix point is located to the right of the first bit of the mantissa. Hence, the value of the mantissa will be at least 1, and at most, almost 2. The normalization rule for the IBM system is the same as for the DEC system: the left-most digit of the mantissa must be nonzero. However, the base of the IBM system is 16, so this rule means that the left-most four bits of the mantissa can assume any value from 0001 to 1111. Thus, the mantissa in the IBM system can be anything from 1/16 to almost 1.

The format used by the three different systems is shown in Figure 1.6. The DEC and the IEEE number system have the same basic format: one bit is used for the sign, 8 bits for the exponent, and 23 bits for the mantissa. Both systems use the "hidden bit" technique; that is, since the left-most bit of the mantissa *must* be a "1," there is no reason to store the bit. Hence, it is assumed to exist, and is "hidden" behind the exponent. Thus, although only 23 bits are available for the mantissa in the 32-bit representation, the mantissa is actually 24 bits long. As explained above, the mantissa for the DEC representation is a fraction somewhere between ½ and almost 1, while the IEEE mantissa is a number between 1 and almost 2. The IBM representation calls for a sign bit, 7 bits for the exponent, and 24 bits for the mantissa. The IBM representation cannot use the hidden bit technique, since the most significant bit of the mantissa can be either a "1" or a "0."

The exponent of these floating point numbers is stored in an excess format. The exponent of the DEC system is stored in excess 128 format; the exponent of the IEEE system is stored in excess 127 format; and the exponent of the IBM system is stored in excess 64 format.

With that much explanation about the makeup of the floating point number systems, let's represent the following numbers in each format: 900,000, 9, and 0.009. The first step in the process is to create representations for the numbers, in binary for the DEC and IEEE systems, and in hexadecimal for the IBM system. Use the algorithms of the previous section to show that:

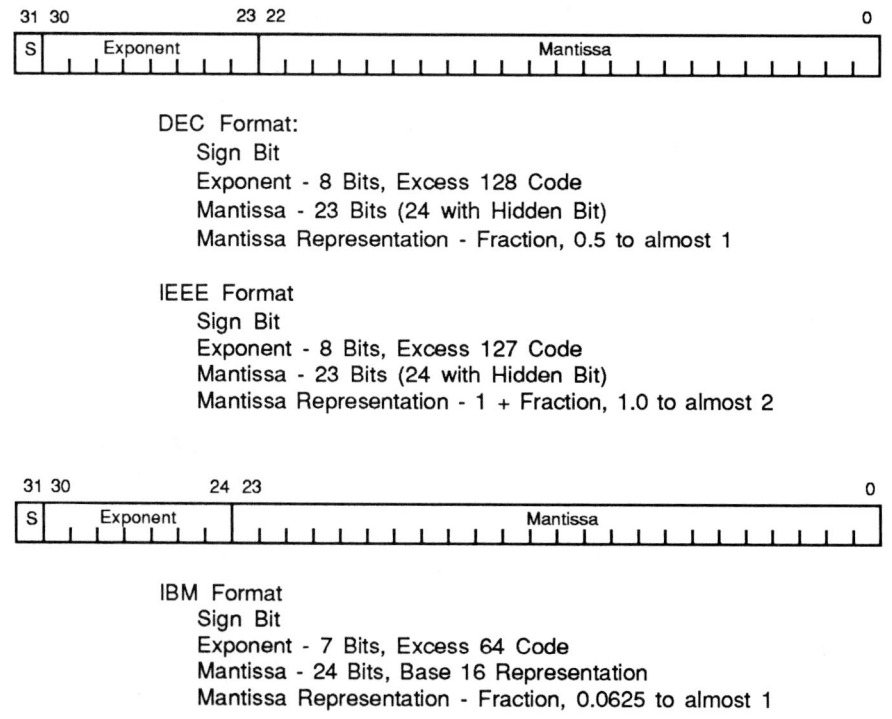

Figure 1.6. Bit Orientation in Floating Point Number Systems.

Number Base 10	Representation Base 2	Representation Base 16
900,000	11011011101110100000	DBBA0
9	1001	9
0.009	0.0000001001001101110100010111100	0.024DD2F

The representations for the integers are exact. The representation for the fraction is correct to the number of bits presented.

The DEC representation calls for a fraction between $\frac{1}{2}$ and 1, so the number 900,000 will be represented as $0.11011011101110100000 \times 2^{20}$. With a value of 20, the excess 128 representation of the exponent is 148, or 10010100_2. Therefore, the representation of 900,000 in DEC 32-bit floating point is 01001010010110111011101000000000. By a similar process, the number 9 is represented as 0.1001×2^4. The exponent of 4 has an excess 128 representation of 132, or 10000100_2. This gives a pattern of 01000010000100000000000000000000 for the number 9. The number 0.009 will have a negative exponent. The value is represented as $0.100100110111010010111100 \times 2^{-6}$. The exponent -6 is represented as 122 in excess 128 code, which has a binary pattern of 01111010. Therefore, the decimal fraction 0.009 has a 32-bit pattern of 00111101000100110111010010111100.

The IEEE representations will proceed in the same manner as the DEC representations above, except that the mantissa will be a number between 1 and 2. Therefore, the number 900,000 will be $1.10110111011101 \times 2^{19}$. The IEEE system uses an excess 127 representation, so the representation for the exponent is 146, or a binary pattern of 10010010. This gives a 32-bit representation of 01001001010110111011101000000000. The number 9 will be represented as 01.001×2^3. With the exponent in excess 127, the exponent pattern is 10000010. This leads to the following pattern for the decimal number 9: 01000001000100000000000000000000. Finally, the fraction is represented as $1.00100110111010010111100 \times 2^{-7}$. The exponent is stored as an 8-bit representation of 120, which is just 01111000. Therefore the IEEE 32-bit pattern for 0.009 is 00111100000100110111010010111100.

The IBM system represents the same values in base 16. The number 900,000 in base 16 is just DBBA0. The system calls for a fractional mantissa, so this representation becomes $0.DBBA0 \times 16^5$. With an excess 64 code for the exponent, the bit pattern for the exponent is 1000101. Therefore, representing the value as a binary pattern to compare it with the previous results, this becomes 01000101110110111011101000000000. The number 9 is very simple in this system, since it is merely 0.9×16^1. The exponent pattern is 1000001, so the whole thing becomes 01000001100100000000000000000000. Finally, the fractional value 0.009 is merely $0.24DD2F \times 16^{-1}$. The resulting 32-bit pattern is 00111111001001001101110100101111.

These results are summarized in Table 1.1. It is interesting to compare the various representations of Table 1.1. The DEC and the IEEE representations are the same, except that the bits of the exponent field are different. The reason for this is that the DEC system has exponents which are one higher than the IEEE system, at the same time that it has an excess which is one higher. This leads to a difference of two in the exponent for the same values. Also, since the IBM system is a base 16 system, the representations for the values will be different for that system. Note, for example, that the representations for 0.009 in the base 2 systems have two more bits of significance than the base 16 representation of the same number. Why?

Table 1.1. Coding of Numbers in Floating Point Systems.

Value	System	Representation Base 2	Representation Base 16
900,000	DEC	01001010010110111011101000000000	4A5BBA00
	IEEE	01001001010110111011101000000000	495BBA00
	IBM	01000101110110111011101000000000	45DBBA00
9	DEC	01000010000100000000000000000000	42100000
	IEEE	01000001000100000000000000000000	41100000
	IBM	01000001100100000000000000000000	40900000
0.009	DEC	00111101000100110111010010111100	3D1374BC
	IEEE	00111100000100110111010010111100	3C1374BC
	IBM	00111111001001001101110100101111	3724DD2F

Since the system ranges are not the same, the values which are representable are different for each system. Nevertheless, it is useful to consider the problem of converting between values which are within the range of each system. Converting between the DEC and the IEEE systems is relatively easy. All that is needed is an adjustment to the value of the exponent field by two, as shown by the patterns of Table 1.1. However, what steps are needed to convert between a representation of a number in a base 16 system and a number in a base 2 system? To demonstrate this, convert the number which has the representation 452CD000 in the IBM system to the representation of the number in the IEEE system. The given pattern has a sign of 0, so the number is positive. The exponent is 45_{16}, but this is an excess 64 system, so the value of the exponent is 5. Therefore, the number is $0.2CD_{16} \times 16^5$. We can easily convert the hexadecimal fraction to a binary fraction: $0.2CD_{16} = 0.001011001101_2$. So, the problem is reduced to dealing with the exponent and normalizing the mantissa. This is accomplished in the following steps:

$$0.001011001101_2 \times 16^6 = 0.001011001101_2 \times (2^4)^5$$

$$= 0.001011001101_2 \times 2^{20}$$

$$= 1.1011001101_2 \times 2^{17}$$

First, convert the base 16 representation for the exponent to a base 2 representation, which is always possible. Then, normalize the mantissa, and adjust the exponent accordingly. Converting the exponent to an excess 127 representation and placing all the parts of the number together results in a 32-bit pattern of 01001000001100110100000000000000, or 48334000. Converting between a number represented in the IEEE system to a number in the IBM system is the reverse of these steps. The exponent needs to be converted to an equivalent base 16 exponent. Since not all powers of 2 are also powers of 16, there will be two parts to the exponent when the process is finished. One part will be the base 16 part, the exponent of which is the exponent of the resulting number. The other part is a base 2 part, the exponent of which indicates by how many bit positions the mantissa must be shifted to create the proper base 16 normalized fraction.

The creation of appropriate binary patterns for floating point numbers is accomplished by finding the proper bit patterns for the number in question, then shifting the radix point until the representation consists of a normalized mantissa and an appropriate exponent, then combining sign, exponent (in appropriate excess code), and mantissa in a single binary pattern. Conversion between binary patterns is a process of adjusting the exponent and the mantissa in an appropriate fashion, as demonstrated by the preceding example.

Let us now create a small floating point system in order to demonstrate some of the mathematics of floating point numbers. The number system we will use is shown pictorially in Figure 1.7. It consists of a sign bit, 5 bits of exponent, and 10 bits of mantissa, all combined in a 16-bit word. The radix of the system is 2, and the bits of the mantissa represent a base 2 fraction. The mantissa is stored in a hidden bit format, so there are actually 11 bits in the mantissa. The exponent is stored in an excess 16 code. We will consider only normalized numbers and assume that if the bits of the exponent are zero, then the value of the number is zero, regardless of the remaining bits in the word.

One basic question to ask about a system: what is the range? That is, what is the largest representable number, and what is the smallest, nonzero representable number? The largest number representable in the system will be the largest mantissa times the radix of the system raised to the largest exponent. Therefore, the largest positive number will have a pattern of 0111111111111111, which corresponds to a value of $0.1111111111_2 \times 2^{15}$, or almost 2^{16}. The decimal representation of this number is 32,752. The smallest, nonzero representable number is the smallest mantissa times the smallest (in this case, most negative) exponent. This results in a bit pattern of 0000010000000000. Note that the 00001 pattern in the exponent field is the smallest exponent possible for a nonzero value. The remainder of the word is composed of zeros, since the hidden bit technique doesn't store the ''1'' of the smallest legal fraction, ½. This number has a value of $0.1_2 \times 2^{-15}$, or 2^{-16}. The decimal equivalent of this is approximately 1.526×10^{-5}. Two other numbers of interest are the number of legal mantissas and the total number of legal floating point numbers. Since the mantissa can be any combination of 10 bits, prepended by the hidden bit, there are 2^{10}, or 1024 legal mantissas. For each of 31 exponents, all mantissas are legal, and this gives a total of $31 \times 1024 = 31,744$ legal values. To this number is added the representation(s) for zero, so there can be 31,745 legal numbers. (This is positive numbers only; including negative

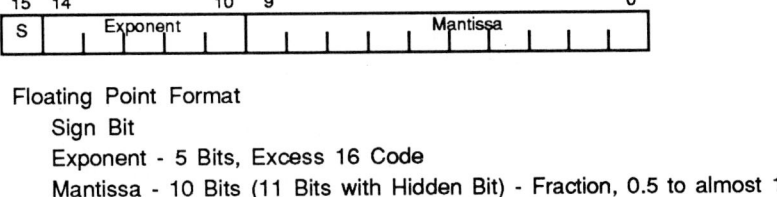

Floating Point Format
   Sign Bit
   Exponent - 5 Bits, Excess 16 Code
   Mantissa - 10 Bits (11 Bits with Hidden Bit) - Fraction, 0.5 to almost 1

Figure 1.7. Configuration of 16-Bit Floating Point System.

numbers doubles this value.) This is only 96.9 percent of the number of legal values in a comparable integer system.

As a final exercise in floating point numbers, consider the problem of doing a simple multiplication with this system. Let the number A be 45, let the number B be 163, and let the result be the number R. All are represented as floating point values. A multiplication problem can be simply stated (for a base 2 system):

$$R = A \times B$$

$$M_R \times 2^{E_R} = M_A \times 2^{E_A} \times M_B \times 2^{E_B}$$

$$= M_A \times M_B \times 2^{(E_A + E_B)}$$

Therefore, the result mantissa is just the multiplication of the mantissas of the two input values, and the result exponent is the sum of the two input exponents. In the system under consideration the number 45 is represented as $0.1011010000_2 \times 2^6$, and 163 is represented as $0.1010001100_2 \times 2^8$. (What are the corresponding bit patterns in the 16-bit system?) If we multiply the two mantissas together in a simple multiplier, we will end up with twice as many bits, and the pattern will be 0111001010011100000000. How is the correct value of the result determined? Looking at this result, we recognize the fact that the mantissa is not normalized; to be a correct mantissa for this system the radix point must be positioned to the left of the first "1" bit. The first "1" bit then becomes the hidden bit, and the next 10 bits form the mantissa to be loaded into the result. With this operation, we have formed the following value: 0.11100101001 1100000000. The break in the number indicates where the floating point number will lose precision, since it cannot retain the bits in the second half. The break point also identifies one of the problems with floating point numbers: How is the determination of the mantissa constructed? The simple method demonstrated here is truncation, where the remaining bits are discarded. If, on the other hand, the system designers determined that some rounding scheme should be used, then instead of simply discarding the extra bits, the rounding algorithm will be applied. Any rounding should be done before the normalization step. The final result is $0.1110010100 1_2 \times 2^{13}$. Note that the sum of the exponents (6 + 8) has been decremented to account for the normalization of the resultant mantissa.

The next floating point operation we will perform is a simple addition. Let $A = 4070$ and $B = 36.75$; then form the sum $A + B$. The result is $R = A + B$, which is (for a base 2 system)

$$M_R \times 2^{E_R} = M_A \times 2^{E_A} + M_B \times 2^{E_B}$$

$$= (M_A + M_B) \times 2^{(E_A - E_B)} \times 2^{E_A}$$

That is, we add to one mantissa a shifted copy of the second mantissa, and the shift amount is determined by the difference in the size of the exponents. If the exponents are equal, no shift is needed. The numbers needed for this example are 4070, which is $0.1111111001 1_2 \times 2^{12}$, and 36.75, which is $0.1001001100_2 \times 2^6$. The difference in the exponents is 6, so the smaller number is shifted 6 places with respect to the larger one, and the addition operation becomes:

```
  0.11111110011
+ 0.00000010010 0110000000
  ─────────────────────────
  1.00000000101 0110000000
```

The result of the addition is a mantissa which must be normalized in the opposite direction as the result of the multiplication, which will increment the exponent. Thus, the result is $0.10000000010_2 \times 2^{13}$. If rounding is included in the unit instead of truncation, this process must be modified accordingly. The system designer must determine at which point rounding will be applied, and the algorithm for the rounding (normal rounding, round-to-zero, etc.).

Addition can also produce results which require normalization to the more significant bit positions, as well as to the lesser significant positions. To demonstrate this, consider the addition of 352 and -343. In the system under consideration, 352 is represented as $0.10110000000_2 \times 2^9$, and 343 is represented as $0.10101011100_2 \times 2^9$. So, the addition operation (subtraction, actually, since the second number is negative) is as follows:

$$
\begin{array}{r}
0.10110000000 \\
-\ 0.10101011100 \\
\hline
0.00000100100
\end{array}
$$

The result must be shifted to the left by five positions to create a normalized fraction. This gives a result of $0.10010000000_2 \times 2^4$. Note that the left shift of the result by five places decreased the exponent by five.

These two operations demonstrate that the addition operation creates results which must be shifted left or right to create the proper mantissa, depending on the position of the most significant digit in the final operation.

1.3. Conclusion

As we have seen, dealing with number systems involves providing interpretations to the patterns of symbols which make up the alphabet of the number systems. In computers, these fundamental symbols are ones and zeros, but they can be grouped together to form octal and hexadecimal numbers. These patterns can form integer values, floating point values, or other systems not considered here. To attach meaning to the patterns, a designer must determine the method by which the pattern is generated or by which a pattern can be interpreted.

This chapter has looked briefly at some of the common mechanisms for attaching meanings to patterns. The two most prominent are integer systems and floating point systems. Additional complexity can be added to integer systems by using excess codes, or by assuming the existence of a radix point and turning the representations into fixed point systems capable of representing fractional values as well as whole numbers. Since humans deal mostly with base 10 numbers, several algorithms were presented for converting between base 10 representations and computer representations, using either simple methods compatible with paper and pencil systems, or by using a calculator to do more bits at a time.

Floating point numbers are also very useful, and this chapter looked at some techniques for dealing with the common representations for floating point numbers. This included conversions to the floating point representations from decimal numbers, and converions between different floating point systems. Also considered were some of the requirements for doing arithmetic in floating point systems.

1.4. Exercises

1.1 Convert the number 1098356_{10} to a base 2 number and a base 8 number.

1.2 Convert the number 7452301_8 to a base 2 number and a base 10 number.

1.3 Convert the number 456239_{10} to a base 2 number and a base 8 number.

1.4 Convert the bit pattern 101100001011 0 to a decimal number.

1.5 Convert the number 0.567_{10} to equivalent base 2 and base 8 representations.

1.6 Convert the number 0.3647_8 to an equivalent base 10 representation.

1.7 Create floating point representations in the IBM and IEEE systems for the following numbers: 1, 100, 100,000, 0.01, 0.000001.

1.8 Create floating point representations in the DEC and IBM systems for the following numbers: 9, 80, 640, 300, 0.07, 0.000009.

1.9 What does the pattern 11001011010010100010100010001000 mean in the DEC 32-bit floating point number system? The IBM 32-bit floating point system?

1.10 What is the decimal equivalent of the smallest representable normalized positive number in each of the three floating point systems discussed in this chapter?

1.5. Additional References

[Bree89] Breeding, K. J., *Digital Design Fundamentals.* Englewood Cliffs, NJ: Prentice Hall, 1989.

[Flet80] Fletcher, W. I., *An Engineering Approach to Digital Design.* Englewood Cliffs, NJ: Prentice Hall, 1980.

[LaFr89] Langholz, G., J. Francioni, and A. Kandel, *Elements of Computer Organization.* Englewood Cliffs, NJ: Prentice Hall, 1989.

[Mano88] Mano, M. M., *Computer Engineering: Hardware Design.* Englewood Cliffs, NJ: Prentice Hall, 1988.

[Poll90] Pollard, L. H., *Computer Design and Architecture.* Englewood Cliffs, NJ: Prentice Hall, 1990.

[Schn85] Schneider, G. M., *The Principles of Computer Organization.* New York: John Wiley & Sons, Inc., 1985.

[Stal87] Stallings, W., *Computer Organization and Architecture.* New York: Macmillan Publishing Co., 1987.

2

Design of Combinational Systems

Systems of digital logic in which the outputs are functions solely of the current inputs are called combinational systems. They are also memoryless, with no feedback, either synchronous or asynchronous, involved in the system. The design of this type of system forms part of a good basic course in digital logic, and several good texts covering this design process are available [Flet80, McCl86, Mano88, Bree89]. In this chapter we describe the design of two different systems, and present alternative design strategies for each of the systems. In designing a combinational system, the designer must determine the relative merits of the factors which influence the final product. The best design, then, is the one which utilizes the critical system resources in the most efficient manner.

2.1. Design Steps for Combinational Systems

Creating a combinational system to perform a predefined function is a challenging task. As mentioned above, the system will operate as depicted in Figure 2.1: each output will be a function only of the inputs to the system and consequently will not be a function of the history of the past inputs or past outputs. From the beginning of the design process until it is completed there are a number of decisions to be made. The manner in which those decisions are made and the strategy behind the decisions determines the type of system which will result. Regardless of the strategy, the steps involved in the process are basically the same from one design to the next. Designers consistently successful in their design efforts will consciously or unconsciously follow these steps:

1. **Understand the requirements.** This may seem like a superfluous statement, but it is a very real requirement. One of the principal difficulties that arise in the creation of a digital system actually stems from the communication of ideas

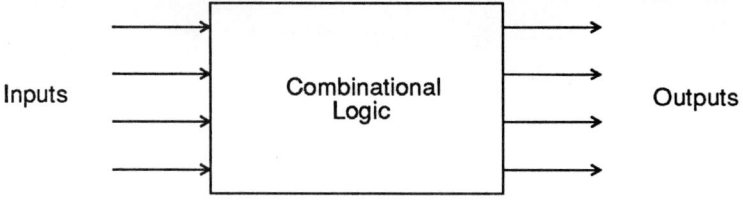

Figure 2.1. Combinational Logic: Outputs Are Functions Only of Inputs.

between those who are active in the definition and use of the system, and those who are responsible for the actual creation of the logic. One person's understanding of what is required to solve a problem will be different from a second person's understanding, and these differences can lead to fundamental flaws in the approach to a solution. It is imperative that the requirements of the system be identified *before* the system is implemented. The system must be fully understood before adequate and correct combinational logic can be designed. This understanding includes the nature and origin of the inputs, the desired behavior and purpose of the logic, and the destination and function of the outputs.

2. **Create a basic block diagram.** This step enables the designer to reinforce the understanding of the problem which was established in step 1. Often during this portion of a design an inadequate problem specification can be identified, or the designer can explore different mechanisms in the process of defining an approach to the design of the result.

3. **Partition the problem as needed.** If the system requires only a small number of outputs which are simple functions of a small number of inputs, then the system need not be partitioned. However, if the system contains several inputs and outputs, then the complexity of the system can be controlled by functional partitioning of the logic into more manageable portions. Again, the method of partitioning and the size of the different portions of the circuit are a function of the design strategy and the available system resources.

4. **Uniquely define the outputs of each portion in terms of the inputs of that section.** The *best* way, and the most comprehensive way, to accomplish this is to create a truth table specifying the correct output for each and every combination of inputs. An alternative method is to use logic equations. *Caution: Don't go straight to equations unless you are a genius.* The creation of correct equations is a complex process, and errors can easily be introduced.

5. **Reduce the truth table or equations to an implementable form.** The creation of logic equations to implement can be done in many ways, as is described in logic texts. One of the most often used mechanisms is to represent the truth table of the function as a Karnaugh map, and then use repeated application of the principle of logical adjacency to create a reduced two-level form. Another method is to use the theorems and relationships of logical analysis to create a final system. Still another is to use a computer aided mechanism to produce the desired result. One such is a program which implements the tabular method for logic reduction (see [Mano79]), one version of which is found in Appendix A. At this point note that the strategy for

implementation and the constraints placed on the circuit will have a direct effect on the final circuit. If the circuit must be fast, and the result must be available in two gate delays (one level for the AND of a Boolean system, another for the OR) then the system must be created accordingly. If, however, the number of gate delays can be increased, then the logic may be reduced, and the ingenuity of the designer can have a large impact on the number of gates involved in the final solution.

6. **Create a logic system to do the work.** In this book this is done by using logic gates, and connecting them together to form the system defined in steps 1–5. If the logic system is to be created in silicon internal to an integrated circuit, then the representation may be different. The point is to create a system which correctly and completely implements the logic required to solve the problem. In order to do this, care must be taken to match the assertion levels of the inputs with the logic, and also to create outputs which satisfy the assertion requirements for the system.

7. **Implement the logic system and debug as necessary.** This process must be performed to ascertain that the system was implemented in a reasonable fashion, and that the resulting implementation solves the problem. A caution is inserted at this point concerning the checkout process. It is sometimes the case that the logic system developed in steps 4, 5, and 6 correctly implements the logic equations involved, but nevertheless, the logic does not solve the problem. Thus, the designer should be aware that the combinational system created reflects the current understanding of the problem. If the logic does not perform the needed service, then the error may not be in the logic system, but rather in the designer's understanding of the system requirements. Thus, the debugging process should not center entirely on the logic system created in steps 4, 5, and 6, but should also include the work done in steps 1, 2, and 3.

Each of the above steps must be completed in a reasonable fashion to be sure that the final system performs as required. It cannot be stressed enough that the final logic system will be a product of the design strategy, and each decision made in the design process will impact on the result. It is often beneficial to iterate through steps 1, 2, and 3 above for at least two reasons. The first is that the designer's understanding of the problem will be enhanced with each exposure to the requirements, and this process can give insight into both the problem and possible solution mechanisms. The second reason is to compare different systems which will result from different design strategies or partitioning mechanisms. The resulting systems will enable a designer to make reasonable decisions about the final design.

In the creation of the final design and its presentation, care must be taken to match the assertion levels for input and output. This will assure that the signals perform as required to meet specifications. The designer must also be careful to use the symbols and names in such a fashion that the design can be communicated to another designer or to the end user. This communication requirement is often not given enough emphasis in designs and systems, resulting in unused or improperly used devices and systems.

2.2. 2-Bit ALU

The object of this exercise is to design a 2-bit ALU. This unit will accept inputs from 2-bit words and create the appropriate output, according to the setting of the control inputs. The first design is done using only NAND gates (and inverters, which can be made from NAND gates). The second design utilizes other types of gates, and a slightly different design strategy. The resulting designs will then be compared to the two-level implementation of the system. This project is large enough to demonstrate some of the techniques which can be used to build larger ALUs; yet it is small enough to be done in a practical fashion.

2-bit ALU: System Definition and Requirements

The first step identified above is needed to specify the action of the system. We first identify the functions to be performed by the ALU, and the relationship between the control lines and the function being performed The ALU itself could be represented by a symbol such as:

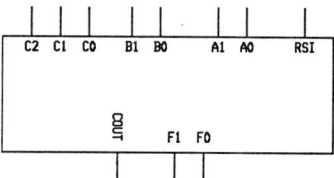

The 8 inputs have the following definitions:

A1, A0

B1, B0

> These inputs form 2-bit numbers to be operated on by the ALU. We assume here that we are working with positive logic, so that the inputs and outpus are asserted high.

C_{IN}

> This is the carry input for the ALU, which is also used as the borrow for subtract operations.

C2, C1, C0

> These three inputs control the output according to Table 2.1. Note that the table includes both logical and arithmetic operations, so that both capabilities must be in the final system.

Table 2.1. Functions to be Provided by the 2-Bit ALU.

C2	C1	C0	ALU Function
0	0	0	0 (zero) passed to outputs
0	0	1	$\overline{A}$ passed to outputs
0	1	0	A · B (logical AND) is passed to outputs
0	1	1	A + B (logical OR) is passed to outputs
1	0	0	A ⊕ B (logical exclusive-OR) is passed to outputs
1	0	1	A minus B (including borrow...) is passed to outputs
1	1	0	B minus A (including borrow...) is passed to outputs
1	1	1	A plus B (including carry...) is passed to outputs

The three outputs include both the number out and carry (borrow) output:

F1, F0

> This 2-bit word will receive the results of the ALU as specified by the control inputs.

C_{OUT}

> This is used in arithmetic operations where a carry (borrow) is needed. It is not defined during the logical operations.

One way of approaching the solution would be to create a truth table with all 2^8 combinations of the inputs, and then create logic for each output from the information contained in the table through whatever reduction mechanism is available. However, different approaches will lead to different implementations. We will solve this problem in two different ways. The first solution is a straightforward approach in which some care is taken to create a system with a reasonably small number of gates, at the same time maintaining a system which has a recognizable solution to the problem. The second solution is one in which the emphasis is on minimizing gate delays in the system.

2-bit ALU: Basic block diagram

The basic approach followed in these implementations is indicated in Figure 2.2. The action of the system is selected by multiplexers, the inputs of which implement the various functions called for in the system definition. The logic of the system, then, must provide the appropriate values to the inputs of the multiplexer, as defined by Table 2.1, and the control lines are connected to the select lines of the multiplexer to select the correct function. Thus, the logic of the system always develops all the functions (sum, OR, AND, etc.), and the control lines, operating through the select function of the multiplexer, select the desired operation for output. With this idea as the basis of the unit, different approaches to partitioning and reduction lead to different implementations.

The logical functions of the ALU can be derived quite simply with minimal gating. The arithmetic functions provide more interesting questions which must be addressed. The basic add/subtract functions can be obtained from truth tables, and the resulting carry/borrow lines forwarded to the appropriate stage. These truth tables are given in Table 2.2. We make the following observations about Table 2.2,

Table 2.2. Truth Tables for Arithmetic Functions.

Inputs			A plus B		A minus B		B minus A	
A	B	C_{IN} †	C_{OUT}	F	C_{OUT} ‡	F	C_{OUT} ‡	F
0	0	0	0	0	0	0	0	0
0	0	1	0	1	1	1	1	1
0	1	0	0	1	1	1	0	1
0	1	1	1	0	1	0	0	0
1	0	0	0	1	0	1	1	1
1	0	1	1	0	0	0	1	0
1	1	0	1	0	0	0	0	0
1	1	1	1	1	1	1	1	1

† C_{IN} is for both carry and borrow
‡ C_{OUT} used for borrow here

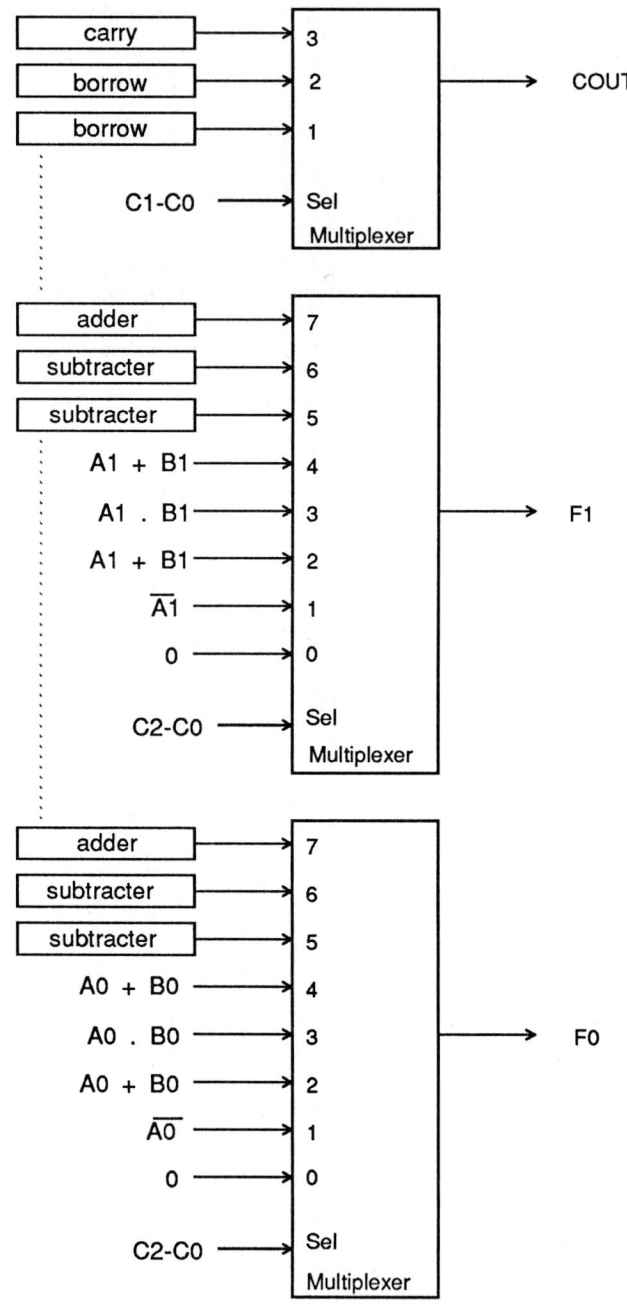

Figure 2.2. Basic Block Diagram for 2-Bit ALU.

and then discuss them more with regard to reduction of the logic. First of all, the output function (F) for addition and subtraction is the same, and is equal to $A \oplus B \oplus C_{IN}$. Also, the form for the carry is the same for each arithmetic operation:

$$F = AB + AC + BC \quad \text{(A plus B)}$$

$$F = \overline{A}B + \overline{A}C + BC \quad \text{(A minus B)}$$

$$F = A\overline{B} + AC + \overline{B}C \quad \text{(B minus A)}$$

The creation of the logic for the ALU must provide correct values for all the required functions. The assumptions made concerning the available basic logic elements, as well as the decisions made in the partitioning process, will have a direct effect on the implementation of the system.

2-bit ALU: Partition for NAND/NOR implementation and define outputs

The basic partitioning for the system has already been indicated, as shown in Figure 2.2. We can also create a table to include this information; this is presented as Table 2.3. The logic of the system must implement the functions listed in Table 2.3, and the manner in which this is accomplished determines the efficiency of the system or the view the designer has of the system resources.

As has been mentioned, one approach to the design is to expand Table 2.3 to include all possible combinations of inputs, and then implement the resulting terms directly. Approaching the generation of the outputs in this fashion will result in 70 separate minterms, which is usually considered unacceptable. Thus, we hope to reduce the overall logic by creating the system as shown in Figure 2.2.

2-bit ALU: Reduction of logic for NAND/NOR implementation

For the first implementation of the ALU we will use only NAND gates (and inverters), and also the ripple-carry mechanism for generating the carry (borrow)

Table 2.3. Logic Equations for the 2-Bit ALU.

$C2$ $C1$ $C0$	C_{OUT}	$F1$	$F0$
0 0 0	-	0	0
0 0 1	-	$\overline{A1}$	$\overline{A0}$
0 1 0	-	$A1 \cdot B1$	$A0 \cdot B0$
0 1 1	-	$A1 + B1$	$A0 + B0$
1 0 0	-	$A1 \oplus B1$	$A0 \oplus B0$
1 0 1	$\overline{A}B + \overline{A}C_{INT} + BC_{INT}$	$A1 \oplus B1 \oplus C_{INT}$ *	$A0 \oplus B0 \oplus C_{IN}$
1 1 0	$A\overline{B} + AC_{INT} + \overline{B}C_{INT}$	$A1 \oplus B1 \oplus C_{INT}$ †	$A0 \oplus B0 \oplus C_{IN}$
1 1 1	$AB + AC_{INT} + BC_{INT}$	$A1 \oplus B1 \oplus C_{INT}$ ‡	$A0 \oplus B0 \oplus C_{IN}$

$$* \; C_{INT} = \overline{A0}\,B0 + \overline{A0}\,C_{IN} + B0\,C_{IN}$$

$$\dagger \; C_{INT} = A0\,\overline{B0} + A0\,C_{IN} + \overline{B0}\,C_{IN}$$

$$\ddagger \; C_{INT} = A0\,B0 + A0\,C_{IN} + B0\,C_{IN}$$

out of the device. Thus, the logic is implemented in a very straightforward fashion. There are three observations we will make which directly impact on the gating implementation. The first is that we will incorporate, as much as possible, the AND-OR requirements of the logic with the AND-OR stages used in the multiplexer, resulting in a reduction in the total logic required for the implementation.

The second observation concerns the creation of the proper output for the arithmetic functions. Note that the generation of the function output (F) for both the least significant bit and the most significant bit are the same: $A \oplus B \oplus C_{XX}$, where C_{XX} is either the carry input or the internal carry. Thus, the same logic gates can be used for all arithmetic functions. Also, if we can find a way to force the carry to be zero, then this same set of gates will also generate $A \oplus B$, which further reduces the requirements for additional gates.

The third observation is that the carry (borrow) logic can also be implemented quite simply, by including the proper terms in a single carry circuit and enabling the terms when necessary. For example, the term AC is used in both the addition and the B minus A subtraction. The appropriate control lines are included with the gate which generates AC to enable contribution from that term only when it is necessary. Treating the terms in this fashion enables a single carry circuit to be used for both addition and subtraction.

Having made these three observations, we will now create logic which will generate the appropriate responses. The logic follows the basic approach depicted in Figure 2.2, combining functions where appropriate, and uses only NAND gates (and inverters).

2-bit ALU: Logic for NAND/NOR implementation

The logic diagrams for the ALU are included as Figure 2.3. The logic for the control and the signal buffers is shown in Figure 2.3(a). As shown in the figure, all inputs are asserted high. Creation of the final results requires both the asserted high and the asserted low signals, so inverters are included to perform this level conversion. An additional stage of inverters creates the levels back again. Thus, A0-H is inverted to create the signal BA0-L, which is in turn inverted to create BA0-H. This adds one more gate delay to the system, but presents a single load at the inputs. Also note that the carry/borrow input is received by a NAND gate configured so that when the exclusive-OR function is needed, the internal carry/borrow line is not asserted. This allows the exclusive-OR capability of the adder to be used for the logical exclusive-OR. The remaining gates in Figure 2.3(a) are used to enable the designated functions. ABAR-H is asserted only when the output should be the inverse of A; AND-H is asserted only when the output should be the logical AND of the A and B inputs; OR-H is asserted only when the output should be the logical OR of the A and B inputs. The remaining signals are used to enable the appropriate terms of the carry for doing the addition and subtraction.

The logic for the least significant bit and the internal carry is shown in Figure 2.3(b). This diagram demonstrates the ideas described above for implementation of the system. One gate (gate 1) is used to provide all the OR function needed in the generation of the least significant bit (F0-H). This is used for both the multiplexing function and the generation of the terms needed for output. The remaining gates for the least significant bit (gates 2–9) are used to create the appropriate logic functions. Gate 2 provides A0 · B0, but only when the control lines call for an AND function. When the control lines call for an OR function, gates 3 and 4 present A0 and B0 to gate 1, where the actual OR is performed. Gate 5 is used to present an inverted version of A0 when the control lines request it. The remaining gates for

Chap. 2: Design of Combinational Systems

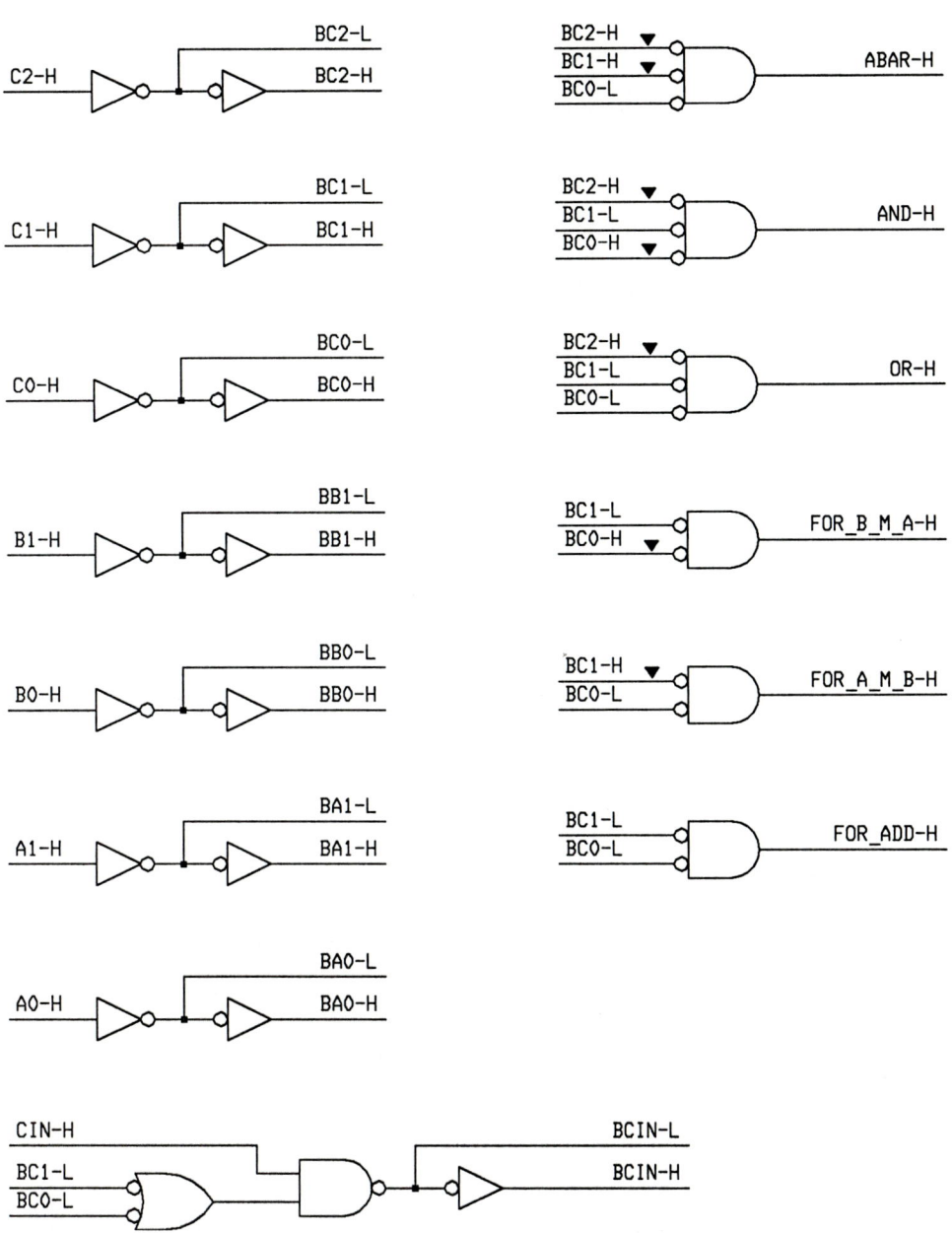

Figure 2.3(a). Logic of ALU: Buffer Gates and Control Signal Generation.

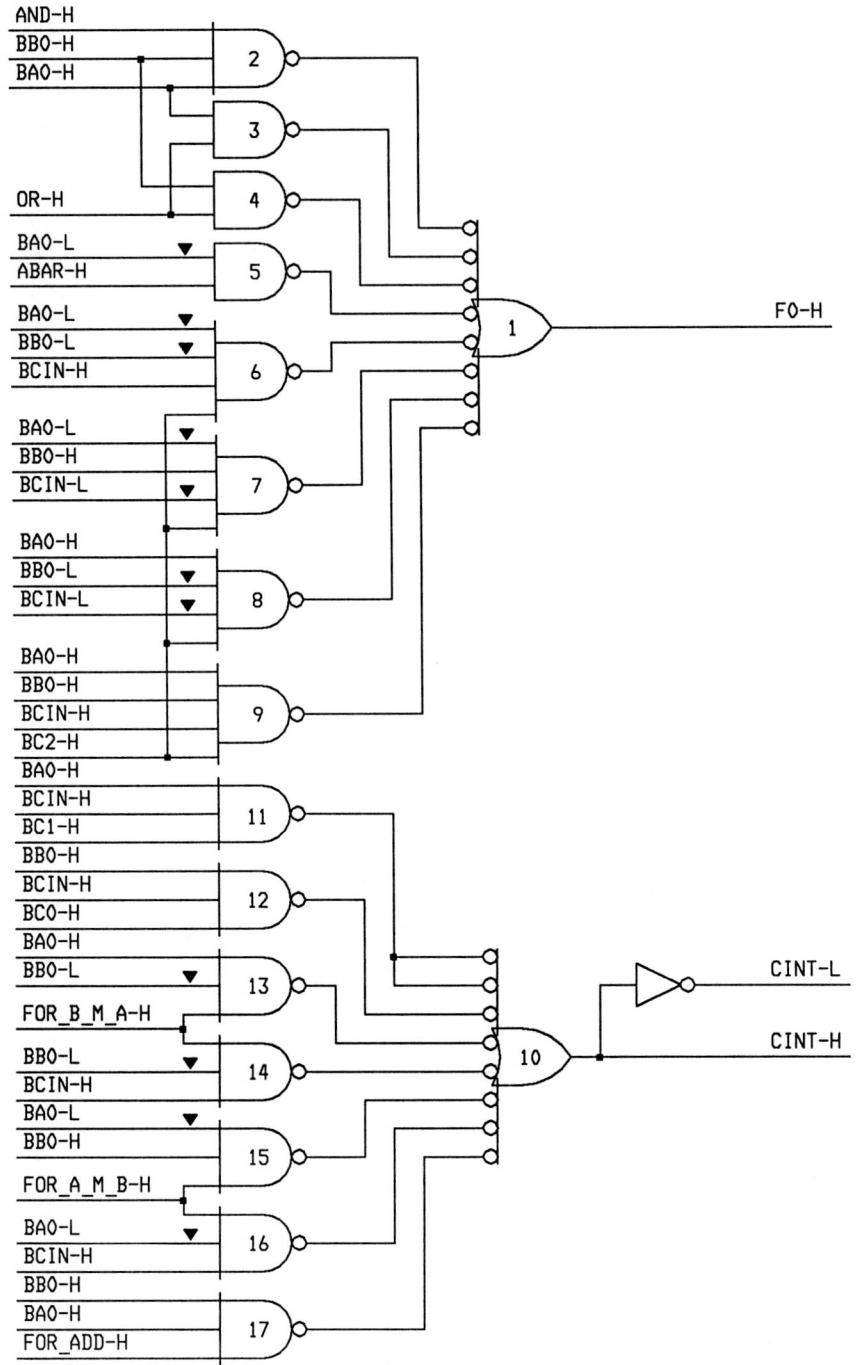

Figure 2.3(b). Logic of ALU: Least Significant Output and Internal Carry Circuitry.

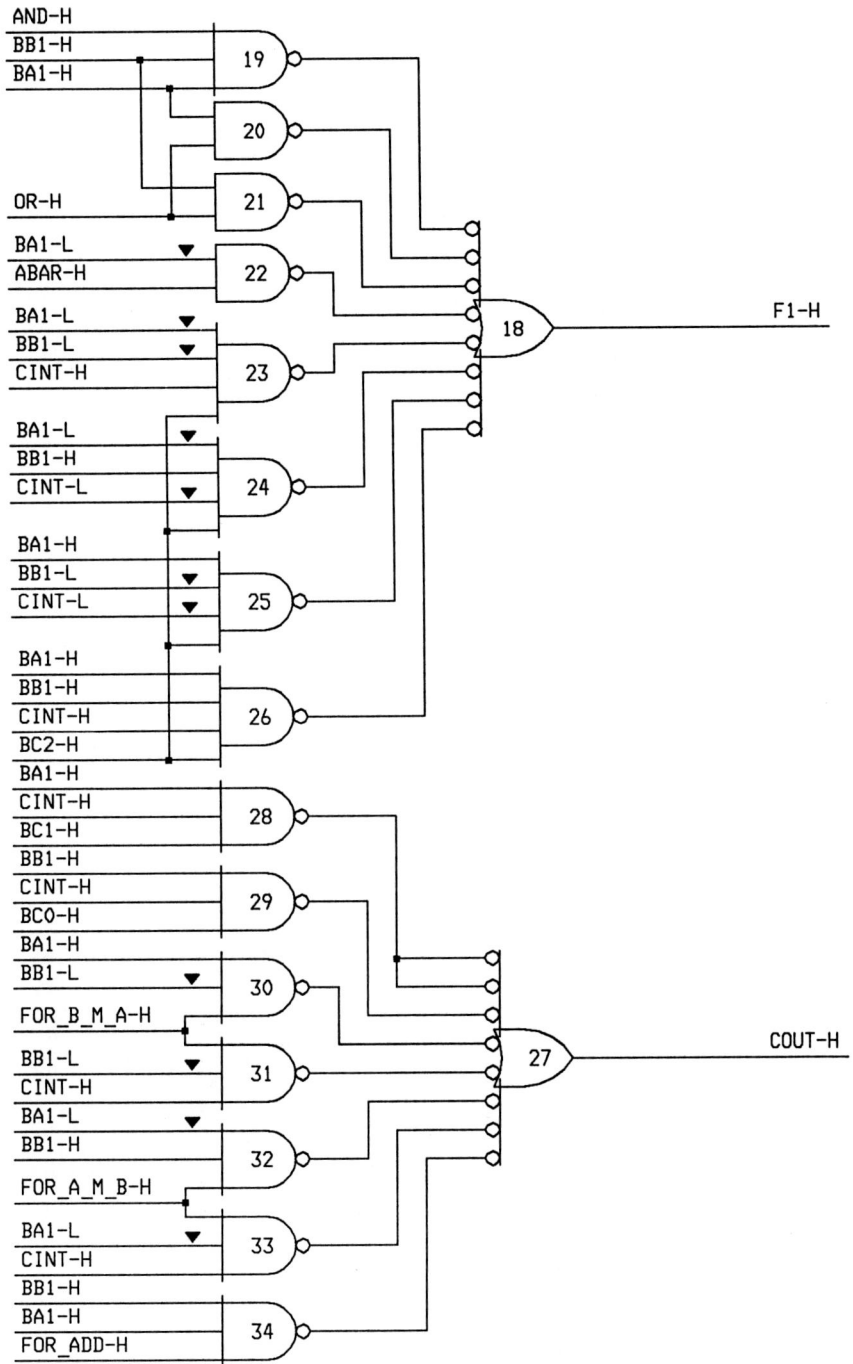

Figure 2.3(c). Logic of ALU: Most Significant Output and Carry-Out Circuitry.

the generation of F0 implement $A0 \oplus B0 \oplus C_{IN}$, which will be used for all arithmetic functions as well as the exclusive-OR function. Partitioning the system as demonstrated here resulted in creation of the F0 signal with a total of nine gates (plus those used for control).

The generation of the internal carry is also included in Figure 2.3(b). Gate 10 is used as the OR for the function, and the AND terms provided by gates 11–17 are enabled only when needed. The $A0 \cdot C_{IN}$ term is needed for both A minus B and A plus B; this occurs whenever control line C1 is asserted. Thus, gate 11 is used to provide this term. Similarly, the $B0 \cdot C_{IN}$ term is needed for both B minus A and A plus B, and this is supplied by gate 12 whenever the control line C0 is asserted. Gates 13 and 14 are enabled for B minus A, and these provide the $A \cdot \overline{B}$ and $\overline{B} \cdot C_{IN}$ terms needed for that function. Likewise, gates 15 and 16 are enabled for A minus B, and provide the terms $\overline{A} \cdot B$ and $\overline{A} \cdot C_{IN}$ needed for that function. Finally, gate 17 provides the $A \cdot B$ function needed during addition. Note that in order to reduce the gating requirements, the control line C2 is not involved in the control of the carry logic. Hence, during the logical operations the carry is undefined.

The logic for the most significant bit (F1-H) and the carry/borrow output (COUT-H) is shown in Figure 2.3(c). This logic has exactly the same form as the logic for the least significant bit and the internal carry. Gate 18 generates the most significant bit from the various terms provided to it. The AND function is provided by gate 19. During the OR request both A1 and B1 are presented (by gates 20 and 21), and gate 18 performs the actual OR function. The inverse of A1 is provided by gate 22 when called for by the control lines. And gates 23–26 provide $A1 \oplus B1 \oplus C_{INT}$ whenever control line C2 is asserted. This implements arithmetic and exclusive-OR functions.

The carry/borrow out (COUT-H) is generated in exactly the same fashion as the internal carry. Gate 27 provides the OR function needed, and gates 28–34 provide the appropriate terms when needed. Thus, control lines C1 and C0 are used, along with FOR_ADD-H, FOR_B_M_A-H, and FOR_B_M_A-H, to enable at the appropriate times the terms needed for the carry function.

The approach shown in Figure 2.3 uses 58 gates (including the inverters) to perform the required logic. A different set of answers to the questions involved in the design process would result in a circuit with different characteristics. Now we will go through portions of the process again, with a different set of decision criteria. This time, we will allow gates other than the NAND/NOR set, and will also request that the carry out be generated with a look-ahead technique.

2-bit ALU: Partitioning for relaxed requirement implementation

The partitioning for this implementation is basically the same as for the previous implementation. The main differences come in the implementation, since we are not limited to NAND/NOR gates, and in the generation of the carry out. Nonetheless, the basic block diagram is still as shown in Figure 2.2.

2-bit ALU: Logic for relaxed requirement implementation

The logic for the new version must implement the same logic as in the first version, but perhaps in a different manner. Thus, the logic equations shown in Table 2.3 maintain their role as definitions of correct behavior. However, we can now use exclusive-OR gates in the implementation, and this will be useful both for the generation of the function bits and the carry/borrow functions.

2-bit ALU: Reduction of logic for relaxed requirement implementation

The obvious reduction in gate complexity comes from the ability to use exclusive-OR gates in the implementation. This allows a reduction in the gating requirement for generation of the function bits (F0 and F1). However, it is also useful for the implementation of the carry/borrow logic. To see how this works, consider the following manipulations of the logic equation for the carry function of the adder:

$$C_{OUT} = A \cdot B + A \cdot C_{IN} + B \cdot C_{IN}$$

$$= A \cdot B + (A + B) \cdot C_{IN}$$

$$= A \cdot B + (A \oplus B) \cdot C_{IN}$$

The last step is not obvious, and, in fact, looks incorrect. However, expansion of the second and third forms yields exactly the same result, and we will use this fact to reduce the logic requirements of the system. The $A \cdot B$ term in the above equations is often called the carry generate term, since there will be a carry if it is asserted, regardless of the incoming carry. The $A + B$ term, or the $A \oplus B$ term, is often called the carry propagate term, since, if this term is asserted, the carry available as an input is passed on to the output.

We have done the above logic manipulations with the addition function, but the same operations can be done with the subtraction functions. That is, the generation of the borrow will be divided into two portions: the "borrow generate" and the "borrow propagate." The generate term will include an inverse of A or B, depending on the subtraction being performed. The carry propagate term will be the inverse of $A \oplus B$. Thus, the carry will be propagated by $A \oplus B$ during addition, and the same signal will be propagated by the inverse of $A \oplus B$ during subtraction. Note how these features are utilized in the implementations of the ALU.

2-bit ALU: Logic diagrams for relaxed requirement implementation

The logic diagrams for the new version of the ALU are shown in Figure 2.4. The buffering action of the inverters and the control logic included in Figure 2.4(a) is almost identical to that included in Figure 2.3(a). The only difference is the addition of an exclusive-OR gate, used to control the level of the carry propagate signal used in the system.

The logic for the least significant output (F0-H) is shown in Figure 2.4(b). The basic approach is the same as before: the OR functions needed will be provided by gate 1, while the AND functions will be supplied by gates 2–6. Gate 2 is used to provide the $A0 \cdot B0$ function. Gates 3 and 4 send A0 and B0 to gate 1 for the OR function, when the control lines request it. And gate 5 is used for the inverse of A0. The principal difference in this implementation shows up with gate 6. In the previous implementation, the exclusive-OR function was provided by a set of NAND gates, and those gates were enabled only when control line C2 was asserted. In this implementation, the exclusive-OR function needed is provided by exclusive-OR gates, and then this logical value is allowed to contribute to F0 only when C2 is asserted. This enabling function is handled by gate 6. The exclusive-OR function is handled in two steps. The first step is the creation of $A0 \oplus B0$, which is performed by gate 8. This will be used by gate 7 to create the $A0 \oplus B0 \oplus C_{IN}$ term needed for logical and arithmetic functions at F0. The $A0 \oplus B0$ term is also used by gate 11 to create P0-H, which is the propagate function for the least

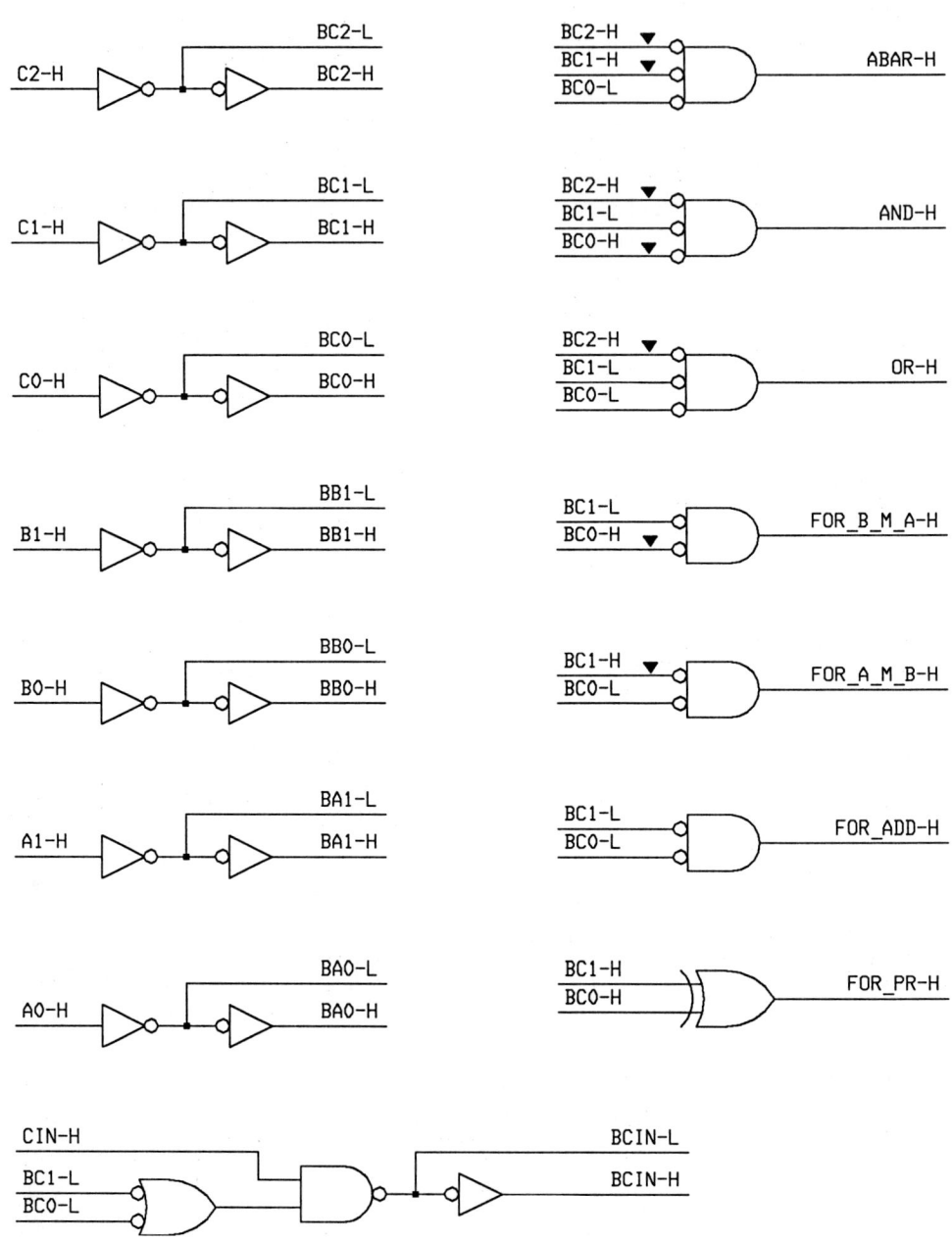

Figure 2.4(a). Logic of ALU: Buffer Gates and Control Signal Generation.

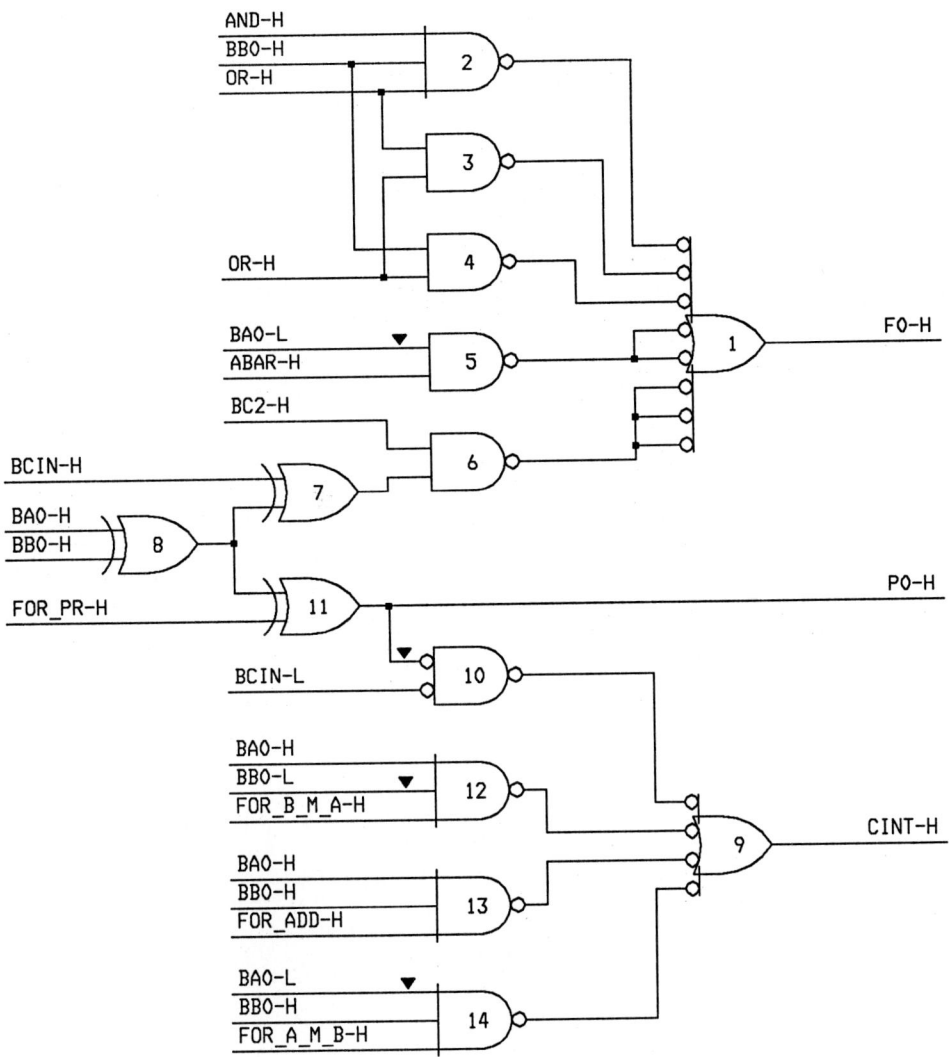

Figure 2.4(b). Logic of ALU: Least Significant Output and Internal Carry Circuitry.

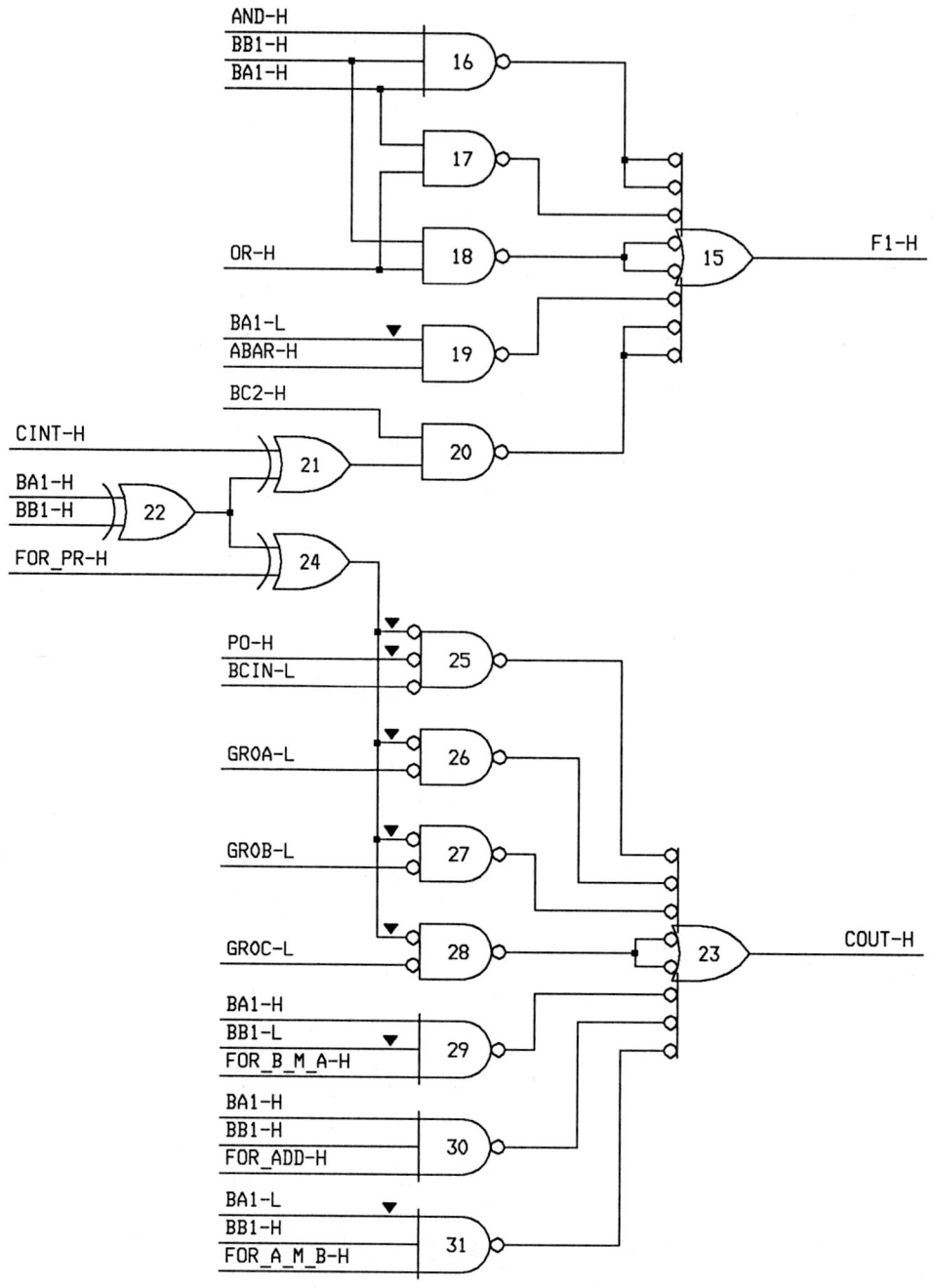

Figure 2.4(c). Logic of ALU: Most Significant Output and Carry-Out Circuitry.

significant portion. For addition, the P0-H function will be $A0 \oplus B0$; for subtraction, gate 11 will create the inverse of $A0 \oplus B0$.

The internal carry (CINT-H) is created as described for the previous implementation. Gate 9 provides the OR function as necessary. Gate 10, which is an OR gate used in an AND function, passes on the carry/borrow as called for by the propagate function (which is available on the P0-H line). Gate 12 is the carry (borrow) generate for the B minus A function; gate 13 is the carry generate for the add function; and gate 14 is the carry (borrow) generate for the A minus B function. These logic elements are used in the generation of the final carry/borrow output.

The most significant bit of the ALU (F1-H) and the carry/borrow output (COUT-H) are shown in Figure 2.4(c). Gates 15–20 generate F1 in exactly the same fashion that gates 1–6 generate F0. And the exclusive-OR gates (21, 22, 24) are used to generate the carry/borrow propagate as well as the function needed by F1. But notice the expansion of the complexity in generating COUT-H, in comparison with the generation of CINT. Gates 29, 30, and 31 are used for the carry/borrow generate functions of the subtractions and the addition, in the same fashion that 12, 13, and 14 perform these functions for CINT. The additional complexity comes from including the carry generate terms from CINT in COUT. That is, the carry/borrow terms will propagate as determined by gate 24. When the output of gate 24 is low, then the appropriate gate (26, 27, or 28) passes on the carry generate from the previous stage. Gate 25 is used to pass on the carry/borrow to the output when the generate signals from both stages are asserted.

The logic shown in Figure 2.4 is very much like the logic shown in Figure 2.3, and uses 54 gates. However, these gates often have fewer inputs than their counterparts in Figure 2.3. The net result is a smaller implementation. The basic question about implementation strategies must be addressed by the system designer. We will make the following observation about the system.

2-bit ALU: Comparison of implementations

There are many different ways to compare designs. The system designer determines which method matches his requirements, based on his application and his utilization of system requirements. At this point we can make some observations about the system, and the use of system requirements. Table 2.4 has been created to specify the use of various system resources by three different implementations. The first implementation is the first one given above, using only NAND/NOR gates. The second implementation is that just completed, with relaxation of the gating utilization and a different method for carry/borrow generation. The third implementation is what would be required for a strictly two-level design.

The number of gate delays shown in Table 2.4 is obtained by counting the number of gates in the longest path from the input to the output. A more correct version would be obtained by using the actual propagation delays, since gates do not have the same propagation delay. In particular, exclusive-OR gates are generally slower than NAND or NOR gates.

The number of gates is obtained by counting the gates on the schematic. Again, this parameter should not be taken by itself in the evaluation of a design. Rather, all the system resources should be considered in making design comparisons.

The last three elements in Table 2.4 will vary somewhat with the method of implementation. For this example, Advanced Low-Power Schottky (ALS) devices were used. Other types of TTL devices will result in different power, but the same

Table 2.4. Utilization of System Resources by ALU Designs.

Parameter	Design 1	Design 2	Design 3
Max gate delays			
C_{IN} to C_{OUT}	6	3	3
C_{IN} to F1	7	6	3
C_{IN} to F0	4	5	3
A0 to C_{OUT}	6	5	3
A0 to F1	7	7	3
A0 to F0	4	6	3
A1 to C_{OUT}	4	4	3
A1 to F1	4	6	3
C0 to C_{OUT}	7	5	3
C0 to F1	8	8	3
C0 to F0	6	7	3
Number of gates	58	55	90
Number of packages	21	18	80
Board space (in^2)	6.72	5.76	25.84
Power (mW)	92.3	105.9	180.5

board space. Implementing these systems directly on silicon would result in a different set of parameters, but the method would be the same: determine the utilization of system resources and compare the results.

The ALU examined in this section demonstrates several of the facets of designing combinational systems. The design strategy is influenced by the application of the system, and this strategy directly impacts the techniques used in the design. The designer can use different techniques to implement a design; however, it is imperative that the final result correctly implement the various functions required of the system. The designer's ingenuity in combining gating systems to implement the function will have a direct impact on the final result.

2.3. Row Reduction Element for Seven Rows

The object of this design is to create a system which performs row reduction. This type of device can be used in high-speed multipliers to reduce the number of rows in a partial product array. In general, a row reduction element with N outputs can accept as input $2^N - 1$ rows. Thus, a full adder is a row reduction element which accepts as input three rows (actually, 3 bits, one each from each of three different rows) and produces a 2-bit output which specifies the number of "1" bits on the input lines. With three inputs, the output must be able to represent 0, 1, 2 and 3, which is the operation of a full adder.

A row reduction unit with three outputs is capable of representing numbers from 0 to 7. Hence, the maximum number of rows it can accept as input is seven. The final result of this design effort, then, must be capable of generating the proper result in a reasonable fashion. Again, we will do the design in two ways: first, the easy way, without worrying too much about gate delays; then a second time, trying to implement the function in 4 or 5 levels of logic.

Row reduction element: System definition and requirements

The word statement of the row reduction element identifies its function: accept up to 7 bits of input and create a 3-bit output representing the number of "1" bits in the input. A symbol for this type of device can be created as follows:

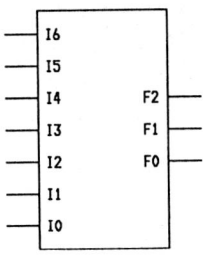

In order for this device to work properly, the 7 bits which form the input must all have the same significance. That is, in a partial product array of a multiplication, all the bits must originate in the same column. The 3 bits of the output will have different significance. The least significant bit of the output will have the same significance as all the inputs. The second bit of the output will be 1 bit more significant, and the third bit will be 1 bit further.

There are no control lines with this function, only the seven different input lines. For the complete definition of the device, we could list all 128 different combinations of the inputs, and generate the proper output for each combination. However, this would tell us little more than we already know, so we will forego that enumeration.

There are a number of ways of approaching this design. The most tedious method is to implement this in two levels of logic, which will result in a system with over 140 minterms. Another method is to relax the gate requirements and try to implement the system in fewer gates, with more gate delays. Finally, the simplest method is to make the 7–3 row reduction element from repeated use of 3–2 row reduction elements. We will proceed first to the simplest, and then to the faster, more complicated method.

Row reduction unit: Basic block diagram and partitioning for simple implementation

The simplest way to make a 7–3 row reduction unit is through repeated use of full adders. A block diagram of this implementation is shown in Figure 2.5. This implementation uses the fact that the process described above can be done in sections. Each full adder will create a 2-bit number identifying the number of "1"s on its inputs, which is the desired function for the reduction element, but with only 3 bits. The results can then be combined to produce the desired final result. In Figure 2.5, the significance of each of the lines is identified by a digit to the side of the line. Thus, the outputs of the first three full adders have the significance of 0 and 1, and the outputs of the final full adder have the significance of 1 and 2. (Or, if you prefer, 1, 2, and 4.)

The implementation technique shown in Figure 2.5 sets forth the partitioning of this solution: the final result is made up of four full adders.

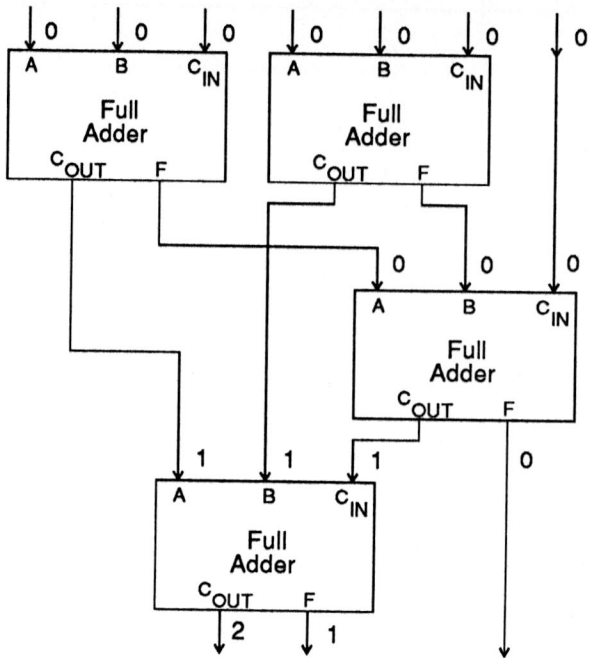

Figure 2.5. Full Adder Implementation of 7–3 Row Reduction Element.

Row reduction unit: Define outputs and reduce

The logic of each of the blocks defined in Figure 2.5 is very simple. The C_{OUT} and F are defined as:

$$C_{OUT} = A \cdot B + A \cdot C_{IN} + B \cdot C_{IN}$$

$$F = A \oplus B \oplus C_{IN}$$

The carry and the function outputs can be combined, as was performed in the ALU above, which results in a full adder requiring only five gates to implement, two 2-input exclusive-OR gates, and three 2-input NAND gates.

Logic for implementation of simple row reduction unit

The logic diagram for the simple row reduction unit is shown as Figure 2.6. Each of the full adders is outlined for reference. Full adder number 1 (FA1) is used to combine inputs I4-H, I5-H, and I6-H. Full adder number 2 is used to combine inputs I1-H, I2-H, and I3-H. The least significant of each of these adders, and the remaining input line (I0-H), are handled by full adder number 3. The least significant bit of full adder number 3 forms one of the outputs (F0-H). The other output of full adder number 3 is combined with the remaining outputs of the other two adders in full adder number 4. The outputs of full adder number 4 are the remaining two bits of the unit (F1-H and F2-H).

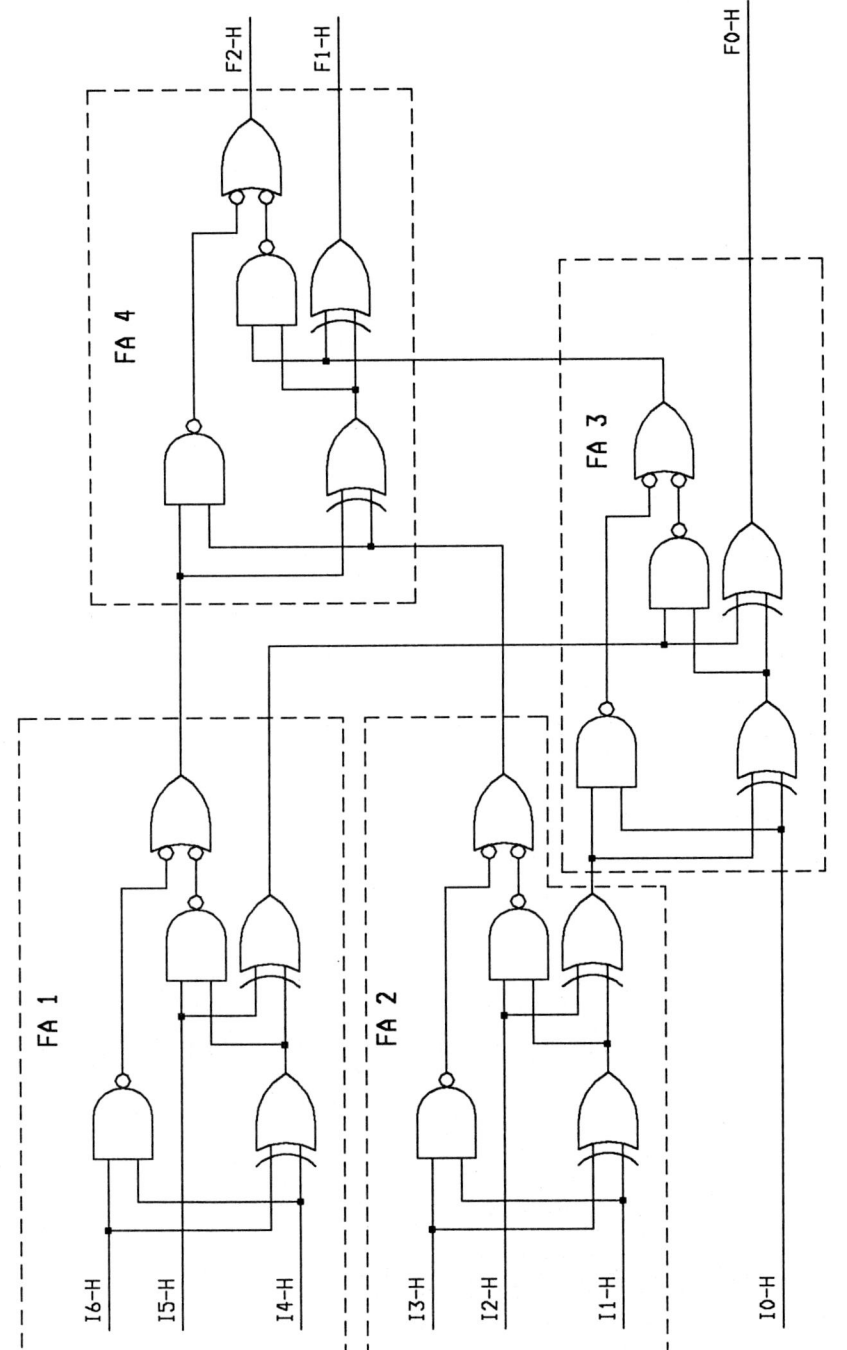

Figure 2.6. Logic for Full Adder Implementation of Row Reduction Unit.

The implementation shown in Figure 2.6 requires only 20 gates, and can be implemented with only five packages (in off-the-shelf technologies). This is contrasted with the a faster implementation, which does not allow the use of exclusive-OR gates and in which the designer strives for high speed.

Row reduction unit: Block diagram and partitioning for NAND/NOR implementation

The implementation which utilizes only NAND/NOR gates will approach the design in a different fashion. Certainly, one could merely replace the exclusive-OR gates in Figure 2.6 with NAND equivalents, but this would greatly increase the delays from input to output. The implementation we will undertake is demonstrated by the very simple diagram shown in Figure 2.7. As indicated by the figure, each of the outputs is implemented directly from the inputs, and the generation of the terms is shared in some way in an attempt to reduce the overall logic requirement. The basic problem we will try to address is how to implement these in a reasonable fashion.

NAND/NOR implementation of row reduction unit: Define outputs and reduce

There are a number of ways to approach the specification and reduction of the logic for this system. Our approach here is to generate truth tables for each variable, but to give the truth tables in the form of a Karnaugh map. This allows graphical representation of the elements, and allows us to see patterns helpful to the process of logic sharing. This specific implementation is certainly not the only way to do it, but we have chosen it to demonstrate the method.

The first output to consider is F0-H, the least significant output. The initial step is to generate the entire truth table, and then map this information onto a Karnaugh map. There are many ways of constructing Karnaugh maps, and a seven-variable map is usually considered too unwieldy. However, such a map can be very useful, and we will use the map shown in Figure 2.8. For our Karnaugh map, we will first take the four variables labeled I0 through I3 and create a four-

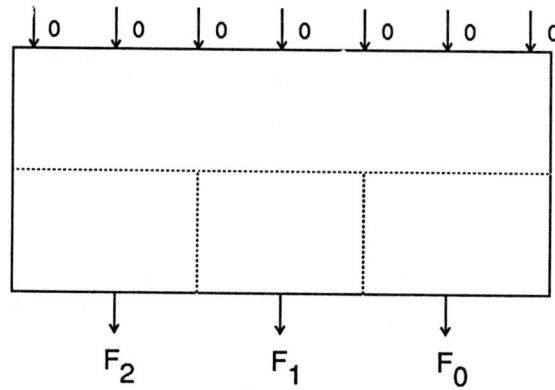

Figure 2.7. Partitioning Method for NAND/NOR Implementation of Row Reduction Unit.

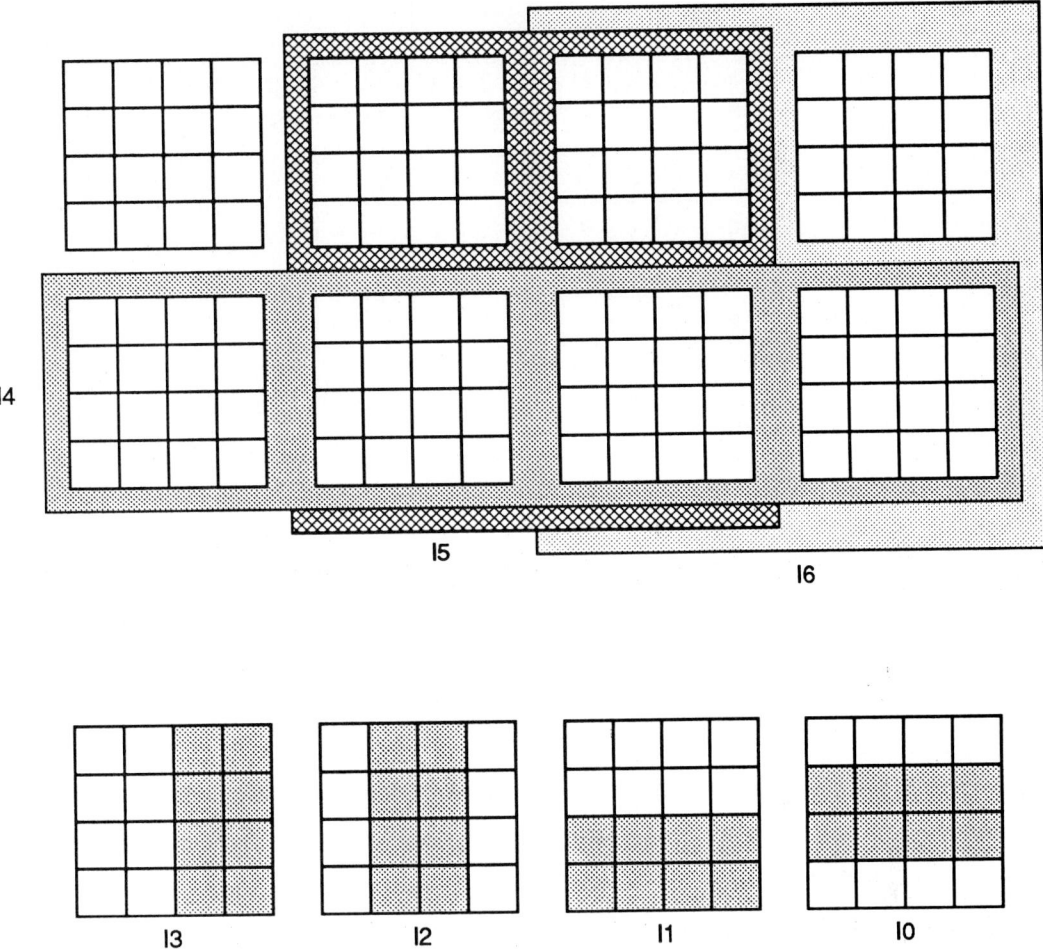

Figure 2.8. Seven-Variable Karnaugh Map for Row Reduction Unit.

variable map. This is shown at the bottom of Figure 2.8, with the shaded areas of the map identifying the assertion areas of each of these four variables. Each of the eight sections in the upper map has this same configuration, so logic terms can be identified. The remaining three variables, I4, I5, and I6, form the three-variable map shown at the top of the figure. Each entry in this map is the four-variable map consisting of the other four variables. This may look confusing at first, but it will be very beneficial in the process which follows.

The map for the least significant bit is shown in Figure 2.9 It is obvious that there is a pattern here, but it is not at all obvious that anything can be done with it. The pattern demonstrated by this map is the exclusive-OR pattern, and the equation for F0 would reduce to:

$$F0 = I6 \oplus I5 \oplus I4 \oplus I3 \oplus I2 \oplus I1 \oplus I0$$

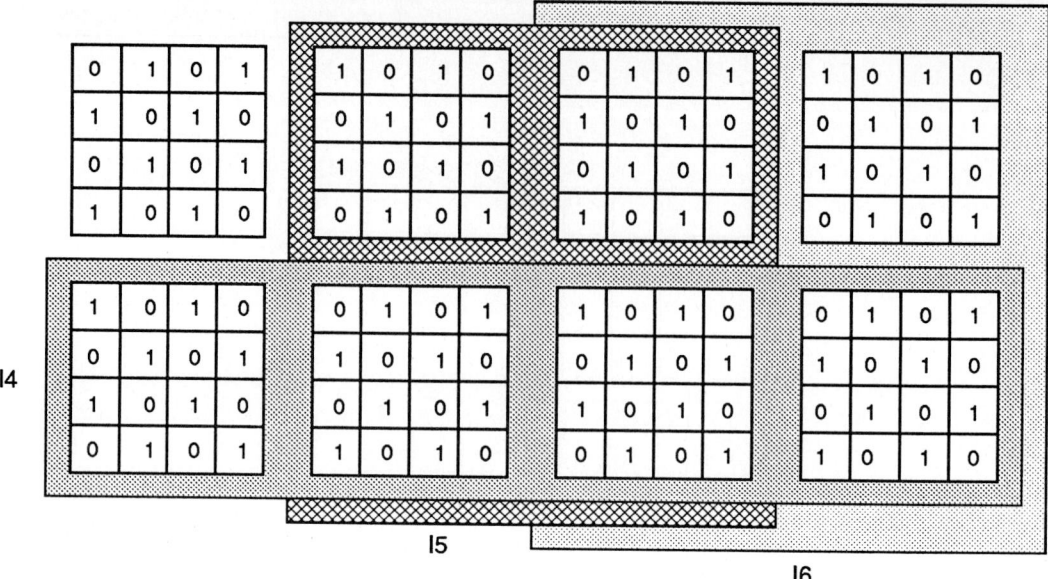

Figure 2.9. Karnaugh Map for F0 of Row Reduction Unit.

However, we stated that we would not use exclusive-OR gates for this implementation. Therefore, we will have to do the same function with NAND gates. The NAND implementation of a 3-input exclusive-OR is relatively straightforward:

$$\text{EX-OR} = \overline{X} \cdot \overline{Y} \cdot Z + \overline{X} \cdot Y \cdot \overline{Z} + X \cdot Y \cdot Z + X \cdot \overline{Y} \cdot \overline{Z}$$

Using this implementation method, we can group the terms in the following fashion:

$$\text{F0} = (\ \text{I6} \oplus \text{I5} \oplus \text{I4}\ ) \oplus (\ \text{I3} \oplus \text{I2} \oplus \text{I1}\ ) \oplus \text{I0}$$

This will not be useful in the generation of the other outputs, but it does provide the needed output for F0.

The map for the generation of F1 is given in Figure 2.10. By grouping patterns together, a number of different implementations are possible. The two-level logic equation for this map has 42 terms in it, but we can do better by allowing more levels of logic. Without further discussion we will give the equation used in this implementation of F1:

$$\text{F1} = (\ \overline{\text{I3}}\ \overline{\text{I2}}\ \overline{\text{I0}} + \overline{\text{I2}}\ \overline{\text{I1}}\ \overline{\text{I0}} + \overline{\text{I3}}\ \overline{\text{I2}}\ \overline{\text{I1}} + \overline{\text{I3}}\ \overline{\text{I1}}\ \overline{\text{I0}} + \text{I1}\ \text{I2}\ \text{I3}\ \text{I4}\ )\ \cdot$$

$$(\ \text{I6}\ \text{I5}\ \overline{\text{I4}} + \overline{\text{I6}}\ \text{I5}\ \text{I4} + \text{I6}\ \overline{\text{I5}}\ \text{I4}\ ) +$$

$$\overline{\text{I6}}\ \overline{\text{I5}}\ \overline{\text{I4}} \cdot (\ \overline{\text{I3}}\ \text{I1}\ \text{I0} + \overline{\text{I3}}\ \text{I2}\ \text{I0} + \overline{\text{I3}}\ \text{I2}\ \text{I1} + \text{I3}\ \text{I2}\ \overline{\text{I1}} + \text{I3}\ \text{I1}\ \overline{\text{I0}} + \text{I3}\ \overline{\text{I2}}\ \text{I0}\ ) +$$

$$\text{I6}\ \text{I5}\ \text{I4} \cdot (\ \overline{\text{I3}}\ \overline{\text{I2}}\ \overline{\text{I1}}\ \overline{\text{I0}} + \text{I3}\ \text{I2}\ \text{I1} + \text{I3}\ \text{I1}\ \text{I0} + \text{I3}\ \text{I2}\ \text{I0} + \text{I2}\ \text{I1}\ \text{I0}\ ) +$$

$$(\ \overline{\text{I3}}\ \overline{\text{I2}}\ \text{I0} + \overline{\text{I3}}\ \text{I1}\ \overline{\text{I0}} + \overline{\text{I3}}\ \text{I2}\ \overline{\text{I1}} + \text{I3}\ \overline{\text{I1}}\ \overline{\text{I0}} + \text{I3}\ \overline{\text{I2}}\ \overline{\text{I1}} + \text{I3}\ \overline{\text{I2}}\ \overline{\text{I0}}\ )\ \cdot$$

$$(\ \overline{\text{I6}}\ \overline{\text{I5}}\ \text{I4} + \overline{\text{I6}}\ \text{I5}\ \overline{\text{I4}} + \text{I6}\ \overline{\text{I5}}\ \overline{\text{I4}}\ )$$

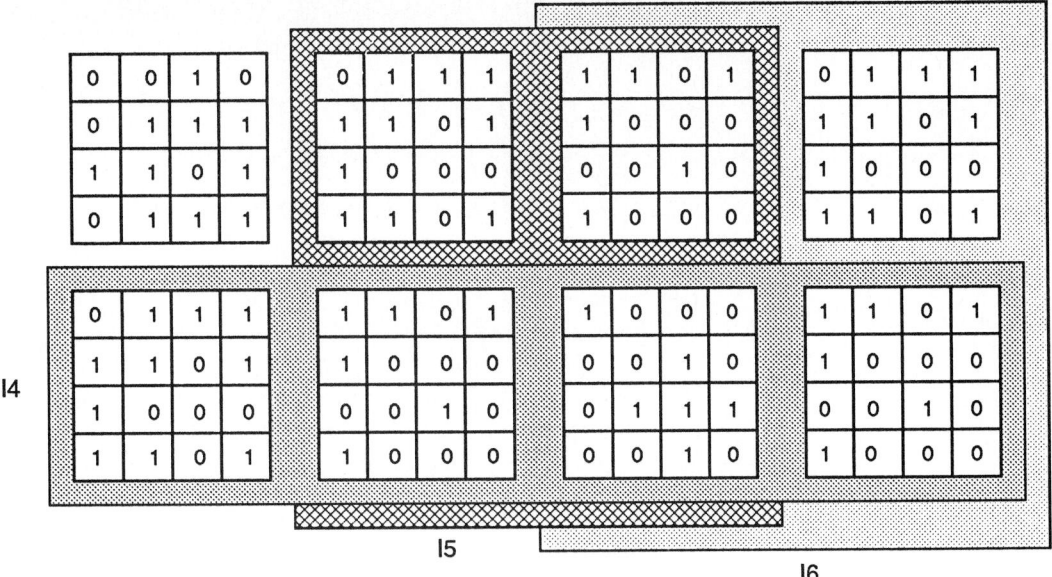

Figure 2.10. Karnaugh Map for F1 of Row Reduction Unit.

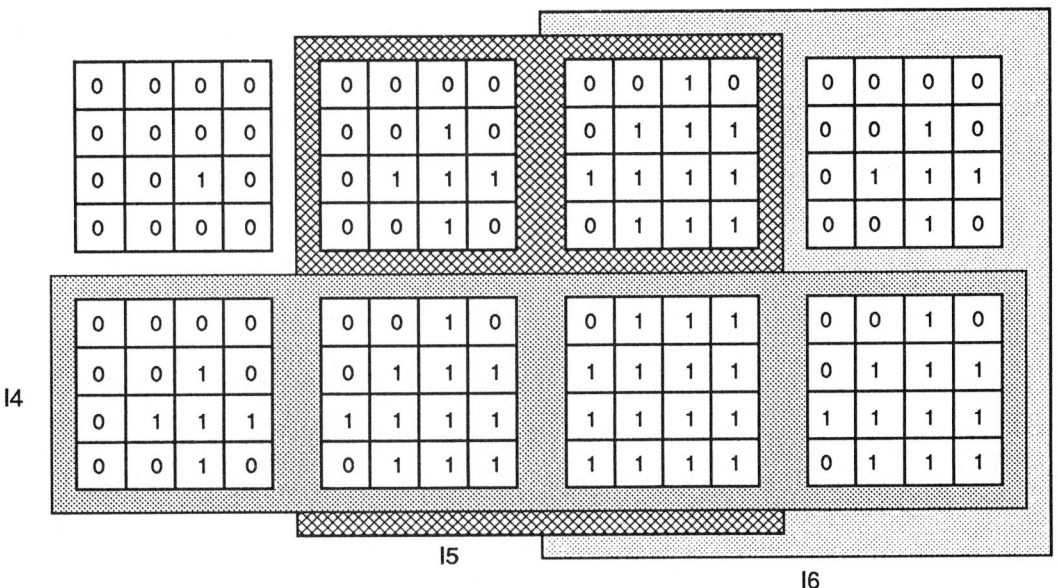

Figure 2.11. Karnaugh Map for F2 of Row Reduction Unit.

This organization of the logic is by no means the only, or even the best, reduction of the map given in Figure 2.10. But the equations demonstrate one method of combining the logic: look for patterns in the Karnaugh map, and group the patterns together to take advantage of their repetitive nature. The same will also be true of the logic for the most significant bit of the row reduction unit.

The map for F2 is given in Figure 2.11. Examination of the map reveals additional patterns which can be utilized in the implementation of the row reduction unit. The logic for this term can be represented as:

$$F2 = (\ I3\ I2\ I1\ +\ I3\ I1\ I0\ +\ I3\ I2\ I0\ +\ I2\ I1\ I0\ )\cdot(\ I6\ +\ I5\ +\ I4\ )\ +$$

$$(\ I3\ I2\ +\ I1\ I0\ +\ I2\ I0\ +\ I3\ I0\ +\ I3\ I1\ +\ I2\ I1\ )\cdot(\ I6\ I5\ +\ I5\ I4\ +\ I6\ I4\ )\ +$$

$$I3\ I2\ I1\ I0\ +\ I6\ I5\ I4\ \cdot(\ I3\ +\ I2\ +\ I1\ +\ I0\ )$$

Again, this logic has been created by observing patterns which exist in the map, and then implementing the logic in such a way that the patterns are utilized. Try to match the various members of the above equations with the existing patterns in the maps of Figures 2.9 to 2.11.

The actual implementation of the logic of the equations is shown in Figure 2.12. The logic for the least significant bit is shown in Figure 2.12(a). Note that the groupings of the bits match the equation as shown above. In the first stage, the exclusive-OR of I6, I5, and I4 is generated, as is the exclusive-OR of I3, I2, and I1. Then, the exclusive-OR of these two results and I0 is produced by the final stage.

The logic for the second bit is shown in Figure 2.12(b). Like the system of Figure 2.12(a), this logic contains two levels of AND-OR gates. That is, the first set of gates is a set of NAND gates used in an ANDing function, followed by a set of NAND gates used in an ORing function. The third set of gates is again a set of NAND gates used in an ANDing function, and the final gate is a NAND gate used in an ORing function. It is instructive to go back to the equations and to the maps and identify the terms implemented by each of the gates in the diagram shown in Figure 2.12.

The logic for the third bit of the row reduction unit is shown in Figure 2.12(c). As before, there are two levels of AND-OR functions implemented with NAND gates. Again, compare the logic with the equations and the maps to ascertain how this is done. Also, look for ways to improve the logic shown. Note that terms used in the generation of F2 are used for the generation of F1, and that terms used for the generation of F1 are used in F2.

Row reduction unit: Comparison of implementations

A comparison of the three different designs is given in Table 2.5, where various parameters are listed. The first design is the first one done here, with four full adders arranged as shown in Figure 2.6. The second design is that just completed, the NAND implementation with 4 or more gate delays. The third design listed in Table 2.5 is one in which each output is generated with two levels of logic (one AND level, followed by an OR level). The three gate delays listed include the inverter needed to create a correct inversion of the input signals.

One interesting note is that the first design appears to be better from a number of different aspects, since it is much smaller, consumes less power, and requires fewer gates than the other designs. However, some of the gates counted in the gate delay column are exclusive-OR gates, which have measurably longer times than NAND or NOR gates. When the author was simulating the designs to verify correctness, the times required for design 1 were considerably longer than the times required for the NAND-only implementation. Therefore, even though the number of gate delays involved appears to favor the first design, the actual times involved need to be considered for a real implementation.

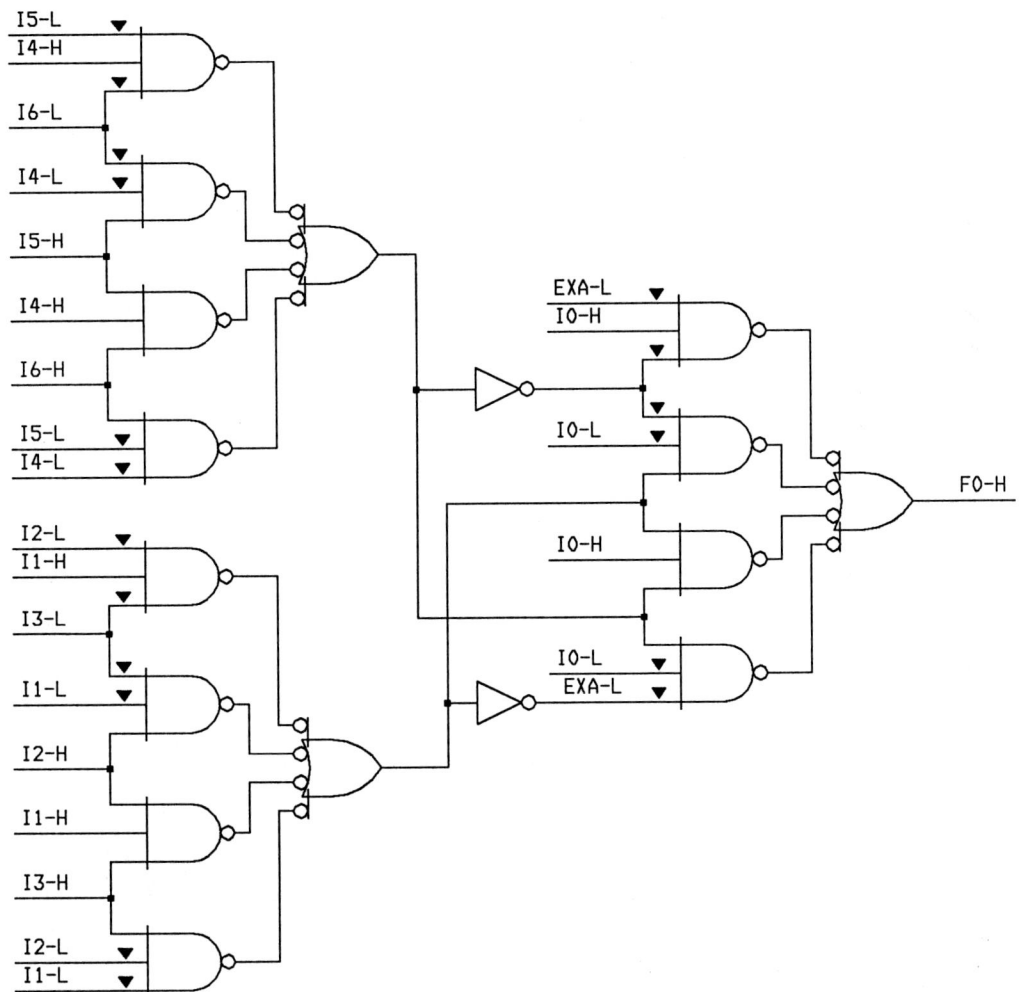

Figure 2.12(a). Logic for F0 of Row Reduction Unit.

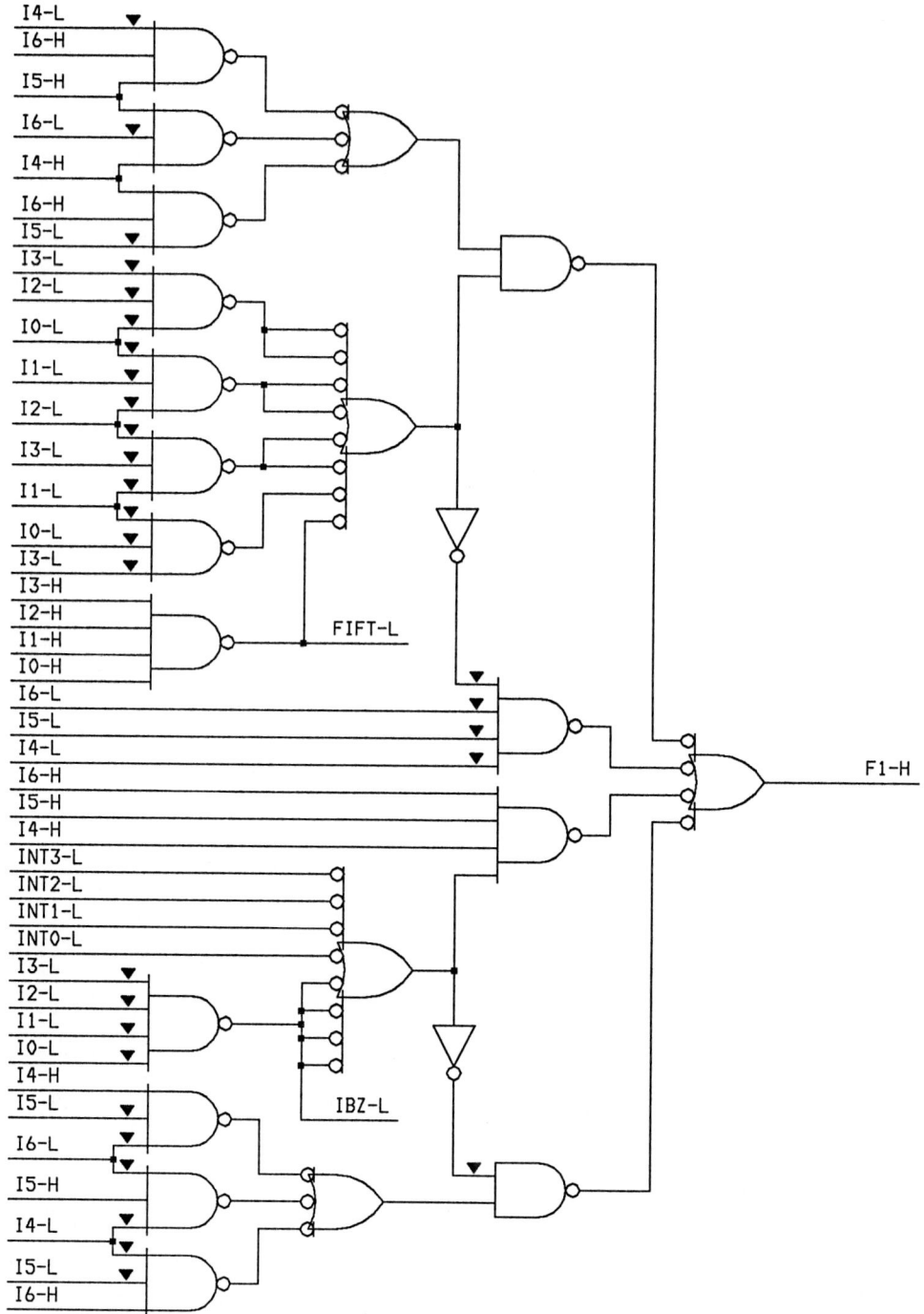

Figure 2.12(b). Logic for F1 of Row Reduction Unit.

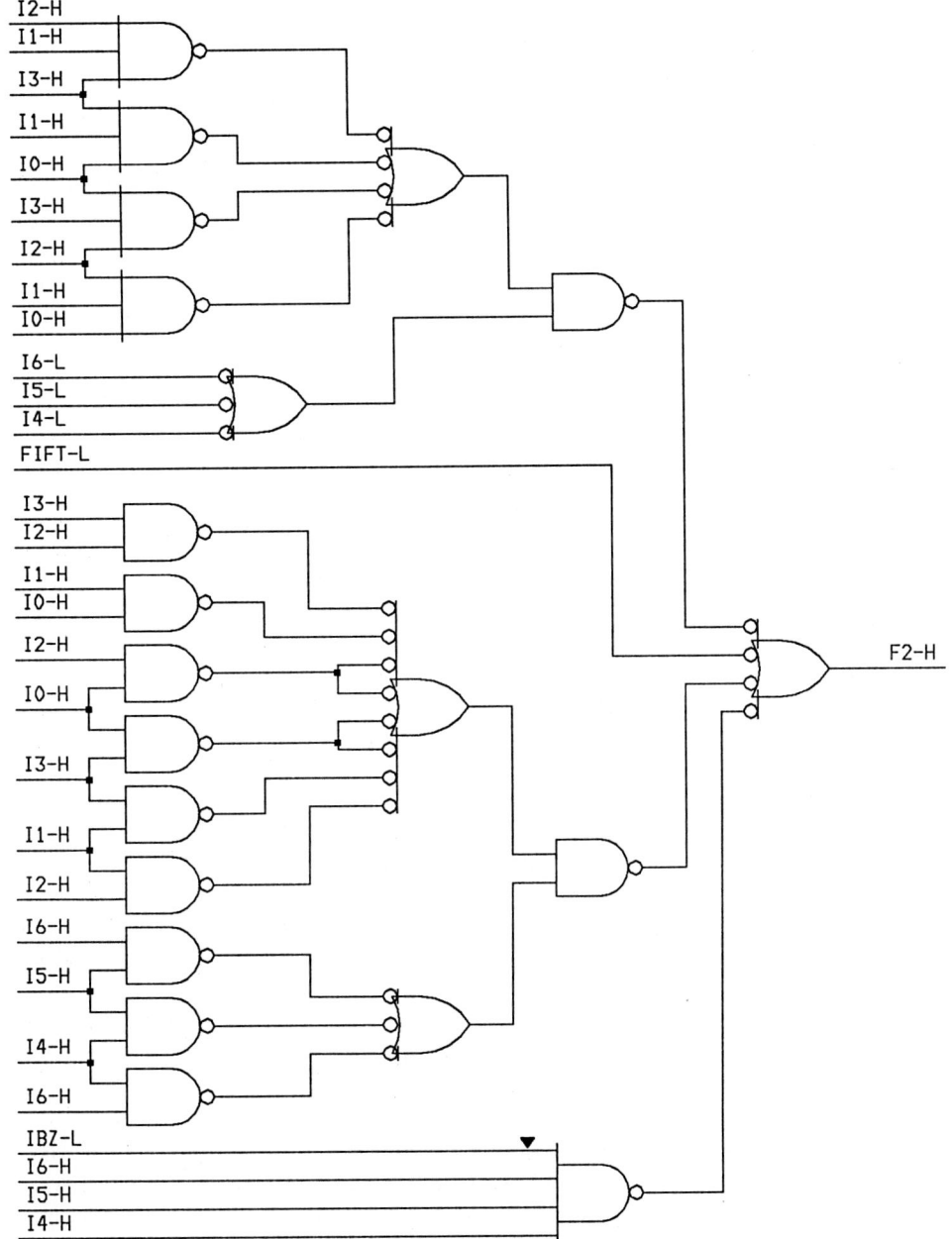

Figure 2.12(c). Logic for F2 of Row Reduction Unit.

Table 2.5. Utilization of System Resources by Row Reduction Designs.

Parameter	Design 1	Design 2	Design 3
Max gate delays			
Ix to F2	7	5	3
Ix to F1	5	6	3
Ix to F0	4	6	3
Number of gates	20	61	160
Number of packages	5	24	139
Board space (in^2)	1.6	7.68	44.96
Power (mW)	31.0	88.025	292.125

2.4. Summary

The creation of a combinational system requires attention to details, at the same time keeping in mind the application and needs of the system. This chapter examined some of the ideas involved with combinational systems. The creation of the actual logic needed follows the implementation details included in basic texts on digital logic. Using those concepts, we have created alternative designs for two different systems by following these steps:

1. Understand the requirements.
2. Create a basic block diagram.
3. Partition the problem as needed.
4. Uniquely define the outputs of each portion in terms of the inputs of that section.
5. Reduce the truth table or equations to an implementable form.
6. Create a logic system to do the work.

Systems created with these guidelines should perform the needed functions in their application systems. The term ''should'' is used instead of the more positive ''will,'' because occasionally such a system does not function as expected. When this happens, it is imperative that all the steps in the process be reviewed, especially the understanding of the function required, the inputs available, and the outputs to be generated. Only when the understanding of the designer correctly matches the application of the system will a correct result be possible.

The ingenuity of the designer will have a direct impact on the implementation methods and the final product of the process. Much can be learned by alternative designs of the same system. Much can also be learned by examining similar designs performed by others. For example, before commencing on a design project for an ALU, one can learn from the design strategies of others by analyzing the gate-level representations given in a data book for a '181 or a '381, or a similar arithmetic device. Having examined strategies used by others in the fabrication of their systems, a designer can then devise his own strategy based on the available system resources.

Each design should be evaluated with respect to its utilization of system resources. This chapter identified the number of gates, the power used by an ALS implementation, and the area of a board needed to implement the function as three possible metrics which can be used. This same method can be extended to silicon area, number of transistors, or any other metric which makes sense in a system. By evaluating designs in light of the use of system resources, a design can be selected which is not only clever and innovative, but which produces the correct result making best use of the system resources.

2.5. Exercises

2.1 Design a 5-input majority gate. That is, the system has 5 inputs and 1 output. If the majority of the inputs are high, then the output is high. If the majority of the inputs are low, then the output is low.

2.2 Design a comparator capable of less-than, greater-than, and equal determination between two 4-bit quantities.

2.3 Modify the design of exercise 2 to include cascading inputs and outputs (less-than, greater-than, equal inputs which accept the corresponding outputs of the previous stage).

2.4 Compare the design of exercise 3 with the '85, and comment on the diffference in the designs (levels of logic, number of gates, number of gate delays in producing outputs, and so forth).

2.5 Create a system which accepts two 2-bit numbers and creates an output which is the product of the 2 input numbers. Thus, the output will have 4 different lines, so that the number 9 can be represented.

2.6 Design a system which accepts a 4-bit binary number and produces a number when the value of that number is divisible by 3.

2.7 Design an adder which accepts two 4-bit numbers as input, as well as a carry input. The outputs produced by the system are the 4-bit number, a propagate line, and a generate line. Maximum time delay from any input to any output is 8 gate delays. Also, each input can present only one load to the outside world.

2.8 Design a 2-bit subtracter with look-ahead borrow. That is, the unit should accept two 2-bit numbers and a borrow as inputs, and produce a borrow propagate and a borrow generate as outputs.

2.9 Analyze the gate-level implementation of a '381, and describe the techniques used in the implementation to minimize the total number of gates used.

2.10 Using only NAND gates, design a system which generates odd parity across 8 bits of input.

2.6. Additional References

[Bree89] Breeding, K. J., *Digital Design Fundamentals.* Englewood Cliffs, NJ: Prentice Hall, 1989.

[Flet80] Fletcher, W. I., *An Engineering Approach to Digital Design.* Englewood Cliffs, NJ: Prentice Hall, 1980.

[Mano79] Mano, M. M., *Digital Logic and Computer Design.* Englewood Cliffs, NJ: Prentice Hall, 1979.

[Mano88] Mano, M. M., *Computer Engineering: Hardware Design.* Englewood Cliffs, NJ: Prentice Hall, 1988.

[McCl86] McCluskey, E. J., *Logic Design Principles, with Emphasis on Testable Semicustom Circuits.* Englewood Cliffs, NJ: Prentice Hall, 1986.

[Pros87] Prosser, F. P., and D. E. Winkel, *The Art of Digital Design: An Introduction to Top-Down Design.* Englewood Cliffs, NJ: Prentice Hall, 1987.

[Wilk87] Wilkinson, B., *Digital System Design.* Englewood Cliffs, NJ: Prentice Hall International, 1987.

3

Instruction Execution and RTL Descriptions

Doing work in digital systems involves moving information from one element in the system to another, and in some cases manipulation of the information in the process. The work can be specified in a number of ways, one of which is the use of a register transfer language, or RTL. This chapter examines the use of an RTL to identify the work to be done in a system, and through various examples looks at different techniques used to implement digital systems.

There are two basic purposes for the descriptions included in this chapter. The first is to demonstrate the manner in which the work can be accomplished. In general, stored program computers perform the tasks needed by the system in a *fetch-decode-execute* process. That is, each instruction is obtained at the appropriate time from the the system memory (the *fetch* portion). Then, the bit pattern of the instruction is interpreted to identify what work is actually called for (the *decode* portion). Finally, whatever action is called for by the instruction is performed (the *execute* portion). This action may move and manipulate data in the system, or this action may deal with the program flow in the system. In any case, when the work specified by the instruction is performed, the system must be left in a state such that the next instruction in line will be able to do its work effectively. And an RTL can be used to specify the action of the instruction.

A principal reason for the transfer specification mentioned above is to identify for the control unit of a system the action which must take place. The design of the control unit can proceed once the order of the transfers is known. Design of the control unit involves creating a functional unit to assert the control lines of the data path in a way that will cause the transfers identified by the RTL to occur. Thus, the RTL specification of work and the design of the control system are intertwined in the creation of a digital system.

The second basic purpose for utilizing RTL descriptions for work done in a system is to provide a basis for analysis. Using RTL descriptions for instruction execution allows a regular investigation into the resources needed for system

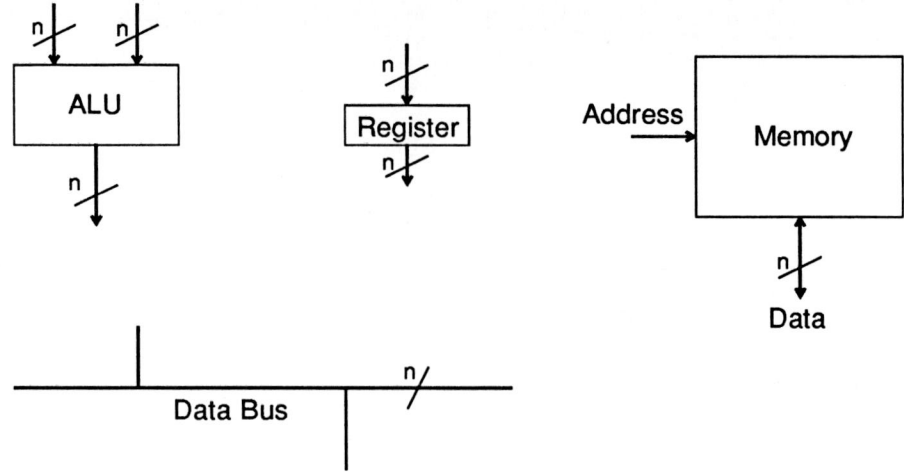

Figure 3.1. Basic Building Blocks for System.

operation. This analysis is performed by creating appropriate RTL descriptions of work and then identifying the resources required to perform the work. A common resource utilized in this process is the time to complete the work, but other system resources could also be included in the analysis, such as the system power consumption, the silicon area required, or the complexity of the control system. Once resource utilization has been determined, then engineering judgments can be made to create a system which will most effectively accomplish the system goals.

The RTL descriptions included in this chapter are relatively simple, and involve the basic elements shown in Figure 3.1. These elements represent only a few of the many building blocks used to perform the work of a system. More complex systems will contain correspondingly more complex building blocks, as well as more buses and other components. Thus, the tool (which in this case consists of the combination of block diagram and RTL descriptions) can be modified to meet the needs of the analysis and the system architecture.

3.1. Definition of the RTL

The register transfer language used use in this chapter to describe the various systems includes capabilities of register transfer specification, parallel transfers, and data manipulation methods. The basic register transfer is indicated with the use of an arrow:

$$PC \rightarrow MAR$$

The above statement indicates that the value stored in a location named PC is to be copied to a location named MAR. If two things are to happen simultaneously, then they are connected with a vertical bar:

$$\begin{matrix} SP & \rightarrow & MAR \\ PC+1 & \rightarrow & PC \end{matrix}$$

Connecting the two arrows as shown in the above statements indicates that the two transfers specified will occur simultaneously. Of course, this is possible only when the data paths allow, and we will assume that the activation of the control lines will be appropriately handled by the control system of the unit. Brackets identify a particular location to be used in a transfer:

$$M[MAR] \rightarrow MBR$$
$$MBR \rightarrow R[RAR]$$

The first of the above statements indicates that the contents of the location in memory (the single capital M is used to denote memory) identified by the address in MAR are moved to the register named MBR. The second statement specifies that the contents of MBR are moved to the register identified by what is in RAR. In a similar manner, a field of the value stored in a register, or otherwise available on a bus or whatever, will be identified by using pointed brackets. The field can be identified by name or by directly specifying the limits of the range:

$$IR<adr> \rightarrow MAR$$
$$MBR<3:0> \rightarrow STUS$$

The first of the two statements above indicates that the address portion ($<adr>$) of the instruction register (IR) is to be moved to the location named MAR. The second statement indicates that the bits in the range from 3 to 0 are to be placed in a location named STUS. This would imply that the location STUS was capable of storing only these 4 bits; otherwise, the location of the field within STUS would need to be identified.

Another provision provided with the RTL is the capability of specifying conditional action. This is accomplished by using a syntax very similar to the if statement of the C language:

```
if ( ... condition ... ) {
        work if condition is true
} else {
        work if condition is not true
}
```

This mechanism allows identification of the assorted register transfers involved with the implementation of conditional branch instructions, or the specification of work performed by a state machine.

We will also allow the specification of repetitive events with C-like constructs. The most appropriate is the while loop:

```
while ( ... condition ... ) {
        work if condition is true
}
```

This mechanism calls for execution of the work so long as the condition remains true, at which time the action will proceed to the statements which follow the while loop.

These capabilities permit specification of most of the work associated with a computer system. Extensions of the language are included when necessary to describe an action needed in a system. For example, transfer of control to a specific RTL statement can be accomplished by including a label (*label_name:*) and a directive to transfer the control (*goto label*). As mentioned above, the RTL needs to be flexible enough to meet the needs of both analysis and design, and hence should be adjusted as necessary to provide solutions to system problems.

These tools can be very useful for a study of techniques for doing work in a computer system. The RTL allows a computer architect to identify the number and type of transfers needed to do the work required by an instruction, and also to assign a time penalty to each of the transfers. Using these times, or any reasonable metric which indicates how system resources are being used, comparisons can be made between alternative structures. Let us now examine assorted structures and study the effects of adding ALUs and buses to the overall system. The approach shown here is to propose a system organization, indicated by a block diagram, and with each organization we will give RTL representations for an add instruction (ADD), an instruction for bit-by-bit inversion (INVERT), a subroutine jump instruction (JMS), and a return from subroutine (RTN). These instructions give some insight into system behavior, but a more detailed examination is needed for an in-depth understanding of the capabilities and limitations of the system.

3.2. Common Bus Implementation of a Single Address Machine

For our first example let us look at a very simple machine, one which uses an instruction set specifying a single address in the instruction word, hence the name "single address" machine. This type of a machine has a single register which is used to store the results of arithmetic operations; in most systems of this type this register is called the accumulator (ACC). The block diagram of the system to be considered in this section is shown in Figure 3.2. The memory is used to store data and programs which are to be used by the system. Information is transferred from the memory to the remainder of the system through the memory buffer register (MBR). In systems using a destructive readout memory technology, such as ferrite core memories, this register is required to save the contents of the memory as it is

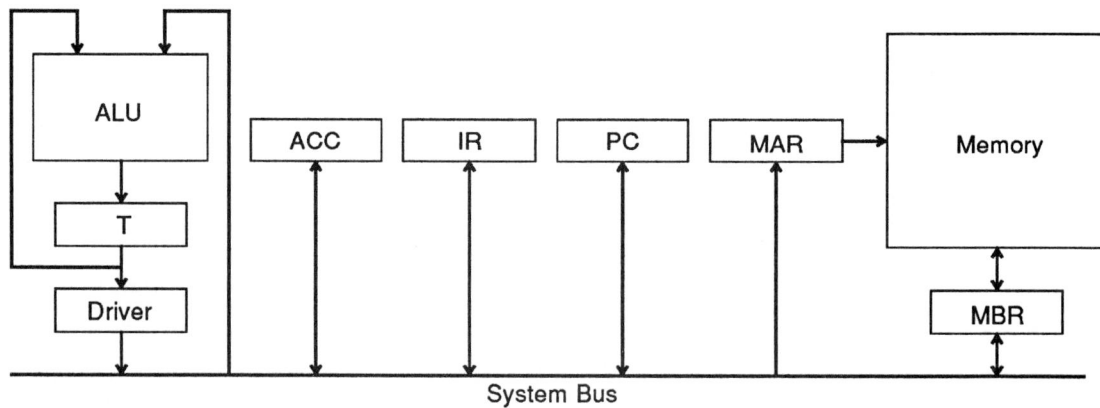

Figure 3.2. Block Diagram for a Single Address Machine with a Single Bus.

Chap. 3: Instruction Execution and RTL Descriptions

accessed, so that it can be returned to the location to maintain the value for the next access. Most semiconductor memories do not require this external register for their operation.

The address at which a memory transaction is to occur is identified by the memory address register (MAR). Transfers to the MAR, as well as transfers to and from the MBR and the other elements of the system, take place on a bus common to all the assorted system elements. Also connected to this bus is the accumulator register (ACC), used to store the results of the arithmetic and logical instructions, and the program counter (PC), which is used to specify the location of the next instruction to execute. Note that in this implementation the ACC is not directly associated with the arithmetic/logic unit. Instead, we will make the assumption that this system has only one element capable of doing any arithmetic operation, and that is the ALU. Thus, any operation requiring arithmetic interaction must be configured to use the ALU shown. Hence, the program counter is not really a counter, but merely a register. The tradeoff being investigated in this case is whether to minimize the hardware complexity of the system, which will result in more involved register transfers and control to accomplish the work.

Before we can accurately create RTL statements describing the actual transfers, we need to know some basic information about the elements of the system, as well as the expected mode of operation. For example, the system of Figure 3.2 shows an ALU, and to make use of the ALU we need to know what its capabilities are. For this problem we will assume a very simple ALU, a unit capable of four functions: pass the bus to the temporary register, add the bus to the temporary, NAND the bus and the temporary, and increment the temporary register.

In addition to the capabilities of the system components, we also need to know the expected behavior of the system itself — in this case, the expected mechanism for subroutine linkage. For this example we will assume a linkage mechanism like the PDP8 by Digital Equipment Corporation, and other small systems of the time. That is, a subroutine will leave an empty location as the first location of the subroutine, and the JMS instruction will place in that location the address to which control should be passed at the completion of the routine. This assumes that the return address can be uniquely represented in the location provided. In general, this is not a good linkage mechanism, since it prevents the use of recursive subroutines or shared code. The other system assumption we will make is that all operations utilize one location of memory, both for instruction storage and data storage.

The various portions of Figure 3.3 contain RTL representations of the instructions mentioned above. Figure 3.3(a) lists the RTL transactions involved in

	fetch:	These register transfers are common to all instructions.	
1		PC $\rightarrow$ MAR	Get program counter to MAR
		PC $\rightarrow$ T	and to T at the same time.
2		M[MAR] $\rightarrow$ MBR	Instruction moves to MBR.
		T + 1 $\rightarrow$ T	Increment program counter.
3		T $\rightarrow$ PC	Updated program counter value returns to PC.
4		MBR $\rightarrow$ IR	Instruction moves to IR.
	decode	System decodes instruction	
	execute:	and the execute portion performs the work.	

Figure 3.3(a). RTL for Single Address Machine: Fetch Portion of Instruction.

instruction fetch, which is common to all the instructions. The first transfer moves the contents of the program counter to the memory address register and to the temporary register. This is possible in a data sense, inasmuch as the bus containing the value is accessible by both destinations. In order for this to be a real possibility, we will assume that the control system can assert the control lines associated with both registers simultaneously to receive the value. The second RTL statement also involves two simultaneous transfers. The first is to place into the memory buffer register the contents of the memory location identified by the address in the MAR. The second transfer results from incrementing the contents of the temporary register. The transfer identified in RTL statement number 3 places the updated value for the program counter back in the PC. Finally, RTL statement number 4 identifies the transfer of the instruction to the IR. After these transfers, the system must decode the instruction, then begin execution. The RTL statements involved in the execution of the instructions are included in the other portions of Figure 3.3.

The time required to complete the instruction fetch will be the sum of the times required for the individual parts of Figure 3.3(a). In order to assign reasonable times to the transfers, one must know something about the actual implementation. For this system we will assume that register transfers can be accomplished in 40 nsec, that arithmetic operations require 80 nsec, and that memory transfers require 200 nsec. Using these times, we can ascertain the execution time of the instructions.

The transfers of RTL statement number 1 of Figure 3.3(a) take different amounts of time. The first is strictly a register transfer, and incurs a penalty of 40 nsec. The second transfer utilizes a path through the ALU, and hence requires an additional 100 nsec. Certainly a controller can be constructed that will allow these transfers to occur as soon as possible. However, we will assume that the time required is the longest of the times. This results in a controller with a reduced complexity, but longer times. Thus, RTL statement number 1 requires 140 nsec. Statement number 2 requires a memory access, which is 200 nsec. The increment of the temporary register requires 140 nsec, which is totally hidden by the 200 nsec memory access time. Statement number 3 is a register transfer, and can be accomplished in 40 nsec. And statement number 4 is another register transfer, with a time requirement of 40 nsec. Thus, the fetch portion shown in Figure 3.3(a) incurs a time penalty of 420 nsec. To this will be added the times for the other transfers required for completion of the individual instructions.

With a single address machine, dyadic instructions find one of the operands in the accumulator, and the other operand in the location identified by the single address included in the instruction. We are assuming that the complete address is found in the instruction, and that no further indexing or other address modification is needed. With that assumption, the RTL statements for implementation of an ADD

	fetch:	
	decode	
5	execute: IR<adr> → MAR	Move address to MAR.
6	M[MAR] → MBR	Data moves to MBR.
	ACC → T	At same time, ACC goes to T.
7	MBR → T	Add in value from MBR.
8	T → ACC	Sum goes to ACC.

Figure 3.3(b). RTL for Single Address Machine: ADD Instruction.

instruction are included in Figure 3.3(b). The first action of the execution portion of transfers is to place the address of the desired location into the MAR. This is accomplished with transfer number 5. Two actions are specified by transfer number 6. The first is to place the desired data, the address of which is located in the MAR, into the MBR. The second action is to place the contents of the accumulator into the temporary register. These actions can take place at the same time since they do not use the same set of resources. The two quantities which are to be added by the instruction are now available in the temporary register and the MBR. These values are summed in transfer number 7 of Figure 3.3(b), and the result is placed in the temporary register. Finally, the value is returned to the ACC in transfer number 8.

The time required by the ADD instruction is the 420 nsec of the fetch portion, plus whatever is required by the transfers of Figure 3.3(b). Statement number 5 is a register transfer, which takes 40 nsec. The two transfers of statement number 6 can be accomplished in 200 nsec. The addition of statement number 7 requires a register transfer time plus the ALU time of 100 nsec, for a total of 140 nsec. Finally, the sum is returned to the ACC in the transfer of statement number 8, which takes 40 nsec. Thus, the ADD instruction requires a total of 840 nsec.

The RTL transfers for the execution portion of the INVERT instruction are shown in Figure 3.3(c). The bits to be inverted are contained in the ACC, so transfer number 5 moves the contents of the ACC to the temporary register. In transfer number 6 the contents of the ACC are again found on the bus, this time to be combined with the bits of the temporary register in a NAND operation which results in a logical inversion of every bit in T, the results being returned to T. Finally, transfer number 7 moves the bits back to the ACC.

The timing of the INVERT instruction can be obtained by examination of the transfers of Figure 3.3(c). Statements 5 and 6 both require a register transfer and an ACC operation, for a total of 140 nsec. The transfer of statement 7 can be accomplished in 40 nsec. Thus, the total time of the instruction is the 420 nsec of the fetch, plus 320 nsec in the execution phase, for an execution time of 740 nsec.

The transfers involved in the implementation of the JMS instruction are shown in Figure 3.3(d). During the fetch portion, the PC was adjusted to point to the next instruction in the instruction stream. As mentioned above, the action of the JMS instruction should place this value at the first location of the subroutine, then start execution at the second location of the subroutine. This process is initiated by transfer number 5, which moves the address of the subroutine to the memory address register. In addition, transfer number 5 moves this value to T, where it can be incremented to identify the second location of the routine, where execution is to begin. There are two transfers identified by statement number 6. The first moves the current contents of the PC to the MBR, so that it can be written to the memory. The second transfer of statement number 6 increments the contents of the temporary register, which results in the address of the first instruction of the

	fetch:		
	decode		
5	*execute:*	ACC → T	Copy ACC to T.
6		ACC NAND T → T	This will invert on a bit by bit basis.
7		T → ACC	Move back to ACC.

Figure 3.3(c). RTL for Single Address Machine: INVERT Instruction.

```
        fetch:
        decode
5       execute:  IR<adr>  ⊤→  MAR         Move address to MAR.
                  IR<adr>  ⌐→  T           Also to T.
6                     PC  ⊤→  MBR          This is the return address.
                    T+1  ⌐→  T            Increment to point at instruction of subroutine.
7                   MBR  ⊤→  M[MAR]        Return address to first location of subroutine.
                      T  ⌐→  PC            Instruction address to PC.
```

Figure 3.3(d). RTL for Single Address Machine: JMS Instruction.

subroutine. The final RTL statement of Figure 3.3(d) also identifies two transfers. The first transfer involves only the memory, and places the return address into the first location of the subroutine, the address of which had previously been moved to MAR. The second transfer moves the address of the first instruction of the subroutine to the PC, leaving the system ready for the initiation of a new instruction.

The times for the JMS instruction are obtained from the transfers of Figure 3.3(d). The transfers of statement number 5 require 140 nsec. The transfers of statement number 6 also require 140 nsec. Finally, the transfers of statement number 7 can be accomplished in 200 nsec. The total is then the 420 nsec required by the fetch portion of the instruction execution, plus the 480 nsec for the register transfers of Figure 3.3(d), for a total of 900 nsec.

The return from subroutine instruction must adjust the program counter to contain the next instruction to execute in the calling routine. This address is stored in the first location of the subroutine, as indicated by the RTL statements of Figure 3.3(d). Thus, the return instruction must fetch this return address from the initial location of the subroutine, and transfer it to the program counter. One set of RTL statements to accomplish this is shown in Figure 3.3(e). The address of the initial location of the subroutine is obtained from the IR by the transfer specified by RTL statement number 5. This address is placed in the MAR. The contents of the initial location of the subroutine are retrieved by the transfer of statement number 6, which places this value in the MBR. Thus, when this transfer has been completed, the MBR contains the address of the next instruction to execute. Finally, the transfer of RTL statement number 7 moves this address to the program counter, and the system is ready to continue execution of the calling program.

The timing for the RTN instruction is determined in the same fashion as the timings for the other instructions of Figure 3.3. The transfer of statement number 5 requires 40 nsec. The memory access of statement number 6 takes another 200 nsec. And the transfer of statement number 7 is completed in 40 nsec. Thus the

```
        fetch:
        decode
5       execute:  IR<adr>  →  MAR         Move address to MAR.
6                 M[MAR]   →  MBR         This gets address to MBR.
7                    MBR   →  PC          This action returns to calling routine.
```

Figure 3.3(e). RTL for Single Address Machine: RTN Instruction.

RTN instruction is completed in 700 nsec, 420 nsec for the fetch portion and 280 nsec for the execution portion.

This section has examined a single bus implementation of a single address machine and identified the register transfers needed to perform the work required by some of the instructions. This can be used as a basis for further analysis of the instruction implementation techniques, instruction timings, and machine complexities. That is, by continuing the analysis to include other instructions, we can determine the average instruction execution time for a particular instruction mix. We can also determine the effect that different system components will have on overall instruction time. For example, if the ALU is replaced with a unit which is twice as fast, how much would the execution time decrease? Or, if only single transfers are permitted, not the parallel transfers exemplified by RTL statement number 1 of Figure 3.3(a), how much longer will the execution times be? These questions, as well as many, many others, can be investigated by using the RTL statements and their associated block diagrams. The next section uses the same techniques to look at a two address machine.

3.3. A Two Address Machine Implemented with a Single Bus

Although many single address processors have been created, it is far more common to use a two address scheme. The two address format identifies two operands (for dyadic operations), and, using those operands, performs the specified function. The result then takes the place of one of the two operands. To prevent the length of the addresses from using too many bits, this technique is most often used with a register set. Thus, addresses contained in the instruction refer to locations in a set of registers which forms a part of the basic machine. A block diagram of the system we will use for examining some of the two address techniques is shown in Figure 3.4.

The memory of Figure 3.4 is used for the same purposes as the memory of the previous section, storage of program and data. However, this system does not

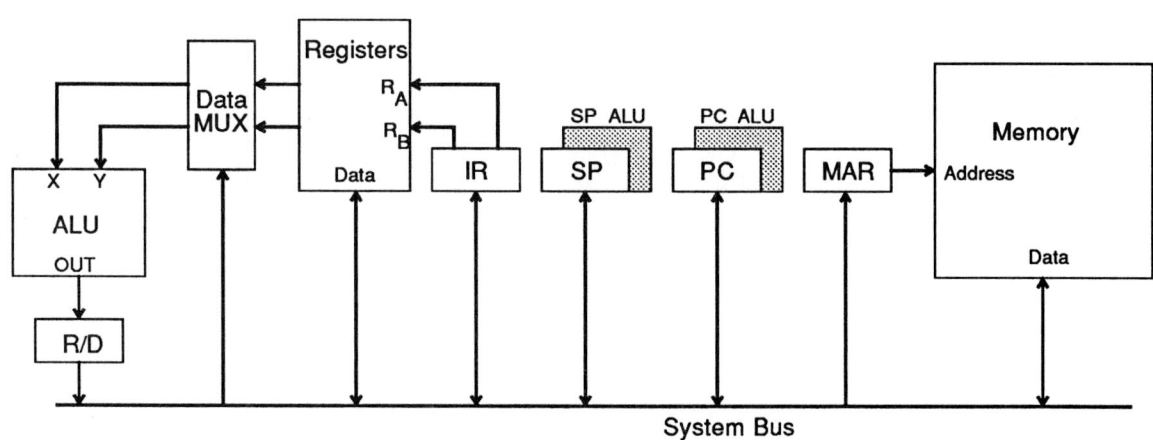

Figure 3.4. Block Diagram for a Two Address Machine with a Single Bus.

contain a memory buffer register. The information is transferred directly to and from the addressed location in the memory over the bus. Many memories contain an electrical buffer to minimize the loading effect of the memory, and some systems also incorporate a storage element as part of the memory system itself. But for our purposes here, we will assume that the transfers take place without a register. The address at which the memory transaction occurs is contained in the memory address register, MAR.

The program counter (PC) of Figure 3.4 provides the same function as the program counter of the previous section. The contents of the PC identify the next location to be accessed in the instruction stream. However, as part of the program counter of the system, we will incorporate an ALU which can be used as needed. The principal use of this ALU is to increment the program counter as needed. For the single address machine we made the simplifying assumption that the program counter needed to be incremented by one to point to the next value to extract from the instruction stream. Here, we will allow the program counter to increment by whatever amount is needed, depending on the instruction itself. We will denote the length of the instruction as Ilen.

A new element is available to the system, as shown in Figure 3.4. This is the stack pointer register, or SP. This register is used to create a stack in the memory. The contents of SP point to a location in the memory which can be used to store information. When information is pushed onto the stack, the required action is to first decrement the SP, then transfer the data value to the location identified by SP. When information is popped off of the stack, the value which is stored in memory at the location identified by SP is extracted from memory, and then the contents of the SP are incremented to point to the next available value. The increment and the decrement operations are performed by the ALU associated with the SP in Figure 3.4.

The instruction register (IR) shown in Figure 3.4 performs the same task as the instruction register of Figure 3.2; it stores the instruction to be executed. Since the instruction register will be used to hold instructions which are more complicated than the previous system, let us state the assumptions that we are making for the system. For our analysis we will limit ourselves to four kinds of instructions. The first is dyadic instructions, which have the general format OPER A, B. This instruction identifies an operation that requires two operands. The operation will be specified by the OPER field, and the two operands are found by using the registers A and B of the register set. The operands could be found in the registers themselves, or the registers can be considered pointers to the actual operands. The interpretation of the fields is associated with the bits which we are grouping with OPER.

The second type of instruction is that which requires only one operand, such as a clear or an increment. These instructions will have the general format OPER B. Again, the OPER bits identify the function, and the operand is found by using register B of the register set. The third type of instruction which we will utilize is a control instruction with an address, exemplified by the JMS instruction. We will assume that this requires two elements, the bits to identify which control instruction it is, and then the address itself. The length of the field for the identification of the function we will denote as Ilenid, and the length of field for the address we will denote Ilenad. Finally, there is the control instruction which requires no address; this is exemplified by the RTN instruction.

The IR will hold the instruction for the purpose of decoding it, and it will present the A and B fields as needed to the register set to specify the appropriate registers. These registers can be used to hold addresses or data. We will further assume that information can be extracted from one or two registers in a single operation, or information can be placed into a register in an operation, but that these two operations cannot proceed concurrently. If the registers are to be written into in the same cycle in which they are accessed, then additional parts are needed in the system. It will be very useful for us to include with the register set a register for the temporary storage of operands. We will assume that such a location does exist, and that it is referred to as Rtemp.

The data multiplexer (Data MUX) in Figure 3.4 is used to select appropriate operands to send to the ALU. We will assume that this unit is appropriately directed by the system controller, and that it can provide the needed values to the X and Y inputs of the ALU. Thus, it could choose two registers, a single register, zero, or the system bus as needed to perform the operation of the system.

The final functional unit shown in Figure 3.4 is the ALU. We will assume that this unit is capable of performing an addition of the two inputs, incrementing one of the inputs, passing the Y input, and NANDing the two inputs together. Associated with the ALU is a combination register and driver labeled R/D. This unit is used to store the results from the ALU until they can be transferred to their final destination.

We will now look at RTL implementations of the four simple instructions of the previous section: ADD, INVERT, JMS, and RTN. The additional assumption needed to proceed is to specify the action associated with calling and returning from subroutines. In this example, we will use the stack to pass the address. That is, the subroutine call will place the address of the next instruction to execute in the calling routine onto the stack, then transfer control to the subroutine. The return from subrouting is caused by popping the address from the stack and placing it in the PC.

The RTL implementation for the fetch portion of the instruction is shown in Figure 3.5(a). Note that in this implementation there are only two transfers. The first loads the PC with the address of the instruction. The second transfers the instruction directly to the instruction register. The program counter is not updated, since the length of the instruction is not known until the instruction has been identified. Another difference between the RTL statements of Figure 3.5(a) and those of Figure 3.3(a) is the lack of a memory buffer register in the two address system. In order to compare the two systems on times alone, and not on which has the higher speed technology, we will use the same times for the fundamental operations: register transfers require 40 nsec, arithmetic operations require 100 nsec, and memory transfers require 200 nsec. With these assumptions, the fetch portion of an instruction requires 240 nsec to complete. This is a little more than half the time required for the fetch of the previous system. Why?

The RTL statements for two different implementations of the ADD instruction are included in Figure 3.5(b). The first is for a register-to-register add, where the operands are located in the register memory. For this block diagram, this is a very simple operation. RTL statement number 3 for the direct addressing mechanism identifies two separate operations. The first performs the addition and leaves the results in R/D. The second operation increments the program counter to identify the

	fetch:	This is the fetch portion of an instruction.
1	PC $\rightarrow$ MAR	Move PC contents to MAR.
2	M[MAR] $\rightarrow$ IR	Move instruction to IR.

Figure 3.5(a). RTL for Two Address Machine: Fetch Portion of Instruction.

Direct addressing system (operands in registers)

fetch:
decode

3	*execute:*	RA + RB	→ R/D	Do the add and leave results in R/D.
		PC + Ilen	→ PC	Increment the program counter.
4		R/D	→ RB	Return results to RB.

Indirect addressing system (operands in memory)

fetch:
decode

3	*execute:*	RA	→ MAR	Contents of RA to MAR.
4		M[MAR]	→ Rtemp	Transfer data to Rtemp.
		PC + Ilen	→ PC	Increment the program counter.
5		RB	→ MAR	Contents of RB to MAR.
6		M[MAR] + Rtemp	→ R/D	Do the add and leave sum in R/D.
7		R/D	→ M[MAR]	move result to memory.

Figure 3.5(b). RTL for Two Address Machine: ADD Instruction.

next instruction to execute. RTL statement number 4 moves the result of the addition to the register set. The total time required for these two statements is 180 nsec, 140 nsec for the addition step, and 40 nsec for the transfer. Thus, a register-to-register addition can be performed in 420 nsec. Note that more than half of this time is required by the instruction fetch portion of the instruction.

The second set of RTL statements included in Figure 3.5(b) implements an addition with indirect addressing. That is, the registers identified in the instruction contain not the operands, but the addresses of the operands. The first task is to transfer the address of the first operand to MAR, which is accomplished by RTL statement number 3. RTL statement number 4 accomplishes two things. The first is to transfer the operand to the temporary register. The second is to increment the program counter. Since incrementing the program counter requires 140 nsec, and accessing memory requires 200 nsec, both operations can be completed within 200 nsec. The address of the second operand is transferred to MAR in RTL statement number 5, and the addition itself performed as shown in statement number 6. Note that the times must be added for the work specified by statement number 6, totaling 340 nsec. Finally, statement number 7 transfers the result to the destination, which is still pointed to by the contents of MAR.

The time required for the indirect version of the add instruction will be much longer than the direct version, since three additional memory accesses are needed. Statements 3 and 5 require 40 nsec each, and statements 4 and 7 require 200 nsec each. The most time is consumed by the work of statement number 6, which takes 340 nsec. Thus, the work portion of the indirect add of Figure 3.5(b) consumes 820 nsec, and the entire instruction is completed in 1060 nsec. Thus, the operand placement will have a direct impact on the speed of execution of the instruction.

Figure 3.5(c) contains the RTL statements needed for direct and indirect implementations of the INVERT instruction. The direct version is almost identical to the implementation of the ADD instruction. RTL statement number 3 does the inversion by NANDing the value with itself. This assumes that the data multiplexer of Figure 3.4 can supply the proper value to both inputs of the ALU. At the same

INVERT with direct addressing.

fetch:

decode

3	*execute:*	RB NAND RB	→ R/D	Do the inversion and leave results in R/D.
		PC + Ilen	→ PC	Increment the program counter.
4		R/D	→ RB	Return results to RB.

INVERT with indirect addressing.

fetch:

decode

3	*execute:*	RB	→ MAR	Contents of RB to MAR.
4		M[MAR] NAND M[MAR]	→ R/D	Transfer results to R/D.
		PC + Ilen	→ PC	Increment the program counter.
5		R/D	→ M[MAR]	Move result to memory.

Figure 3.5(c). RTL for Two Address Machine: INVERT Instruction.

time, the program counter is incremented. Statement number 4 then places the results back in the appropriate register. The total time for this instruction is identical to the direct addressing ADD, which is 420 nsec.

The second set of RTL statements in Figure 3.5(c) is for the indirect version of the INVERT instruction. Statement number 3 transfers the address to the MAR. Statement number 4 does the work, both the work of inversion and the work of incrementing the PC. The work of inversion calls for transferring the operand out of memory onto the bus, then through the data multiplexer and to the ALU. This assumes that the data multiplexer can direct the bus value to both inputs of the ALU at the same time. The result is left in R/D. The result is transferred by statement 5 back to memory, at the same location where the operand was found.

This second set of transfers takes more time than the first, mainly because of the memory interactions. Statement number 3 requires 40 nsec. Statement number 4 requires a memory interaction time, an ALU time, and a register time, for a total of 340 nsec. The incrementing of the PC occurs concurrently with the other action, so the 140 nsec required for the PC interaction is hidden. Finally, statement number 5 incurs a time penalty of 200 nsec. These times total to 580 nsec, which results in a total execution time for the INVERT instruction.

fetch:

decode

3	*execute:*	PC + Ilenid	→ PC	Increment the program counter.
4		PC	→ MAR	Move to MAR; points to address of subroutine.
5		M[MAR]	→ Rtemp	Get address of subroutine to Rtemp
		PC + Ilenad	→ PC	Adjust PC for instruction of calling program.
		SP - 1	→ SP	Adjust SP to receive return address.
6		SP	→ MAR	Transfer stack address to MAR.
7		PC	→ M[MAR]	Transfer return address to stack in memory.
8		Rtemp	→ PC	Move address of subroutine to PC.

Figure 3.5(d). RTL for Two Address Machine: JMS Instruction.

The RTL for the JMS instruction is shown in Figure 3.5(d). As explained above, we are assuming that the subroutine linkage will use the stack in memory. Once the fetch and decode portions of the instruction are complete, the program counter will be incremented to point to the address portion of the instruction. That is, we are assuming that the address is part of the instruction, and that once the op code has been decoded, the address must be obtained from the instruction stream. Thus, statement number 3 of Figure 3.5(d) is used to adjust the PC to point at the address, and this adjusted value is moved to the MAR in statement number 4.

Statement number 5 of Figure 3.5(d) identifies three separate actions which should take place. The first is obtaining the address of the subroutine and placing it in the temporary register. This is the only one of the three actions which utilizes the system bus. The second action is to bump the PC to point at the next instruction. This is where the system should return after the execution of the subroutine. The third action is to decrement the stack pointer to identify the location at which the stack operation should take place. Statement number 6 moves the SP to the memory address register, where it will be used to identify the stack location in memory. The transfer of statement number 7 moves the PC to memory, placing the return address onto the stack. Finally, the transfer of statement number 8 moves the address of the subroutine from the temporary register to the program counter. This leaves the system ready to fetch instructions from the subroutine.

The timing of the JMS instruction implementation of Figure 3.5(d) is obtained from the RTL statements. Statements 4, 6, and 8 are pure register transfers, each of which takes 40 nsec. Statement 3 requires an increment with the transfer, which takes 140 nsec. Statements 5 and 7 each require 200 nsec for the memory interaction. The adjustments of the PC and the SP of statement number 5 require less than 200 nsec, and hence are hidden by the memory access time. The total time of the execution portion is thus 660 nsec. To this must be added the time for the fetch portion of the instruction, which is 240 nsec. The total time is therefore 900 nsec.

The return from subroutine instruction is very simple in this machine, as indicated by the RTL statements of Figure 3.5(e). The assumption to be made at this point is that the return address is on the top of the stack. Thus, the stack pointer currently points at the value to be placed in the PC. The transfer of statement 3 moves the stack pointer to the memory address register to initiate the needed action. There are two transfers identified in statement number 4. The first gets the top of stack value to the PC, and the second increments the SP to point to the next value on the stack.

The return from subroutine instruction of Figure 3.5(e) is quite rapid in its execution. Statement number 3 requires only 40 nsec, and statement number 4 can be accomplished in the memory transfer time of 200 nsec, so the execute portion requires only 240 nsec. Thus, the return instruction requires only 480 nsec to complete.

fetch:

decode

3	*execute:*	SP → MAR	Move SP to memory address register.
4		M[MAR] → PC	Get address from stack to PC.
		SP + 1 → SP	Update stack pointer

Figure 3.5(e). RTL for Two Address Machine: RTN Instruction.

The timings presented in this and the previous section are easily derived with the assumptions which have been made. A more complete set of times can be obtained by examining the RTL implementations for more instructions. This information can be used for a number of different comparisons. For example, comparisons between the single address machine and the two address machine can be made based on a code segment, such as evaluation of a set of Fortran arithmetic statements, or some other task. Another example involves the complexity of the hardware of the system. Each of the memory accesses required by the various parts of Figure 3.5 called for a transfer of the address to the MAR. How much time could be saved in the implementation of the instructions if the address of the memory was obtained from a multiplexer, and the inputs to the multiplexer were the MAR, the PC, and the SP? Would this time decrease justify the added complexity and board space/silicon area required to implement this technique? Although the answers must be based on the system goals and the judgment of the system architect, the timings and other information available from block diagrams and RTL statements provide valuable information on which to make reasonable engineering judgments.

3.4. Extended Addressing Modes

Different approaches to the creation of a computer architecture carry with them different costs as far as the location of operands is concerned. Historically, newer architectures contained more elaborate addressing mechanisms, allowing the operands to be found in ever greater possible locations. This trend has been reversed in recent years by the introduction of RISC machines, or reduced instruction set computers. However, complex instruction set computers (CISC machines) are still used in a number of applications. We will now examine, with the use of RTL statements, some of the many mechanisms used for specification of operand location.

The machine we will use for examination of the addressing modes is the same one used for the previous section, the block diagram of which is shown in Figure 3.4. We will continue to use the two address format, and will demonstrate various addressing techniques with the ADD instruction. We will, however, change some of the assumptions concerning the architecture of Figure 3.4. We will assume that the basic ADD instruction is 2 bytes long, which will allow specification of the addressing modes for the two operands. If registers are involved, they are identified within the 2 bytes of the instruction. Additional addresses or constants required by the instruction are found in the locations immediately following the instruction. The addresses utilized by the system are byte addresses. We will assume that the register and data paths of the system are designed for 32-bit operations, and that the addition being performed is a 32-bit addition.

In the preceding section, we used operands located in registers or operands located in memory. The two methods for location specification were register direct (the operand is in the identified register) and register indirect (the operand is in memory at the address contained in the identified register). Many different options are possible, but we will examine only a few. For each ADD instruction, the fetch portion of the implementation is shown in the transfers of Figure 3.5(a). Thus, execution portion must appropriately adjust the PC, both to identify quantities used in the current instruction and to leave the PC pointing at the next instruction.

One mechanism for operand access can be called register indirect differed. That is, stored in the contents of the register identified in the instruction is an address, and at that location in memory is stored another address; the second address identifies the location of the actual operand. Figure 3.6(a) gives the RTL implementation of an ADD instruction which specifies the destination with register indirect differed addressing.

As can be seen from the transfers called out in Figure 3.5(b), this process just adds one more level of indirection to the register indirect operation. The transfer of statement number 3 copies the address from RB to MAR. Statement number 4 identifies two transfers. The first copies an address (assumed to be 32 bits) to MAR for the next indirect step. The second transfer is the increment of PC; the increment value is 2 since that is the length of the instruction. The transfer of statement number 5 obtains another address from memory, and places that address directly into the MAR. The work of the instruction is contained in the transfer and addition specified by statement number 6. The value is obtained directly from memory and sent to the ALU, along with the contents of the MAR. The sum is left in R/D. The final transfer is shown in statement number 7, which indicates that the result is moved back to memory. Note that the contents of MAR still contain the appropriate address in memory.

The time required for this instruction starts with the 240 nsec needed by the fetch portion, and then adds the appropriate amounts from the RTL statements of Figure 3.6(a). The transfer of statement number 3 is the only one requiring the 40 nsec register penalty. Statements 4, 5, and 7 incur the memory penalty of 200 nsec, and the work of statement 6 requires 340 nsec. Thus the total for this instruction is 1.22 μsec.

Another mechanism which can be used for identification of operand location is direct addressing. That is, the address of the operand is contained in the instruction itself, and the system refers to the information by extracting the address from the instruction stream and using it appropriately. Figure 3.6(b) includes the RTL statements for implementation of an ADD instruction where both operands are identified by addresses carried in the instruction stream.

Since both operands are located in memory, the first task which must be performed is to obtain those addresses. Thus, the execute portion of the instruction begins with an appropriate increment of the PC. Statement number 3 indicates that the PC is incremented by two, at which time it will point at the address which needs to be fetched. Statement number 4 specifies two transfers, the first of which moves the address of the first operand to MAR, and the second of which adjusts PC to point to the next address needed. The memory fetch indicated in statement number 6 obtains the first operand, and this information is moved to the temporary register. The location of the address of the second operand is moved to the MAR by the transfer of statement number 7. Two transfers are shown in statement number 8; the first moves the address of the second operand to the MAR, and the second

	fetch:		
	decode		
3	*execute:*	RB → MAR	Contents of RB to MAR.
4		M[MAR] → MAR	Transfer first address to MAR.
		PC + 2 → PC	Increment the program counter.
5		M[MAR] → MAR	Transfer second address to MAR.
6		M[MAR] + RA → R/D	Do the add and leave sum in R/D.
7		R/D → M[MAR]	Move result to memory.

Figure 3.6(a). RTL for ADD with Register Direct, Register Indirect Differred Addressing.

```
      fetch:
      decode
   3  execute:        PC + 2  → PC          Increment PC to identify first address.
   4                     PC  → MAR          Move PC contents to MAR.
   5              M [ MAR ]  ⇁ MAR          Transfer address of first operand to MAR.
                    PC + 4  ⅃→ PC           Increment PC to identify second address.
   6              M [ MAR ]  → Rtemp        Transfer first operand to Rtemp.
   7                     PC  → MAR          Move PC contents to MAR.
   8              M [ MAR ]  ⇁ MAR          Transfer address of second operand to MAR.
                    PC + 4  ⅃→ PC           Increment PC to identify next instruction.
   9   M [ MAR ] + Rtemp  → R/D            Do the add and leave sum in R/D.
  10                    R/D  → M [ MAR ]   Move result to memory.
```

Figure 3.6(b). RTL for ADD with both Operands in Memory,
Addresses of Operands in Instruction Stream.

transfer bumps the PC to point to the next instruction to execute. Statement number 9 does the work, fetching the second operand from memory and adding it to the first operand (located in the temporary register), then placing the results in the R/D register. Finally, the transfer indicated in statement number 10 moves the result to memory at the same location where the second operand was located.

The fetch portion of the instruction requires 240 nsec, to which must be added the times for the other statements in Figure 3.6(b). The increment of the PC in statement 3 requires 140 nsec. Statements 4 and 7 require 40 nsec each. The memory transactions of statements 4, 7, 8, and 10 each incur a penalty of 200 nsec. And the work statement (number 9) needs 340 nsec to complete. Thus, the total time is 1.6 μsec for this instruction.

The next operand specification mechanism to be considered can be called instruction stream addressing, or immediate operand specification. That is, the operand is considered to be a constant, and it is stored in the instruction stream, so it is ''immediately'' available for use. The instruction must extract the data from the instruction stream and use it in the calculation of the desired result. Obviously, this type of operand specification cannot be used for the destination of the result. Therefore, the instruction shown in Figure 3.6(c) adds the immediate value to the contents of RB, leaving the result in RB.

The work of the instruction is initiated by the increment of statement number 3, which adjusts the PC to point at the immediate operand. This address is then moved to MAR by the transfer of statement number 4. Statement number 5 specifies two tasks. The first is the actual work of the instruction, obtaining the

```
      fetch:
      decode
   3  execute:         PC + 2  → PC      Increment PC to identify first operand.
   4                      PC  → MAR      Move PC contents to MAR.
   5         M [ MAR ] + RB  ⇁ R/D       Do the add and leave sum in R/D.
                     PC + 4  ⅃→ PC       Increment PC to identify next instruction.
   6                     R/D  → RB       Move result to RB.
```

Figure 3.6(c). RTL for ADD with First Operand in the
Instruction Stream, Second Operand in RB.

immediate operand from memory and adding it to the contents of RB. The result of this addition is left in R/D. The second operation bumps the program counter to point to the next instruction to be executed. The result of the addition is returned to RB in the transfer specified by statement number 6.

Since not as many transfers are required by this instruction, its execution time will be smaller than the previous instructions of this section. Statement number 3 requires 140 nsec, while the transfer of statement 4 is accomplished in 40 nsec. The memory fetch and addition of statement 5 incur a penalty of 340 nsec. The result is returned to RB (statement 6) in 40 nsec. Thus, the total time (including the fetch portion) is 800 nsec.

A mechanism sometimes used to identify a location in memory is called PC relative. That is, instead of using an absolute address to point to a spot in memory, the location is addressed as an offset from the current contents of the program counter. In that way, the program can be loaded anywhere in memory and execute properly, or identical copies of the program can execute and refer to different data sets. This method is demonstrated by the RTL statements of Figure 3.6(d), which shows an ADD instruction in which both operands are accessed with register indirect addressing, at the same time offsetting the accesses by the contents of PC.

The execution portion of the instruction begins by adding the current contents of PC to the value contained in RA, as shown in statement number 3. Also in statement number 3 is the instruction to keep a copy of the PC in Rtemp. This value is available on the bus, and it can be stored in the temporary location to save some time later. The value in R/D is the address of the desired operand, and statement number 4 moves this to the MAR. Thus, the first transfer of statement number 5 fetches the first operand from memory, and this is placed in the temporary register. At the same time, the current value in Rtemp, which is a copy of PC, is being added to the value in RB. This is the address of the second operand, and it will be placed in the R/D register. The transfer of statement number 6 moves this value to the MAR. Two tasks are identified by statement number 7. The first is the work of the instruction, which takes the value stored in memory at the location just identified, and to it adds the value stored in the temporary register. This sum is stored in R/D. The second operation is to increment the program counter to point to the next instruction in the instruction stream. The final transfer required by the instruction is completed in statement number 8, which moves the sum to memory.

```
   fetch:
   decode
 3  execute:        RA+PC  ⌐→ R/D       Address in RA is offset by PC.
                       PC  ⌐→ Rtemp     Keep copy of PC for later use.
 4                     R/D  → MAR        This is actual address of operand.
 5            M[MAR]   ⌐→ Rtemp     Get first operand to Rtemp.
              RB + Rtemp  ⌐→ R/D       Address in RB is offset by PC.
 6                     R/D  → MAR        Address of second operand moves to MAR.
 7     M[MAR] + Rtemp  ⌐→ R/D       This is work of instruction; result in R/D.
                    PC + 2  ⌐→ PC        Bump PC to point to next instruction.
 8                     R/D  → M[MAR]  Move result to proper destination.
```

Figure 3.6(d). RTL for ADD with Both Operands Accessed by Register Indirect Addressing, Including a PC Relative Mode.

The execution time for this instruction is 1.2 μsec. The first 240 nsec are required by the fetch portion of the instruction. Statements 4 and 6 are simple transfers, each requiring 40 nsec. The work of the instruction is accomplished in statement 7, which takes 340 nsec to complete. And statements 5 and 8 incur a memory access penalty of 200 nsec each. Thus, the total is 1.2 μsec.

One of the accesses often required by programs is to obtain a specific element in an array or other data structure. One method which can be used for this is called indexing, where one address (the starting address of the array, for example) is indexed by, or added to, another value, which can then be considered the offset into the array. The transfers of Figure 3.6(e) can be considered immediate indexing, since the base address for the indexing operation is located in the instruction stream. The second address involved in the indexing operation is located in a register. Thus, the instruction shown in Figure 3.6(e) is an addition, with the first operand identified in an immediate index mode and the second operand located in a register, RB.

The transfer of statement number 3 increments the contents of PC to identify the location of the immediate value, which is the base address used for the indexing operation. Statement number 4 moves this value to the MAR so that the access can be made. The value is extracted from memory and added to the contents of RA in the first operation of statement number 5. This results in the address of the desired operand, and this address is stored in R/D. The second task of statement number 5 is to increment the PC to point to the next instruction. Statement number 6 moves the address of the operand to the MAR. The work of the instruction is performed by the action specified in statement number 7. The desired operand is extracted from memory and added to the contents of RB, and the result is left in R/D. Finally, the transfer of statement number 8 moves this result to RB, where the results of the instruction should be placed.

The increment of the program counter indicated in statement 3 of Figure 3.6(e) requires 140 nsec. The register transfers of statements 4, 6, and 8 each take 40 nsec. The two remaining statements, 5 and 7, each incur 340 nsec, 200 nsec to get information from memory, and 140 nsec to do the addition. Therefore, the total time required by this instruction is 1.18 μsec, 240 nsec for the fetch portion, and the remainder for the transfers of Figure 3.6(e).

Let's take a quick look at how this method can be changed slightly to save some time. The intent here is to do the same operation as provided in Figure 3.6(e), but to do it faster. The change is to use a base address located in a register, rather than a base address extracted from an instruction stream. The expectation is that many of these accesses will be done at the same time, and rather than incur a

	fetch:		
	decode		
3	*execute:*	$PC + 2 \rightarrow PC$	Increment PC to identify address.
4		$PC \rightarrow MAR$	Move PC contents to MAR.
5	$M[MAR] + RA \rightarrow R/D$		Add index base value to RA; sum to R/D.
	$PC + 4 \rightarrow PC$		Increment PC to identify next instruction.
6		$R/D \rightarrow MAR$	Move address of first operand to MAR.
7	$M[MAR] + RB \rightarrow R/D$		Do the add and leave sum in R/D.
8		$R/D \rightarrow RB$	Move result back to RB.

Figure 3.6(e). RTL for ADD with First Operand Accessed in Immediate Index Mode, Second Operand in Register RB.

memory fetch penalty with each instruction to fetch the base address, assume that a previous instruction has stored the base address in some *known* location. We will identify this know location as Rbase. The reason for assuming that the register is automatically known by the instruction is so that the two register addresses specified by the instruction, RA and RB, can be used in the normal activities of the system. Figure 3.6(f) repeats the operation of Figure 3.6(e), but using register indexing, rather than immediate indexing.

As seen by the RTL transfers of Figure 3.6(f), fewer transfers are required for the register index mode. The indexing step, which adds the two addresses together, is performed in the first operation of statement number 3. The second operation is the increment of the program counter to point at the next instruction. The address of the desired operand is moved to the MAR by the transfer of statement number 4. The work of the instruction is accomplished in statement number 5, and the result placed in RB by the transfer of statement number 6.

This instruction requires only 800 nsec, as opposed to the 1.18 μsec of the immediate indexed instruction. This time comes from the 240 nsec of the fetch, plus two register transfers at 40 nsec each (statements 4 and 6), plus 140 nsec to generate the address (statement number 3), plus 340 nsec to do the actual work (statement number 5). Thus, if indexing is to be done many times, register indexing (in this architecture) saves 380 nsec for each indexing operation.

The final addressing mode to be demonstrated in this section is the register indirect autoincrement mode. In this mode, operands are located in memory and are accessed in a register indirect fashion. However, after the address has been used, it is incremented to point at the next value. Thus, with a single instruction, not only is the destination operand changed, but also the addresses used in the operation. Figure 3.6(g) contains RTL statements for an addition operation where both the source operand and the destination operand are identified in a register indirect autoincrement fashion. This would be very useful in a loop which was used to add two arrays together.

The transfer of statement number 3 of Figure 3.6(g) moves the address of the first operand to the MAR. Statement number 4 identifies three separate actions. The first is to retrieve the operand from memory and place it in the temporary register. The second action is to increment the address used for the first operand. This assumes that either the ALU can increment by four, or that the Data MUX can provide the value as an input to the ALU. The final action of the statement is to increment the program counter to point at the next instruction. The transfer specified by statement number 5 moves the updated address back to RA. Two operations are identified by statement number 6. The first is to place a copy of RB in the MAR. The second is to increment the value in RB by 4 and place the updated value in R/D. At the completion of this step, MAR contains the address needed for

	fetch:			
	decode			
3	*execute:*	RA + Rbase $\rightarrow$ R/D		Add index base value to RA; sum to R/D.
		PC + 2 $\rightarrow$ PC		Increment PC to identify next instruction.
4		R/D $\rightarrow$ MAR		Move address of first operand to MAR.
5		M[MAR] + RB $\rightarrow$ R/D		Do the add and leave sum in R/D.
6		R/D $\rightarrow$ RB		Move result back to RB.

Figure 3.6(f). RTL for ADD with First Operand Accessed in Register Index Mode, Second Operand in Register RB.

3	*execute:*	RA → MAR	Move first address to MAR.
4		M[MAR] → Rtemp	Get operand to Rtemp.
		RB + 4 → R/D	Increment address in RA.
		PC + 2 → PC	Update PC.
5		R/D → RA	Return incremented address to RA.
6		RB → MAR	Copy RB to MAR.
		RB + 4 → R/D	Bump contents of RB by 4; result in R/D.
7		R/D → RB	Move updated address back to RB.
		M[MAR] + Rtemp → R/D	Do the work of instruction; sum to R/D.
8		R/D → M[MAR]	Return result to memory.

Figure 3.6(g). RTL for ADD with Both Operands Accessed in Register Indirect Autoincrement Mode.

both the second operation and the result, and R/D contains the updated address which needs to be returned to RB. The return of the address value is the first action of statement number 7, while the second action is to do the actual work of the instruction. This involves extracting the value from memory, adding it to the value in Rtemp, and placing the result in R/D. The transfer of statement number 8 moves this result to the appropriate location in memory.

The timing for this instruction involves the 240 nsec of the fetch, plus the operations of Figure 3.6(g). Statements 3 and 5 are register transfers and incur 40 nsec penalties. Statements 4 and 8 are memory transfers and incur a penalty of 200 nsec each. Statement number 6 takes 140 nsec, and statement number 7 requires 340 nsec. This totals to 1.2 μsec. Thus, this is a relatively long instruction, but it accomplishes many different operations.

Many other types of addressing modes can be derived by combining and extending the methods demonstrated in this section. Each will result in its own time requirements, and the impact of including all the instruction modes can be assessed only by looking at the entire machine and its expected usage. Using RTL statements to specify the action is a tool which can be used in appropriate ways to analyze a portion of the resource utilization of the system. It also provides a basis from which control design for the system can proceed.

3.5. Implementation of a Rudimentary Vector Machine

The preceding sections have presented some RTL implementations for some fairly standard instruction set architectures. This section presents RTL implementations of a rudimentary vector machine. As explained above, in a stored program computer the work to do is specified by instructions stored in a program in the machine, and the work is accomplished by extracting the instructions from program memory and performing the transfers and manipulations which are specified there. One way in which the data portion of the machine can be more productive is to specify more work with each instruction. This is the basic premise behind vector machines. A vector is an organized set of data, and a vector instruction performs the same operation on all elements in the set, or in other words, on every element of the vector. In this way, the system minimizes the overhead required for doing arithmetic operations.

A block diagram for a portion of the data section of a vector machine is shown in Figure 3.7. The diagram shown is used in this section, but it was not created to mimic a particular real machine. Rather, this diagram was created to demonstrate mechanisms which could be used to describe complex data operations. The system was created in such a way that the various components can all perform their assigned function in a specific period of time, T_{CYC}. Thus, the system will achieve optimal performance when as many elements of the system as possible are utilized in each cycle. One of the jobs of the system architect is to define/design a system in which that can be accomplished.

In the system shown in Figure 3.7, the vector data is stored in the vector memory. This element can store the vector elements that are needed, and can both provide information and accept information in the same cycle. In any cycle, two operands can be made available, one on source bus A and the other on source bus B. The element placed on source bus A is identified by vector address register A (VARA). The element placed on source bus B is identified by vector address register B (VARB). If a value is to be written to a vector location, the address of the location is identified by vector address register C (VARC), and the data comes over the destination bus. The three vector address registers are made with counters so that they can be loaded or incremented. Also, they can be used to provide one address, and incremented at the end of the cycle to provide the next address for the next cycle.

The address registers are used to identify locations for data transfers. These locations can be within the data memory area of the machine (not shown), or they can be in the vector memory. Address registers can be loaded from or provided to

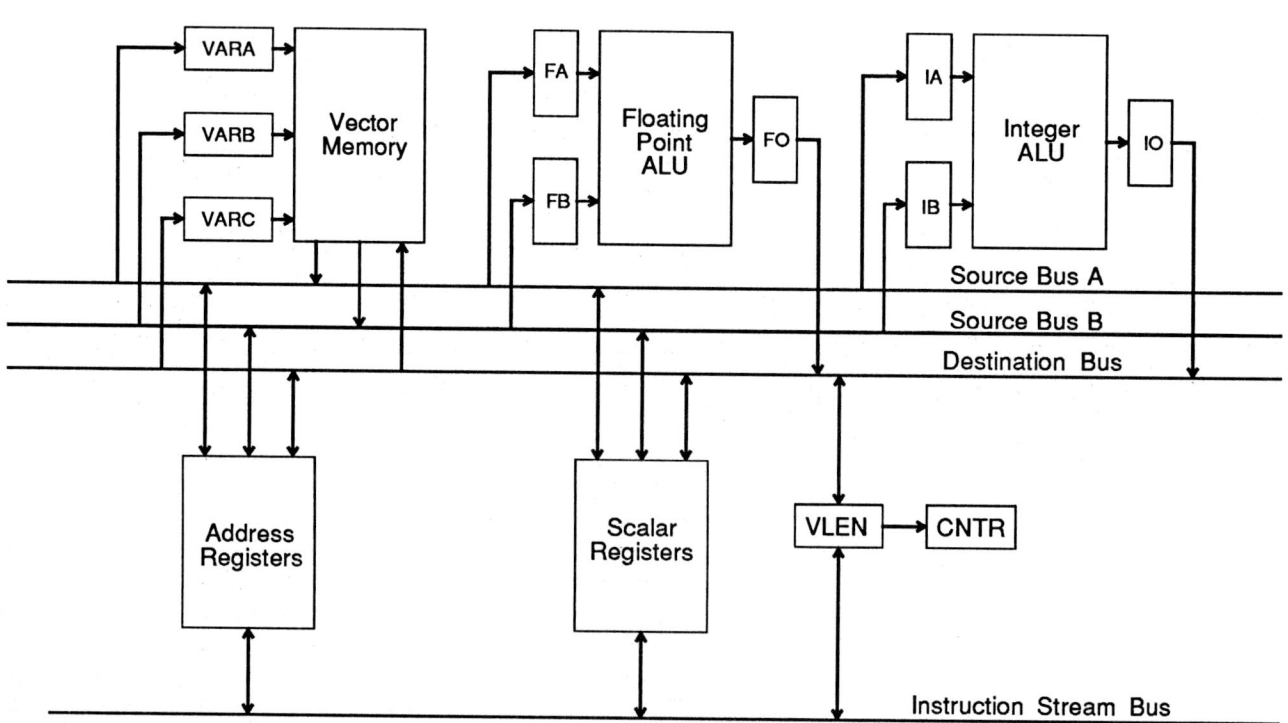

Figure 3.7. Partial Block Diagram for Data Path of Vector Machine.

any of the three data buses shown, as well as connected to the control portion of the system. This latter connection is shown simply as the instruction bus in the figure.

The scalar registers are used to hold scalar values needed in the calculation of values in the system. Like the address registers, the scalar registers can provide information to and receive information from the data buses, as well as the instruction bus.

The floating point ALU shown in Figure 3.7 is configured to allow pipelining of operand flow. The basic premise of the system is that all operations can occur in T_{CYC}, so the floating point ALU can accept two operands and provide a result in a single T_{CYC}. The operands are obtained from the two input registers FA and FB, and the result is left in the output register, FO. Thus, three cycles are required to perform any work in the floating ALU, one to fill FA and FB, one to perform the requested operation and place the result in FO, and one to write the result back to the appropriate destination. However, these operations are independent, and for independent data sets, the calculations can be overlapped (pipelined), which can provide a new result every cycle.

The integer ALU operates in the same fashion as the floating point ALU. Information is provided over the two source buses, and one value is loaded into input register A (IA) and input register B (IB). The integer ALU operates on the data and provides the result to the output register (IO), which can then be directed to the appropriate destination over the destination bus.

Two other registers are shown in Figure 3.7. The first is the vector length register, VLEN. The length of the vector identifies the number of elements contained in the vector, and hence the number of operations to be completed in a vector instruction. Thus, prior to a vector operation, one of the operations which is part of the the overhead or setup is to load VLEN with the number of elements in the vector. The remaining register is the vector count register, VCNT. This is actually a counter which will decrement with each vector operation performed, and test the result for zero. Vector operations begin by transferring the length of the operation to VCNT, and then it is decremented with each operation until it reaches zero. At that time the vector operation has been completed.

The system shown in Figure 3.7 is not complete, since the system elements involved with the control functions are not shown. That is, we have not included the PC or the SP, or the path(s) to program memory and data memory. The following fragment, therefore, concerns the data path operations only. It is left as an exercise for the reader to propose the control path elements, their interconnections, and appropriate RTL statements to perform the work.

Consider the following code segment to form an inner-dot-product by scaling one vector, then multiplying it by another vector, and then summing the elements. We will assume that the length of the vector is stored in scalar register L, that the scale factor is stored in scalar register S, and that the addresses of the two initial vectors are stored in address registers X and Y. We will also assume that temporary vector storage is available at the vector address stored in Z, and that the final result should be placed in scalar register F, With those assumptions, consider the following code segment:

1	MOV S_L, VLEN	Set up vector length.
2	SVMUL S_S, A_X, A_Z	Scale vector stored at X.
3	VVMUL A_Z, A_Y, A_Z	Multiply two vectors together.
4	VSUM A_Z, S_F	Sum all values, result to S_F.

The first instruction loads VLEN with the length of the vector. The second

1	$S_L \rightarrow$ VLEN	Fill vector length register.

2	$S_S \rightarrow$ FA	Scale factor to FA.
	$A_X \rightarrow$ VARB	Address of first vector to VARB.
	$A_Z \rightarrow$ VARC	Address of temporary vector to VARC.
	VLEN $\rightarrow$ VCNT	Fill counter.

3	VM[VARB] $\rightarrow$ FB	First value in vector to FB.
	VARB + 1 $\rightarrow$ VARB	Increment address in VARB.

4	VM[VARB] $\rightarrow$ FB	Move value to FB.
	VARB + 1 $\rightarrow$ VARB	Increment address in VARB.
	FB $\times$ FA $\rightarrow$ FO	Do multiply, place result in FO.

5	while (VCNT != 0) {	Repeat until count is zero.
	VM[VARB] $\rightarrow$ FB	Move value to FB.
	VARB + 1 $\rightarrow$ VARB	Increment address in VARB.
	FB $\times$ FA $\rightarrow$ FO	Do multiply, place result in FO.
	FO $\rightarrow$ VM[VARC]	Move previous result to VM[VARC].
	VARC + 1 $\rightarrow$ VARC	Increment address in VARC.
	VCNT - 1 $\rightarrow$ VCNT	Decrement counter.
	}	

6	$A_Z \rightarrow$ VARA	Address of first vector to VARA.
	$A_Y \rightarrow$ VARB	Address of second vector to VARB.
	$A_Z \rightarrow$ VARC	Address of result vector to VARC.
	VLEN $\rightarrow$ VCNT	Fill counter.

7	VM[VARA] $\rightarrow$ FA	Move value from first vector to input register.
	VARA + 1 $\rightarrow$ VARA	Increment address register.
	VM[VARB] $\rightarrow$ FB	Move value from second vector to input register.
	VARB + 1 $\rightarrow$ VARB	Increment address register.

Figure 3.8. RTL Statements for Vector Machine Implementation.

instruction is a scalar-vector multiply. That is, each element of a vector is to be multiplied by a scalar, with the result stored in the vector memory at location A_Z. Thus, if the vector length is 100, then instruction number 2 specifies 100 multiplication operations. Instruction number 3 is used to multiply the two vectors together. The first vector is found in the vector memory starting at the location identified by A_Z, the second vector is found in the vector memory starting at location A_Y, and the third vector is the same as the first, starting at location A_Z. This will result in writing over one of the input vectors. This instruction also specifies 100 multiplies. Finally, instruction number 4 sums all the elements of the vector starting at location A_Z, placing the results in S_F.

One set of RTL statements which will accomplish the work of these vector instructions is included in Figure 3.8. In addition to the statements shown in Figure 3.8, there are corresponding control transfers to keep the data path operating as stated in the figure. The control transfers must be included with any implementation of this system. The transfer specified by RTL statement number 1 in Figure 3.8 is used to implement the first statement of the above code segment. That is, the value in scalar register S_L is moved to the vector length register, VLEN. This is the only code segment statement to be accomplished with a single transfer.

The transfers required by the second statement of the code segment are initiated in statement number 2 of Figure 3.8. Since the block diagram allows for numerous paths, the four separate transfers of this statement can be performed

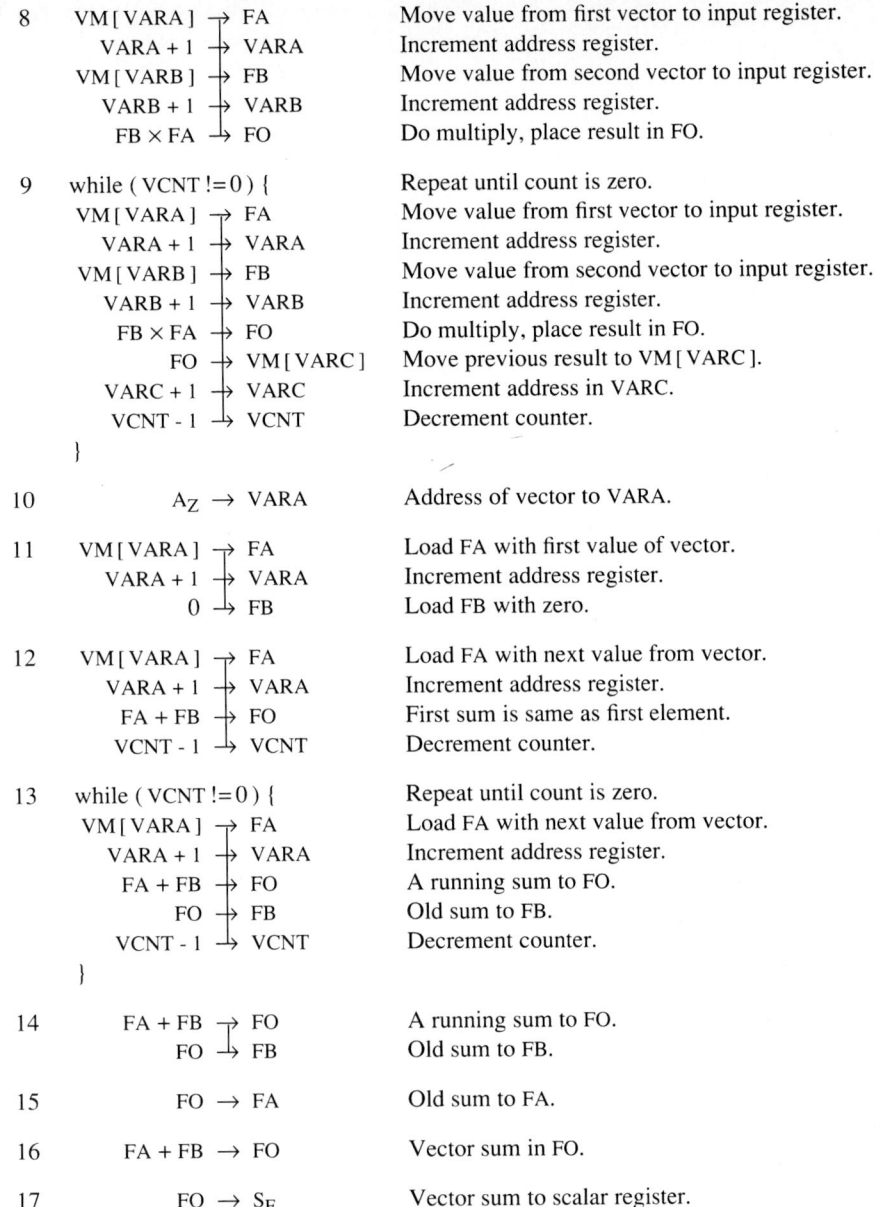

8	VM[VARA] → FA		Move value from first vector to input register.
	VARA + 1 → VARA		Increment address register.
	VM[VARB] → FB		Move value from second vector to input register.
	VARB + 1 → VARB		Increment address register.
	FB × FA → FO		Do multiply, place result in FO.

9   while (VCNT != 0) {   Repeat until count is zero.
      VM[VARA] → FA   Move value from first vector to input register.
      VARA + 1 → VARA   Increment address register.
      VM[VARB] → FB   Move value from second vector to input register.
      VARB + 1 → VARB   Increment address register.
      FB × FA → FO   Do multiply, place result in FO.
      FO → VM[VARC]   Move previous result to VM[VARC].
      VARC + 1 → VARC   Increment address in VARC.
      VCNT - 1 → VCNT   Decrement counter.
    }

10   A_Z → VARA   Address of vector to VARA.

11   VM[VARA] → FA   Load FA with first value of vector.
      VARA + 1 → VARA   Increment address register.
      0 → FB   Load FB with zero.

12   VM[VARA] → FA   Load FA with next value from vector.
      VARA + 1 → VARA   Increment address register.
      FA + FB → FO   First sum is same as first element.
      VCNT - 1 → VCNT   Decrement counter.

13   while (VCNT != 0) {   Repeat until count is zero.
      VM[VARA] → FA   Load FA with next value from vector.
      VARA + 1 → VARA   Increment address register.
      FA + FB → FO   A running sum to FO.
      FO → FB   Old sum to FB.
      VCNT - 1 → VCNT   Decrement counter.
    }

14   FA + FB → FO   A running sum to FO.
      FO → FB   Old sum to FB.

15   FO → FA   Old sum to FA.

16   FA + FB → FO   Vector sum in FO.

17   FO → S_F Vector sum to scalar register.

Figure 3.8. RTL Statements for Vector Machine Implementation (continued).

simultaneously. The first transfer moves a value from the scalar register S_S to FA. The second transfer moves the address of the source vector to be used in this calculation to VARB. The third transfer moves the address of the destination vector to VARC. The fourth transfer moves the length of the vector to the counter (VCNT), where the number of iterations can be controlled.

Two transfers are called out in statement number 3 of Figure 3.8. The first moves a value to FB. The value to be moved to FB is found in the vector storage area, at the address specified in VARB. The second transfer indicated here is the increment of the address in VARB. Notice that these two transfers occur simultaneously, so that the address, which is used for the first transfer, does not change until the data transfer has been completed.

Statement number 4 identifies three transfers required by the process. The first moves another value to FB, where it will be used in the scaling process. The second transfer increments the address at which the information is found in the vector memory. The third transfer is the first of the multiplications to actually be performed. The result is placed in the FO register, from which it must be moved before the process can be complete.

The repetitive part of the process is identified by the transfers of statement number 5. This statement calls for six different transfers, as well as a conditional branch. The condition is identified in the first line, and the construction of the condition indicates that the process will continue until the value in VCNT has reached zero. The first transfer gets another value from the vector memory, just as the two preceding statements have done. And the second transfer is the increment of the address into the memory. The third transfer identifies the load of the output register (FO) with a multiplication result. The fourth transfer identifies the move of the result of the previous iteration to vector memory. This is the first transfer which moves information to a location that is accessible by the user; all transfers prior to this point have been to internal registers not accessible by the user. The fifth of the transfers is the increment of the destination address for the vector placement, and the sixth transfer is the decrement of the amount stored in VCNT. Because of the conditional construct of this statement, the transfers identified by this statement will be executed the number of times identified by the vector length, and hence the correct number of output values will be created. Note that this will result in one extra multiplication (the final value loaded into FO will not be moved to vector memory), and two extra moves from vector memory. These extra operations will have no effect on the final result.

When the iterative transfers specified by statement number 5 have been completed, the work identified by the second instruction in the code segment above has been completed. Thus, the work called out by the third statement of the code segment is initiated in RTL statement number 6 of Figure 3.8. This statement calls for four different transfers. The first three transfers fill VARA, VARB, and VARC. Note that two of these addresses are identical. The fourth transfer fills the counter (VCNT) with the length of the vector. When these transfers have been completed, then the system can initiate the data movement needed for the specified work.

The first transfers required by the vector operation of statement number three in the above code segment are specified in RTL statement number 7 of Figure 3.8. The first and third transfers specified by the statement move data to FA and FB. The second and fourth transfers increment the addresses at which the data was obtained to identify the next elements in the vector memory. Thus, at the completion of the work identified by RTL statement number 7 the first values to multiply have been loaded into the input registers of the floating point ALU.

The transfers of RTL statement number 7 are repeated in RTL statement number 8, along with loading the output register (FO) with the first multiplied value. Also, at the completion of RTL statement number 8, two new values are ready for multiplication, and the addresses have been updated to identify the next input values.

Like RTL statement number 5, RTL statement number 9 is repetitive in nature, and will be executed the number of times identified by the vector length. This value was loaded into VCNT by a transfer in RTL statement number 6. The first and third transfers of RTL statement number 9 move data from the vector memory to the input registers of the floating point ALU. The second and fourth transfers update the pointers for the input transfers. The fifth transfer loads the current value into the output register. The sixth transfer moves the output from the previous iteration to vector memory, and the seventh transfer updates the address at which the output is loaded into the vector memory. The last transfer is the decrement of the value in VCNT. These transfers will be repeated once for every element in the output vector, and hence the proper number of values will be loaded into the result.

The remaining RTL statements in Figure 3.8 are needed to perform the work of the last statement in the code segment above, the vector sum. The desired result is to sum all the elements of a vector, and the instruction identifies the address register which identifies the start of the vector, and the scalar register where the result should be found. The transfer of RTL statement number 10 moves the value from the address register identified to VARA. This identifies the source of the information.

The transfers of RTL statement number 11 continue the addition process. The first transfer moves a data value to one of the input registers (FA) of the floating point ALU, and the second transfer increments the address to point to the next value of the vector. The third transfer loads the other input register (FB) with zero. This assumes that the control system can force a zero value onto the bus, or that it has a direct clear capability on the register.

The transfers of RTL statement number 12 continue the process. The first transfer loads FA with the next value from the input vector. The second transfer updates the address. The third transfer is the first to load the output register. Note that this first value is merely a copy of the first element of the vector, since it was added to the value zero. Also note that FB is not changed by this operation, so the register still contains a zero. The final operation here is to decrement the value in VCNT by one, since the number of values involved here is the number of input values, not the number of output values (which will be one).

The repetitive part of the summation operation is identified by the transfers of RTL statement number 13. Again, this operation will continue until VCNT has reached zero. The first transfer gets a new value to FA, and the second transfer updates the address at which the next input value will be obtained. The third transfer loads the output register (FO) with a sum. Note that the first time through the loop this value will be the sum of the odd numbered elements of the vector, and the second time through the loop this value will be the sum of the even numbered elements. Thus, when the VCNT value reaches zero and we break out of the loop, the work is not completely finished. The fourth transfer performs a feedback, moving the result of the previous iteration to FB. In order for this to occur, the input registers must be able to obtain values from either the destination bus or an input bus. The final transfer of RTL statement number 13 is to decrement the value in VCNT.

When the action moves from RTL statement number 13, the final input value has been loaded into FA, and one of the two running sums has been loaded into FO, the other into FB. The transfers of RTL statement number 14 add the last input value into a running sum, and move the other running sum to FB. The work which remains, then, is to add the value in FO to the value in FB. The transfer of RTL statement number 15 moves the value in FO to FA, so that the final sum can be performed. This is identified by the transfer of RTL statement number 16.

The final action required of this code segment is the transfer of RTL statement number 17, which places the final sum into the appropriate scalar register. It would be hoped that some of the work of statements which follow the four of the code segment above could be accomplished while this final transfer is being performed, and the control system should be capable of identifying this work and initiating the appropriate transfers.

The method of determining the time for this system is the same as for the systems of the previous examples. We have made the assumption that all operations require the same time for completion, T_{CYC}. We will now assume that T_{CYC} is 100 nsec, and attempt to determine the time for the code segment above, and the resulting data rate.

Statement number 1 of the RTL statements of Figure 3.8 is a simple transfer and will be completed in one cycle, or 100 nsec. The same is true of statements 2–4, even though each statement identifies more than one transfer that will occur. Thus, the first four statements will require $4 \times T_{CYC}$ to complete, or 400 nsec. Other statements which incur a time penalty of one cycle include statements 6–8, 10–12, and 14–17.

The fifth RTL statement of figure 3.7 will be repeated once for each value in the vector. Therefore, this statement will require $N \times T_{CYC}$ to complete. This will also be true of statement number 9. Statement number 13 is a repetitive statement, but it will be executed only $N - 1$ times. Thus, the total time required by this system is $(15 + 3 \times N) \times T_{CYC}$, which is 31.5 µsec. In this time period, $2 \times N$ multiplications are performed, and $N-1$ additions. (Actually, more than that are performed, but those are the ones resulting in useful values.) Thus, in 31.5 µsec 299 floating point operations have been performed, for a resulting data rate of 9.49 MFLOPS (million floating point operations per second).

3.6. Conclusion

Representation of work within a digital system by using a register transfer language is a very useful tool. This tool can be used in the analysis of a system, both in terms of function and correctness, and also in terms of performance. The tool can also be useful in the design process of the control section for a digital system,

The language itself should reflect the capabilities of the system, with the language expanding to describe new features as they are made available by the system architect. We have used the *while* construct in this chapter; other constructs are possible, and indeed desirable, as a mechanism to describe the actual or projected behavior of a digital system. The system should not be constrained by the language; the language should be created to describe the system.

Two different uses of RTL statements have been demonstrated here. The first is the use of RTL statements to specify action needed to produce the desired results. If the arithmetic capabilities of a system are not extensive, then care must be taken to assure that all the required action is accomplished in the execution of (or RTL description of) an instruction. In this case, the designer must have a firm understanding of the data transfer requirements of an instruction (or primitive operation in a stand-alone system), and that understanding must be reflected in the transfers specified.

The second use of RTL statements in this chapter is to provide timing information. This information can be very useful in the comparison of two or more different techniques for accomplishing work. The timing information can also be a useful metric to determine the actual speed of execution of a system, or to ascertain the time costs associated with the different portions of doing work in a digital system.

3.7. Exercises

3.1 For a single address machine which follows the block diagram of Figure 3.2, give the RTL implementation of an "increment-and-skip-if-zero" instruction. That is, the value located at the address identified by the instruction is incremented; then, if the resulting value is zero, the program counter is incremented by one, which effectively skips the next instruction.

3.2 For a single address machine which follows the block diagram of Figure 3.2, give the RTL implementation of a jump instruction, which changes the program counter to continue execution at the location identified by the address contained in the instruction.

3.3 For a single address machine which follows the block diagram of Figure 3.2, give the RTL implementation of a program counter relative jump. That is, the address of the next instruction is found at the location identified by the sum of the address found in the instruction and the current contents of the program counter.

3.4 For a two address machine which has a block diagram like Figure 3.4, give an RTL implementation for a two address, indirect, invert instruction. That is, the information found at the location in memory identified by the address in the first register is inverted, and the result is placed in the location in memory which is identified by the address found in the second register.

3.5 For a two address machine which has a block diagram like Figure 3.4, give an RTL implementation for an exclusive-OR instruction. Note that the exclusive-OR function must be created by using the capabilities of the ALU.

3.6 For a two address machine which has a block diagram like Figure 3.4, give an RTL implementation for an indirect, autoincrement move instruction, MOVE $*R_A+$, $*R_B+$. That is, the data is moved from one location in memory to another. The address of the first location is stored in regsiter A, and the address of the second location is stored in register B. After the addresses in the registers have been used, they will be incremented to point at the next value in memory.

3.7 For a two address machine which has a block diagram like Figure 3.4, give an RTL implementation for a PUSH R_N, R_S instruction. This instruction uses the R_S register as a stack pointer to general memory, and at that location places the contents of R_N. The address in R_S is decremented before placing the information on the stack, so that the stack grows down in memory. In addition, give the RTL statements for the implementation of the POP R_N, R_S instruction, which performs the reverse function.

3.8 Create a block diagram similar to Figure 3.2, with the same basic registers which are available there. However, utilize a system bus which is only 8 bits wide. For this system, give the RTL statements for the transfers needed for the invert, add, jump subroutine, and return from subroutine instructions. What is the overall effect on the time of the execution of the instructions?

3.9 For the simplified vector machine of Figure 3.7, create a set of RTL statements for doing a vector inversion instruction.

3.10 Create a block diagram of the vector machine shown in Figure 3.7, using Figure 3.7 as a starting point and adding the other system elements as needed. For this expanded system, give the RTL statements for jump subroutine and return from subroutine instructions. What can be done to improve the performance of the system? That is, in what ways can parallel/concurrent activities be built into the system?

3.8. Additional References

[Bart85] Bartee, T. C., *Digital Computer Fundamentals* (6th ed.). New York: McGraw Hill Book Company, 1985.

[Boot84] Booth, T. L., *Introduction to Computer Engineering: Hardware and Software Design.* New York: John Wiley & Sons, Inc., 1984.

[FoIb85] Foster, C. C., and T. Iberall, *Computer Architecture* (3rd ed.). New York: Van Nostrand Reinhold Co., 1985.

[FuBu77] Fuller, S. H., and W. E. Burr, "Measurement and Evaluation of Alternative Computer Architectures," *Computer.* Vol. 10, No. 10, October 1977, pp. 24-35.

[HaVr78] Hamacher, V. C., Z. G. Vranesic, and S. G. Zaky, *Computer Organization.* New York: McGraw-Hill Book Company, 1984.

[Haye88] Hayes, J. P., *Computer Architecture and Organization* (2nd ed.). New York: McGraw-Hill Book Company, 1988.

[HiSp85] Hitchcock, C. Y., and H. M. B. Sprunt, "Analyzing Multiple Register Sets," *Proceedings of the 12th Annual International Symposium on Computer Architecture.* Silver Spring, MD: IEEE Computer Society Press, 1985, pp. 55-63.

[Kain89] Kain, R. Y., *Computer Architecture, Software and Hardware.* Englewood Cliffs, NJ: Prentice Hall, 1989.

[Lang82] Langdon, G. G., Jr., *Computer Design.* San Jose, CA: Computeach Press, Inc, 1982.

4

Design of Synchronous Sequential Systems

Systems of digital logic in which the outputs are functions of both the inputs and the current state of the system are called sequential systems. If these systems utilize a clock to synchronize the action of the system, then they are called synchronous sequential systems or clocked sequential systems. Design techniques for synchronous systems are taught in courses on basic digital electronics and in computer design courses. (See, for example, [Flet80, Mano88, Bree89], and [Poll90].) This chapter illustrates the design process by presenting several different designs for two units. The process of doing these designs identifies different techniques which can be used to create clocked sequential systems. The different techniques are beneficially utilized in different situations, and each technique has aspects which will lead to applications in different situations.

4.1. Design Steps for Clocked Sequential Systems

The behavior of a clocked sequential system is determined by a combination of the inputs and the state of the system. This is depicted in Figure 4.1. The inputs are obtained from events external to the system, and are presented to the sequential system as logical values. The state of the system is represented by the collection of storage elements internal to the system. That is, all the bits stored in registers and flip-flops in the sequential system jointly define the current state of the system. This is represented in Figure 4.1 as a single storage element called the Present State Register (PSR). The next state, the state which the system will assume after the next system clock, is determined by the next state logic. This logic determines from the present state and the current inputs what the next state should be. Thus, a large part of the design of sequential systems is a correct determination of the next state logic.

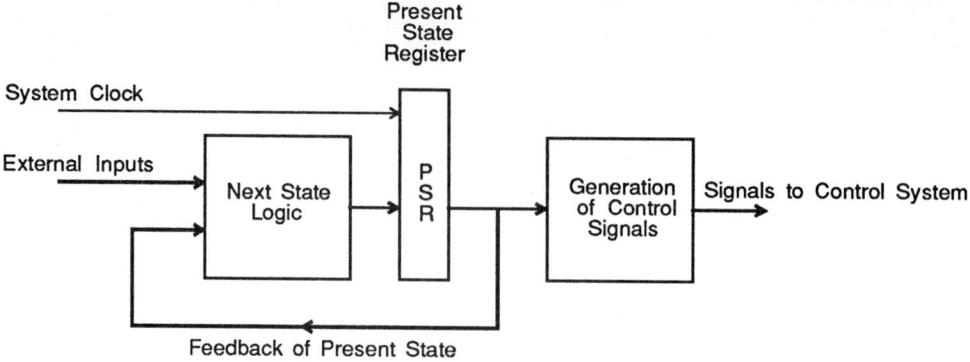

Figure 4.1. Clocked Sequential System: Outputs Are Functions of Inputs and Current State of System.

Most sequential systems do not exist to merely proceed from one state to another, but rather the systems are created to perform work in these states. Work in this sense is performed by activating control signals. Thus, in Figure 4.1 there is a block responsible for creation of signals which will control some other portion of the system. The diagram shown in Figure 4.1 indicates that the generation of the control signals is accomplished by decoding the present state. This method is the most common and the most easy to control, but systems can be created which generate the control signals based not only on the present state, but also on the inputs. Generation of control signals using both input signals and present state values can lead to systems which have glitches on the control lines and unpredictable behavior. Hence, this type of unit is not demonstrated in this text.

Since sequential systems do not stand alone, it is imperative that the creation of the sequential system be considered in the context of the system which it is to control. Once this system has been determined, then the sequential system can be correctly designed. Thus, in the creation of a system with a data processing function and a sequential controller, several distinct steps are required. These steps are summarized as follows:

1. **Understand the processing requirements.** As with the combinational design steps, this seems to be superfluous. However, before any design can be made, it is imperative that the requirements of the system be known. This includes a detailed understanding of the control signals provided from sources of information, as well as any control or feedback needed by those sources of information. It also includes a detailed understanding of the data requirements: the formats used, the transfer mechanisms available, any byte ordering problems, etc.

2. **Understand the processing algorithms.** This differs from the processing requirements, since identifying that a function is to be performed does not identify the manner in which the function is implemented. That is, if two numbers are to be multiplied, a direct multiplication method could be used, or an iterative multiplication method, or any of a number of other techniques. Thus, the designer must not only understand the processing required by the problem (step 1), but also the mechanisms involved in obtaining the results.

3. **Create a conceptual level data path block diagram.** In this step, the requirements identified in step 1 and the mechanisms identified in step 2 are merged into a single block diagram. This block diagram identifies the major computational sections of the system and the paths required for transfer of information in the system. This includes the transfer paths to/from external devices and systems, as well as the paths needed in the system itself. It is in this step that a designer utilizes his ingenuity to create a system which will use appropriate mechanisms (those identified and understood in step 2) to solve the problem to be addressed by the system (which was identified and understood in step 1). Also included at this step are any statements needed, either in a Register Transfer Language (RTL) or in a natural language, to describe the transfers which must take place to perform the needed work.

4. **Create a detailed data path block diagram.** This step actually has two parts. The first is to assign to the sections identified in step 3 the appropriate parts to store and manipulate the data of the system. This requires that the designer be intimately familiar with the available parts, their capabilities, and also their limitations. The second part is to identify the control signals of the parts which will be needed to perform as required by the algorithms, and to assign appropriate names to those control signals. If this naming is done in a mnemonic fashion, the names will indicate the function to be performed with the assertion of the signal. The name will also identify the assertion level of the signal, which is very beneficial in the design portion of the project as well as the checkout portion. At the completion of this step, the designer has available a list of the signals which must be controlled by the sequential part of the system. This list is very useful in the next step of the process.

5. **Create a state diagram describing the sequential nature of the system.** This is another process which ideally will have two parts. The first part is the creation of a state diagram describing the basic behavior of the system; this state diagram is created in conjunction with the preliminary block diagram. That is, as a designer is experimenting with alternative arrangements of system components, the manner in which those components interact to accomplish the work of the algorithm should also be tested with preliminary state diagrams. In this way, when a proposed arrangement of the principal parts of the system is considered, the manner in which the algorithm is implemented on that set of parts is also considered. The second part of the creation of a state diagram occurs when the detailed block diagram is available. At this point, the designer can create a state diagram describing the desired behavior of the system in detail. This detail includes identifying the control signals to be asserted, the order of their assertion, and other characteristics of the control system. The state diagram, then, is the key to understanding the action which is to take place to control data flow in the unit, and to interact with other functional units in the system.

6. **Design the logic of the control section.** Once the detailed block diagram of the data section is available, as well as the state diagram which describes the correct behavior of the control system, the actual logic of the control system can be designed. This requires combinational design skills, since each of the bits of the next state is determined by a logic system which uses as inputs the present state of the system and the current system inputs. The design of this system can take one of many forms, depending on the desired system objectives.

7. **Create the logic design for the entire system, implement and debug.** The final step for the system is to combine the required elements of the data path section with the control section into a single set of logic diagrams or schematics. These will provide the necessary information to implement the entire system, and check it out. If all steps have been done correctly, the checkout phase will be primarily a verification that the unit functions as it should.

The steps listed above, if followed conscientiously, result in a system which performs the needed functions to do the work required by the unit. As with any algorithm of this nature, iteration through the mechanisms will result in systems with better characteristics, where the determination of better characteristics depends upon the metric chosen by the designers.

It should also be noted that the designer's understanding of the initial problem, the algorithms for problem solution, and the logic system created to solve the problem, will change and grow as more work is done on the system. Sometimes the correct understanding of what is really needed to solve the problem is not obtained until one solution has been created and implemented. Then, in the unit checkout process, the designer notices that the problem is not properly solved, even though the hardware operates as envisioned in the design process. If this occurs, the designer must alter his understanding of the desired system behavior until the proper results can be obtained. Then the digital system can be changed appropriately.

The process described above can best be understood by doing, so let us now apply the steps to the design of two different arithmetic processes. Other types of functional units, such as I/O interfaces or device controllers, can be approached in exactly the same fashion.

4.2. Floating Point Multiplier: Method 1

In this section we will design a small floating point multiplier. This system is not as large or complex as a multiplier which conforms to the IEEE floating point standard, but it does contain a sufficient amount of complexity to provide a good vehicle to observe the steps described in the previous section. Once the basic problem is understood, and a solution mechanism proposed, we will examine different methods for implementing the control section.

Floating point multiplier: Understanding the problem

In this exercise we will design a Floating Point Multiplier (FPM), and the first task is to understand the requirements and specification of the system. The object of the exercise is to create a system which will get its input values from registers filled by an external source, and provide the result to a bus which will be read by an external source. In this way, we will not be concerned with the protocols needed for transfer of information to and from our unit. Also, when this nebulous external device requests a result, our device will extract the current contents of the registers, perform the floating point multiplication process, and make the answer available. Finally, the multiplier unit must also make available to the bus, upon demand, the current contents of the two input registers. This has two effects in the system. First, it forces the system to have more parts than it otherwise might, since tri-state drivers are required to give the information to the bus. Also, the input registers cannot be used in the performance of the algorithm, since that action would destroy the contents of the register. The second effect is a beneficial one from the standpoint of system checkout and preventive maintenance. Since the input registers are observable from the system bus, they can be interrogated from an external source to verify correct behavior and to check the system for errors. This is a very useful feature, and can be incorporated in many devices to improve testability of the system.

The floating point format we will utilize, strange as it may seem, is a 16-bit normalized format. The radix of the system is 2, there are 11 bits in the mantissa, the mantissa is stored in fractional format (with a hidden bit technique, so there are actually only 10 bits used for storing the mantissa). The exponent contains 5 bits, and is stored in an excess 16 format. The number itself is stored in a sign-magnitude format. The arrangement of the bits in the word is as follows:

$$\text{S} \quad \text{EEEEE} \quad \text{MMMMMMMMMM}$$

The "S" bit is for the sign of the number, the "EEEEE" bits are the exponent, and the "MMMMMMMMMM" bits are the mantissa. As mentioned above this format utilizes the "hidden 1" bit for the most significant bit of the mantissa; thus, although the format shown shows only 10 bits for the mantissa, there are actually 11 bits involved in the representation of the mantissa. Results are only single precision; that is, no double precision results are required from the device.

The FPM unit must contain registers for the multiplier and the multiplicand, but the signals for loading these registers will be supplied by an external source. The action of loading the multiplier register will also set a flag which will indicate to the FPM that data is available and the operation should commence. When the answer is available, we will set a flag to indicate that the answer is available. The data to be input to the multiplicand and multiplier registers is available on a tri-state bus, and the output must be available to a tri-state bus.

The task at hand is to to design the FPM. An additional constraint is that the multiplication will be done by a shift-and-add algorithm, and that any result will be rounded with normal rounding. Also, the unit must give a maximum exponent (all ones) for all overflow results and must give a zero (exponent bits all zero) for underflow results, or when one or both of the input arguments is zero. Thus, there are no status bits to provide with this simple unit; if underflow or overflow occurs, the result is to set the exponent to zero or maximum value.

Floating point multiplier: Understanding the processing algorithms

Now that we know what the device is supposed to do, let us consider how it is going to accomplish this task. First of all, we know basically that we need to add the exponents and multiply the mantissas. That is, if A and B are floating point numbers, then A is actually $M_A \times 2^{E_A}$, where M_A is the mantissa of A and E_A is the exponent of A. And B is actually $M_B \times 2^{E_B}$, where M_B is the mantissa of B, and E_B is the exponent of B. Thus,

$$A \times B = (M_A \times 2^{E_A}) \times (M_B \times 2^{E_B})$$
$$= (M_A \times M_B) \times 2^{(E_A + E_B)}$$

Thus, if M_R is the mantissa of the result, and E_R is the exponent of the result, then:

$$M_R = M_A \times M_B$$

and

$$E_R = E_A + E_B$$

This seems simple enough until consideration is given to the possible values for the mantissas and the exponents. Let us consider the two elements of the result in that order.

The minimum value for a legal mantissa is 0.10000000000_2, which is just 0.5_{10}. The maximum value for a legal mantissa is 0.11111111111_2, which is 0.99951171875_{10}. Therefore, the multiplied values will range from a minimum of 0.01_2, to a maximum of 0.1111111110_2. The reason for making this observation is that the final output will either be correctly positioned (with respect to the significance of the input bits), or a post-normalization of one digit position will be needed. This will require the hardware to provide for a 1-bit position shift, if necessary, with a corresponding adjustment of the exponent. That is, when post normalization is needed, the exponent must be decremented by one.

The creation of the the resultant exponent, E_R, seems straightforward enough, since $E_R = E_A + E_B$. However, E_A and E_B are stored in an excess 16 format, which is formed by adding 16 to the actual value of the exponent. This gives rise to the situation indicated in Table 4.1. The value of the bit pattern of the exponent, taken as an absolute binary number, needs to be reduced by 16 to ascertain the actual exponent value. Also, when the bit pattern of the exponent is zero, the value of the number is zero, regardless of the mantissa bits or the sign bit.

In considering the multiplication process from the aspect of the exponents, we make the following observations. If the exponent bits of either the multiplier or

Table 4.1. Exponents for the Floating Point Format.

Pattern	Exponent Value	Pattern	Exponent Value
00000	special†	10000	0
00001	-15	10001	1
00010	-14	10010	2
00011	-13	10011	3
00100	-12	10100	4
00101	-11	10101	5
00110	-10	10110	6
00111	-9	10111	7
01000	-8	11000	8
01001	-7	11001	9
01010	-6	11010	10
01011	-5	11011	11
01100	-4	11100	12
01101	-3	11101	13
01110	-2	11110	14
01111	-1	11111	15

† value of number is zero,
regardless of other bits in word.

the multiplicand are all zero, then the value of that number is zero and the value of the result will be zero, and no further processing is necessary. If, however, both multiplier and multiplicand contain exponents which have nonzero patterns, then further processing is required. The two exponents need to be added together, and their sum will identify the final exponent. In absolute binary, the exponents can range from 1 to 31; the actual values represented by these patterns are −15 to +15. The sum of these bit patterns will then range from 2 to 62 (−30 to +30.) We can make a number of statements to describe the various possibilities, but perhaps the easiest way to visualize the result is to place these results in another table, shown as Table 4.2. The logic of the system must be capable of detecting underflow and overflow and of setting the exponent bits appropriately. This will also have an impact on the control system, since if an underflow or an overflow condition is detected, further processing is unnecessary, and the correct result can be made available immediately.

The description of the system indicated that the actual multiplication of the mantissas was to be accomplished by a shift-and-add algorithm, so consideration must be given to the desired method of performing this function. There are a number of different shift-and-add algorithms, some of which operate on the most significant portions of the partial product array first, while others work with the least significant elements in the partial product array first. The advantage derived by working with the most significant bits first is that the multiplication is over as soon as all the remaining bits of the multiplier are zero. Thus, the time required for multiplication is data dependent, but only that time actually required by the significant bits of the multiplier is needed. Thus, in general, this type of multiplication will be faster. The price for this speed is an adder and product register which is as wide as the final result.

The choice made for this system is a shift and add algorithm which starts with the least significant elements of the partial product array. This will be slower than the method just described, but it has other advantages. When not all of the final product is needed, which is the case for the floating point multiplier, the product register need only be as wide as the number of bits needed to form the final result. Also, the adder involved need only be as wide as the rows of the partial product array; thus, the number of parts required for formation of the product is reduced.

There are, therefore, a number of mechanisms either established by the specifications or required by the solution mechanism. First, we need some mechanism for detecting the occurrence of a zero-valued input; this identifies one of the conditions for a zero-valued output. Second, some logic is needed to detect the conditions identified in Table 4.2 for underflow and overflow. If either of these conditions is detected, then the process is finished, and we can correctly set the result register. If neither input is zero, and the calculation passes the initial underflow/overflow test, then the product will be created by a shift-and-add algorithm. The final step of the process is to do the rounding and any action required by post-normalization. The data path must then be capable of performing all of these operations.

Floating point multiplier: Conceptual level data path block diagram

A block diagram of this approach to the floating point multiplier is shown in Figure 4.2. This diagram shows the major parts of the system needed to perform the work, but not all the details. For example, no indication is given concerning the parts or methods to be used to do the exponent addition. Also, we know that the

Table 4.2. Enumeration of Resultant Exponent Possibilities.

The Bit Patterns below represent the possibilities after adding E_A and E_B; the Decimal Equivalent represents the value of the resultant exponent before any adjustments for excess codes; and the E_R's represent the correct bit pattern for the result exponent.

Bit Pattern	Decimal Equivalent	E_R (no PN*)	E_R (with PN*)	Bit Pattern	Decimal Equivalent	E_R (no PN*)	E_R (with PN*)
				100000	0	10000	01111
				100001	1	10001	10000
000010	-30	00000†	00000†	100010	2	10010	10001
000011	-29	00000†	00000†	100011	3	10011	10010
000100	-28	00000†	00000†	100100	4	10100	10011
000101	-27	00000†	00000†	100101	5	10101	10100
000110	-26	00000†	00000†	100110	6	10110	10101
000111	-25	00000†	00000†	100111	7	10111	10110
001000	-24	00000†	00000†	101000	8	11000	10111
001001	-23	00000†	00000†	101001	9	11001	11000
001010	-22	00000†	00000†	101010	10	11010	11001
001011	-21	00000†	00000†	101011	11	11011	11010
001100	-20	00000†	00000†	101100	12	11100	11011
001101	-19	00000†	00000†	101101	13	11101	11100
001110	-18	00000†	00000†	101110	14	11110	11101
001111	-17	00000†	00000†	101111	15	11111	11110
010000	-16	00000†	00000†	110000	16	11111‡	11111
010001	-15	00001	00000†	110001	17	11111‡	11111‡
010010	-14	00010	00001	110010	18	11111‡	11111‡
010011	-13	00011	00010	110011	19	11111‡	11111‡
010100	-12	00100	00011	110100	20	11111‡	11111‡
010101	-11	00101	00100	110101	21	11111‡	11111‡
010110	-10	00110	00101	110110	22	11111‡	11111‡
010111	-9	00111	00110	110111	23	11111‡	11111‡
011000	-8	00000	00111	111000	24	11111‡	11111‡
011001	-7	01001	01000	111001	25	11111‡	11111‡
011010	-6	01010	01001	111010	26	11111‡	11111‡
011011	-5	01011	01010	111011	27	11111‡	11111‡
011100	-4	01100	01011	111100	28	11111‡	11111‡
011101	-3	01101	01100	111101	29	11111‡	11111‡
011110	-2	01110	01101	111110	30	11111‡	11111‡
011111	-1	01111	01110				

* PN = Post Normalization
† Underflow condition
‡ Overflow condition

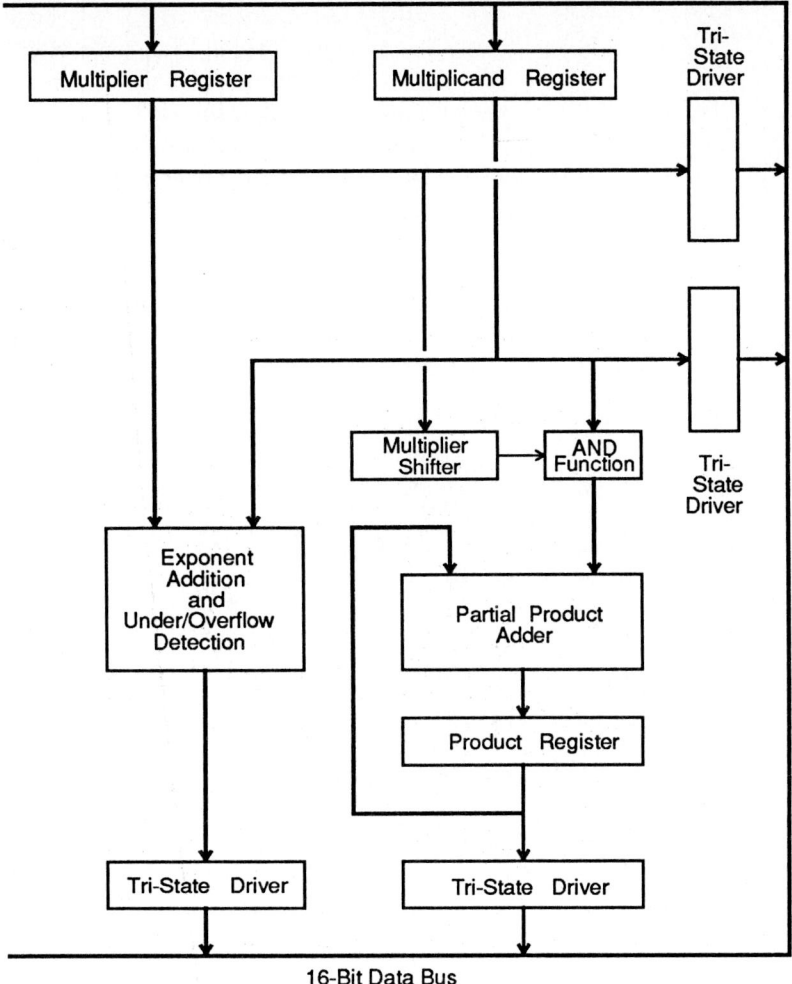

Figure 4.2. Conceptual Level Data Path Block Diagram for Floating Point Multiplier.

exponent must be adjusted for post-normalization, but no mechanism for this is shown in the diagram. Nevertheless, the major functions are represented. There are two input registers for storing multiplier and multiplicand. There are tri-state drivers available to provide the contents of multiplier and multiplicand registers to the bus, as required by the specification. The exponents must be added, and this is alluded to in the figure, as is the need for underflow and overflow detection. The shift-and-add nature of the algorithm is provided by the product register, the multiplier shifter, and the AND function. Finally, the result, in the form of an exponent and a mantissa, is available to the bus through a tri-state driver.

The block diagram as shown in Figure 4.2 does not provide enough detail to allow design of the control system, nor to design the actual data path of the system. What it does provide is a platform the designer can use to experiment with different ideas which can be used in the actual design of the system. We now provide additional details which will be needed to do the arithmetic required in the system.

Floating point multiplier: Detailed data path block diagram

To create a detailed data path block diagram, we work from the basis provided by the conceptual level diagram shown in Figure 4.2. To each block of the conceptual level diagram we will add the detail needed to demonstrate how the system will work, including numbers to identify actual components to be used, and also identify and name the control lines which will be needed. The result is shown in Figure 4.3.

The multiplier and multiplicand registers are formed by '273s; these registers have a clock line and a clear line, but only the clock is shown on the diagram. The clear line is assumed to be tied inactive. The contents of the multiplier and multiplicand registers are available to the bus through '541s, which are tri-state

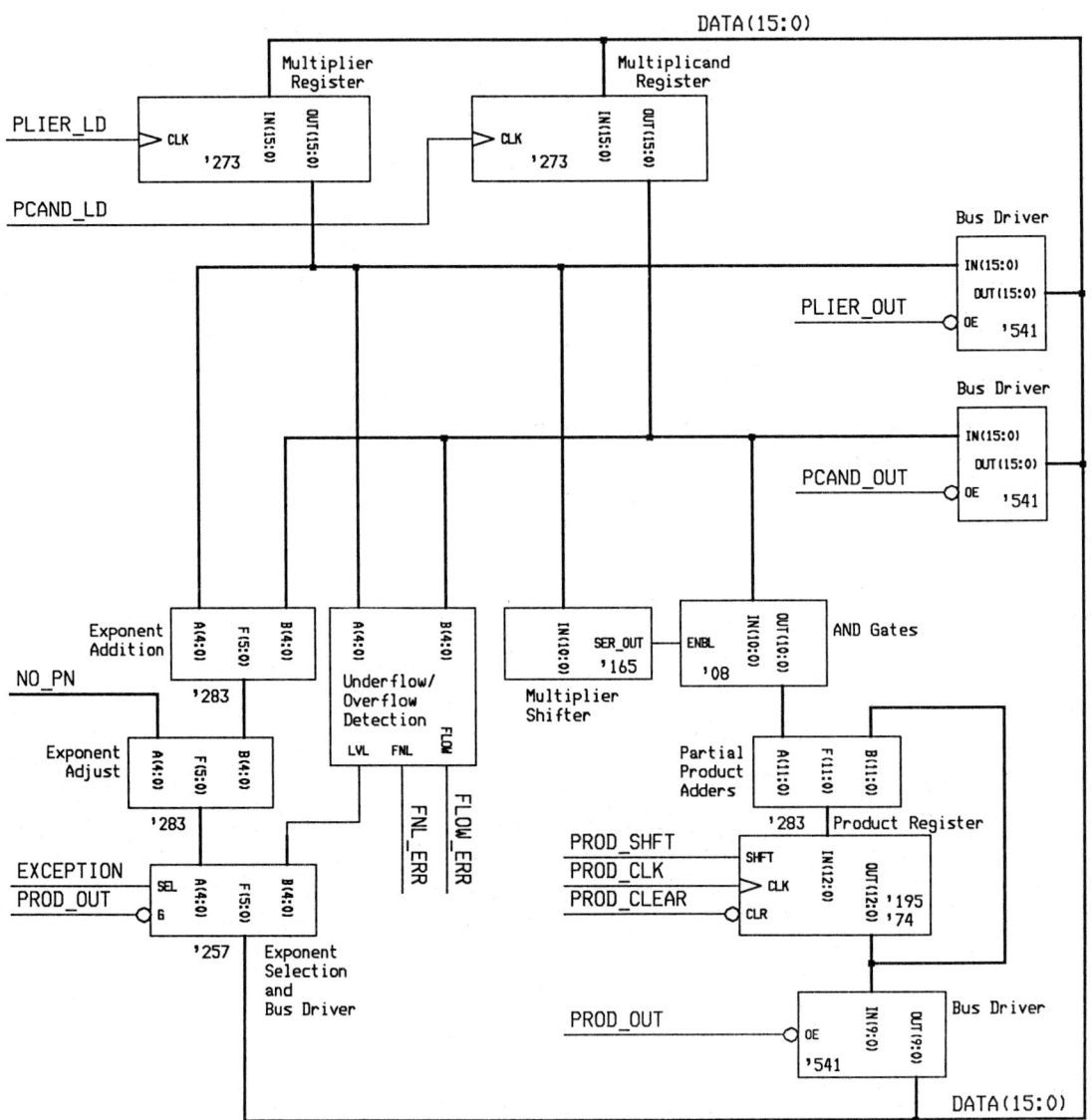

Figure 4.3. Detailed Data Path Block Diagram for Floating Point Multiplier.

drivers. These devices require a low true enable to activate the bus. These control lines, the clocks of the input registers and the output enables of the bus drivers, are identified and named, but control of these lines is the responsibility of the above-mentioned external device. If we were to interface this device directly to a system bus, such as a multibus or a VMEbus, then the control lines could be generated by decoding the appropriate programmed I/O lines.

The exponent of the result is obtained by adding the exponents in a set of binary adders ('283s). By examining Table 4.2, it is evident that the correct final result, the excess representation of the exponent of the result, can be obtained from the sum of the two initial exponents, plus either 16 or 15. If there is no post normalization, then the 16 is added to the sum of the input exponents. If post normalization is needed, then 15 is added to the sum of the input exponents. This exponent adjustment is also handled by a binary adder. The output of the adders is directed to a multiplexer. If an exception condition is detected, then the exponent will be determined by the Underflow/Overflow Detection circuitry, and will be either all ones or all zeros. If no exception is detected, then the exponent will be provided by the exponent adder network. The use of '257s allows the same device to provide both the multiplexing function and the tri-state driver function.

The Underflow/Overflow Detection system consists of the necessary random logic to check for zero input values, as well as to check for exponent sums which, from Table 4.2, will lead to an overflow or an underflow condition. These conditions are reported via control lines (FLOW_ERR, FNL_ERR) to the control section. The control section is then responsible for forcing the appropriate action.

The Multiplier Shifter is constructed from parallel load, serial shift registers. These devices will accept a copy of the mantissa of the multiplier (including the appropriate hidden bit) when the load line is asserted, and then will shift out a single bit each time the clock line is activated. This single bit is used by the AND gates to force that row of the partial product array to be zero, or to be a copy of the mantissa of the multiplicand (with hidden bit). This row is added to the product which is accumulating in the product register, and the result loaded again in the product register. The adder for this operation is a normal binary adder made with '283s. The product register is composed of shift registers ('195), with a 13th bit stored in a single flip-flop. The purpose for doing it in this fashion is that the shift-and-add operation proceeds as expected, except on the last iteration. At that time the MSB of the result is checked. If it is a ''1,'' then the result is as expected, with no post normalization needed, and one last shift is issued to place the result exactly where it should be. If, however, the MSB is a ''0'' after the final addition, then post normalization of one bit position is required. This is accomplished by *not* shifting the result for its final shift. The bits are located as needed in the product register without this final shift, so all that remains is to adjust the exponent accordingly, which is handled by the exponent system. The final result is presented to the bus through a tri-state driver ('541).

Attention to detail at this step of the process will result in a detailed block diagram, which for this project is shown in Figure 4.3. Knowledge of the control lines and the expected behavior of the data operations allows the designer to create a state diagram to describe the expected behavior of the control lines. The designer can also commit the detailed logic of the data path from the knowledge obtained in creating the detailed data path block diagram.

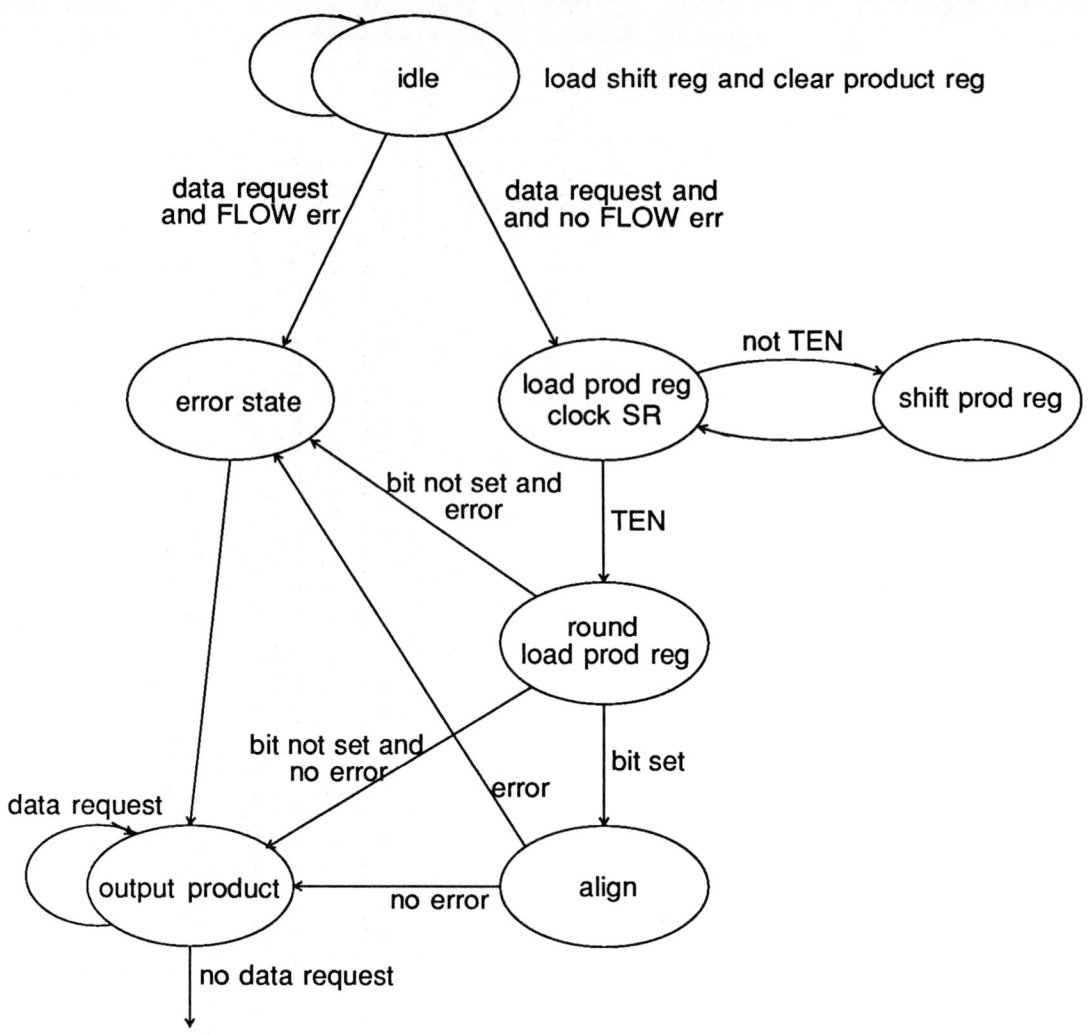

Figure 4.4. Conceptual State Diagram for Floating Point Multiplier.

Floating point multiplier: State diagram

As mentioned above, the development of the state diagram can be done in two phases. The first phase identifies the action which is to happen, and as such is basically a conceptual state diagram. The second phase generates the exact sequence which the hardware must follow to accomplish the work identified by the first state diagram. In many cases, the number of states will be the same between the first and second state diagrams; in some cases there will be more or fewer states in the detailed state diagram than there are in the conceptual state diagram. The preliminary, or conceptual, state diagram for the floating point multiplier is shown in Figure 4.4.

The diagram shown in Figure 4.4 indicates that the control for the system will remain in an idle state until a request is made for an answer. At that time, the controller will leave the idle state for one of two other states, depending on the status of the FLOW error indication. This error condition will exist when the values in the multiplier and multiplicand registers cause one of the error conditions: one of the values is zero (zero result), the sum of the exponents is too small (underflow), or the sum of the exponents is too large (overflow). If one of the error conditions exists, then an error state is visited, which will set an error condition, and then the action goes to the output state. In this state, an output will be made available to the external device; in the case of an error, this will be an exponent of all ones or all zeros.

If there is no detectable error condition, then the state diagram indicates that the SR, or multiplier shifter, should be clocked and that the product register should be loaded. Since this process is to occur 11 times (for the 11 partial products), a counter is set up to identify how many times it has happened. If the count has not progressed far enough, then the product register is shifted. Note that these two states cause the required shift-and-add action to occur; the load occurring in the first state, and the shift occurring in the other. When the load occurs, the appropriate partial product has been added in to the running sum. When the appropriate number of partial products has been added to the product register, then the rounding state is visited, which adds in a value equivalent to half the value of the least significant bit position. At this point, the most significant bit of the product is tested, as well the error condition, which checks for the delayed underflow/overflow check. (See Table 4.2; why must this condition be checked after the mantissa multiplication as well as before?) If the MSB is not set, then no adjustment is needed for post normalization, and the error condition is checked. If the MSB is not set and no error condition exists, we are done. If the MSB is not set and there is an error, then the action moves to the error state, as above. Finally, if the MSB is set, then the alignment state is visited, and the error check made again.

The state diagram shown in Figure 4.4 provides a guide to the action that accomplishes the work of the system. Note that no attempt has been made at this stage to incorporate the control signals which will cause the respective transfers. When a designer considers the transfers required in the system, as indicated by the block diagram and the conceptual level state diagram, then the assertion activity of the control signals to accomplish the work can be ascertained. These signal levels can be incorporated into another version of the state diagram, as shown in Figure 4.5. Note that this state diagram identifies the signals which are to be asserted in each state, and the conditions which will determine the flow of control in the state diagram.

Like any design aid or other tool used to design digital systems, the state diagram is only as useful as the knowledge of the designer. The state diagram is a very useful tool, and can provide guidance for conceptual level decisions and information for detailed implementation information. But it cannot replace the need for a designer to understand the function of the devices which make up the system, and the requirements of the assertion activity on the lines which control the action of those devices. Before a detailed state diagram can be made, the designer must understand how the control lines are to be asserted, and what restrictions are in force concerning those assertions. We will now describe the signals and the states, and provide some of the reasons for the activity which occurs.

The state diagram of Figure 4.5 contains the same number of states and the same state transitions as shown in Figure 4.4. However, the states have been numbered to identify them, and we will refer to them by those numbers. State 0 is the idle state, where the system awaits a signal (DO_MULT) which indicates that a multiply should be performed. In State 0 the signal IDLE_CK is asserted, and its assertion level is low. The purpose of this signal is to cause some work to occur before the actual multiplication process begins. The two things which can occur at this stage are clearing the product register and loading the multiplier shift register.

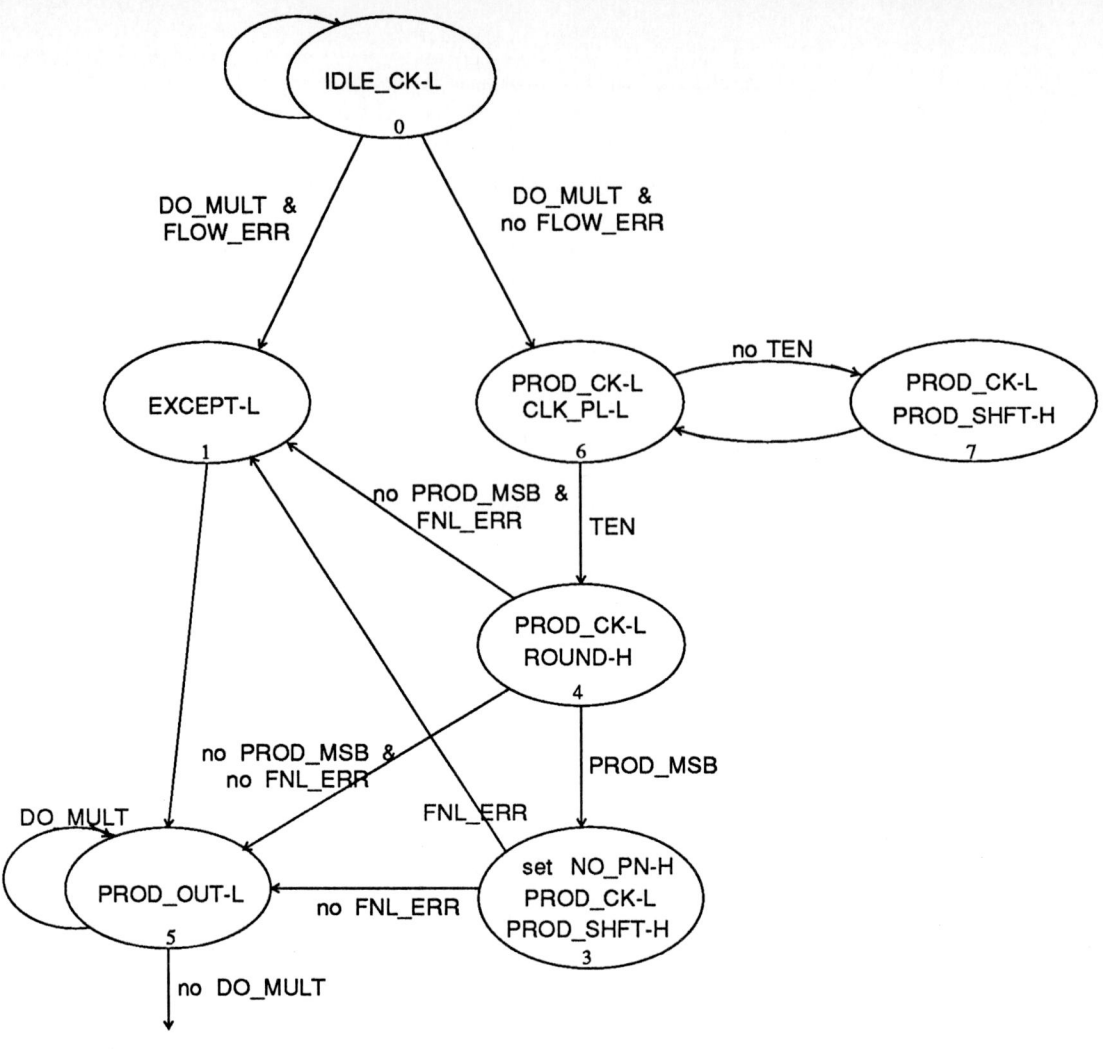

Figure 4.5. Detailed State Diagram for Floating Point Multiplier.

Thus, these two events occur continuously when the system is in the idle state. When the multiply process is to begin, DO_MULT is asserted, and the action of the system leaves the idle state. If an error condition exists, then State 1 is the next state; if no error condition exists, then State 6 is the next state.

If an error condition exists when the DO_MULT signal is asserted, then the action of the state machine moves to State 1. This is also the state which is used if error conditions are detected in State 4 or State 3. The only action here is to assert the exception signal (EXCEPT), which sets a flip-flop to identify the exception condition. The output of this flip-flop (when set) forces the exponent to be supplied by the error system; hence, the exponent will be all ones or all zeroes.

If no error condition exists when the DO_MULT signal is asserted, then the action of the state machine moves to State 6. In this state the product register clock (PROD_CK) is asserted, and the clock for the multiplier shifter (CLK_PL) is also asserted. As we shall see when we examine the circuitry generating these signals, they are asserted for the last half of a cycle only. This has two benefits. The first is that since the signals are not being asserted when the contents of the present state register as well as other control signals are changing, any glitches which might arise because of the race conditions in the logic (either external to devices, or internal to chips such as decoders) will not propagate to the system. The second benefit is to allow control lines such as enables, which do not require glitch-free operation, to be asserted prior to the assertion of the clock lines. One final comment needs to be made concerning these two clock signals (PROD_CK-L and CLK_PL-L). Both signals carry an assertion level indicator of low (the -L suffix to the mnemonic name), yet there is no corresponding logical state indicator on the clock lines of the devices. The reason for this is that the assertion level indicator on the name must match the source of the signal, not the destination, and the logic gates creating these signals does have logical state indicators on the lines which generate these signals.

The action in State 6 is to load the product register with the results of the addition of the current contents of the product register and the next partial product. With 11 partial products in the partial product array, this state should be visited 11 times. The signal generated in this state to shift the multiplier shifter (CLK_PL) is also used to clock a counter, and a control signal derived from that counter (TEN) is tested to determine the proper next state. (Why is the count 10 and not 11?) If not enough iterations have been made, then the next state is State 7; if all the partial products have been added, then the next state is State 4.

As State 6 is used to load the product register, State 7 is used to shift the results to one lesser bit of significance. Hence, the clock for the product register (PROD_CK) is asserted, as is a control line (PROD_SHFT) which indicates that the action of the product register should be a shift rather than a load. The control line (PROD_SHFT) will be asserted through the state, while the clock line (PROD_CK) will be asserted in the last half of the state. The next state (from State 7) is always State 6.

When all the needed additions of partial products have been completed, then the action of the system moves to State 4. The purpose of this state is to do the rounding, which is accomplished by the same adders that do the partial product addition. All the bits of the multiplier will have been shifted out, so the contribution from the multiplicand register is guaranteed to be zero. Thus, a control line is used in this state (ROUND) to assert a bit in the proper position for the round bit, and the product register is loaded again (by PROD_CK). As in State 7, one of the signals is an enable type (ROUND) and will be asserted through the state, while the other (PROD_CK) is a clock signal, and will be asserted only during the last half of the state. The choice of a next state depends upon two factors: the condition of the most significant bit of the product register and the status of the error condition. If the most significant bit of the product register is a zero, then in this implementation of the system, the mantissa bits are aligned properly for output to the bus. Hence, the next state depends on the error condition. If there is an error (FNL_ERR is asserted), then the next state will be State 1; if no error is present, then the next state will be State 5. However, if the most significant bit of the product register (PROD_MSB) is a ''1,'' then a final shift is required before the mantissa will be properly aligned. In this condition, the next state will be State 3.

The purpose of state 3 is to provide one final shift to align the resultant mantissa correctly with the output bus. This will be accomplished by asserting the control line for shifting the product register (PROD_SHFT) and asserting the clock line (PROD_CK), exactly as occurs in State 7. Also, this state sets the proper condition for the sum to be added to the exponent to provide the proper resultant exponent. If this final result ends up in an error condition (checked by FNL_ERR), then the next state will be State 1; if no error condition exists, then the next state will be State 5.

Note that the post normalization of this system is configured differently than a ''normal'' system, since the post normalization in this system is required when

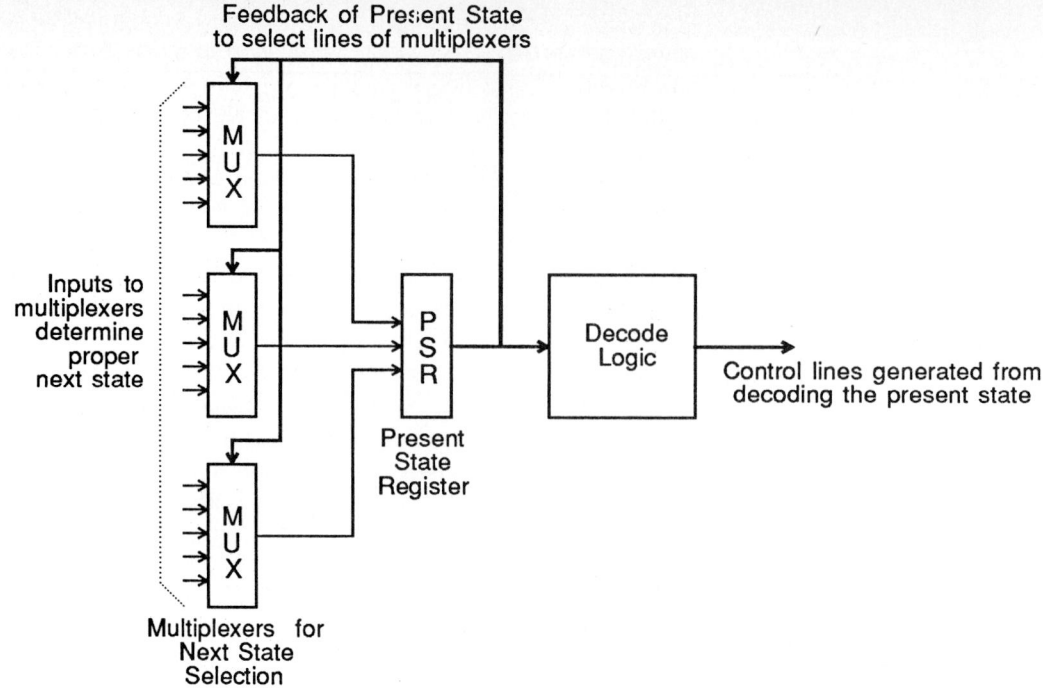

Figure 4.6. Multiplexer Method for State Machine Design.

the most significant bit is a "1," rather than a "0." Thus, the adjustment of the exponent must be done in the same manner that the shifting of the mantissa is done. If no final shift is required, then the value to add to the sum of the input exponents is −15; if a final shift is required, then the value to add to the sum of the input exponents is −16. This is accomplished by the signal NO_PN, which is set in State 3. If State 3 is not visited, then this condition will not be set, and one sum will be provided; if State 3 is visited, then this condition will be set, and a different sum will be provided. Thus, the NO_PN signal is established with a flip-flop, and this condition is cleared in the idle state (by IDLE_CK) and set as appropriate by the calculation.

The output condition is established by State 5, regardless of the condition of the result, whether it consists of an error condition or a normal result. The output is directed to the bus by the PROD_OUT signal, which also indicates to the external device that the desired information is available. The action of the system remains in this state until the external device no longer needs the information, which will be indicated by the release of the DO_MULT signal. At that time, the system returns to the idle state.

The state diagram shown in Figure 4.5 describes in detail the activity which must occur to do the work in the block diagram of Figure 4.3. We will now design the logic to perform the work specified by the state diagram.

Floating point multiplier: Design of control logic and data path

The action of the system has already been defined by the state diagram of Figure 4.5; the next task is to create the logic which will perform as indicated by the state diagram. In order to present different approaches to this task, we will present three different methods for this activity, all of which implement the same state diagram. The difference comes in the manner in which the design is undertaken. The first method is the multiplexer method, which is a very easy method to understand, implement, and check out. It is also beneficial from the standpoint that any needed changes can easily be incorporated in the system.

The multiplexer method to state machine design matches very closely the block diagram shown in Figure 4.1, and a simplified version is shown in Figure 4.6. Figure 4.1 indicates that the next state selection is determined by the present state and the external inputs. The present state is stored in a register called the Present State Register. In the multiplexer method, the next state determination is handled by a combination of multiplexers and external logic. There is a multiplexer for each input of the Present State Register, and the current contents of the Present State register identify which of the inputs of the multiplexer will be selected to form the next state selection. Thus, the appropriate inputs are provided to the multiplexers to follow the desired state diagram. To demonstrate this process, let us consider the state diagram shown in Figure 4.5, and generate the appropriate inputs for the multiplexers.

The state diagram shown in Figure 4.5 has seven states, so 3 bits are needed to represent the present state. Thus, three multiplexers are needed to provide the next state to the Present State Register. To provide the information as to the proper input values for the multiplexers, consider the next state information shown in Table 4.3. The contents of this table are determined from the state diagram by considering each state separately. For example, if the system is presently in State 0, then the next state will be State 0, State 1, or State 6, depending on the input variables. The conditions are indicated on the state diagram in Figure 4.5, and can be represented as follows:

Next State		Condition
0	(000)	DO_MULT not asserted
1	(001)	DO_MULT and FLOW_ERR
6	(110)	DO_MULT but not FLOW_ERR

As can be seen from the enumeration above, the least significant bit is asserted only when both DO_MULT and FLOW_ERR are asserted, hence the corresponding entry in Table 4.3. Also, the two most significant bits are asserted only when DO_MULT is asserted and FLOW_ERR is not asserted, which identifies the other two entries for State 0 in Table 4.3. Other entries for the table are generated in the same way.

Table 4.3. Next State Determination for Floating Point Multiplier.

Present State	Next State		
	MSB		LSB
0	DO_MULT · $\overline{\text{ERR}}$	DO_MULT · $\overline{\text{ERR}}$	DO_MULT · ERR
1	1	0	1
2	$\overline{0}$	0	0
3	$\overline{\text{FNL\_ERR}}$	0	1
4	$\overline{\text{MSB} \cdot \text{ERR}}$	PROD_MSB	1
5	DO_MULT	0	0
6	1	$\overline{\text{TEN}}$	$\overline{\text{TEN}}$
7	1	1	0

The logic for the multiplexer method of control system design can be determined directly from the state diagram and the information in Table 4.3. The diagrams for this logic are shown in Figure 4.7. Figure 4.7(a) shows the connection of the multiplexers and the Present State Register. Note that the present state is connected directly to the select lines of the multiplexers. Thus, when the system is in State 0, each multiplexer is concerned only with the logical value on its 0th input. Also note that aside from the multiplexers, only three gates are required to implement the state diagram of Figure 4.5.

One comment needs to be made concerning the use of the DO_MULT signal in Figure 4.7. This signal is not, in general, synchronous with the system clock, and hence can cause problems. The problems arise when transitions are made on the DO_MULT line at the same time that the system clock is occurring. If this were to happen, the next state logic, which in this case consists of multiplexers and a few gates, will be changing just when it should be stable, which is at the active edge of the clock. When this happens, then the action of the system will not correctly follow the state diagram, and the system behavior will be unpredictable. Thus, steps must be taken to synchronize external inputs with the system clock. One of the simplest ways to accomplish this is to accept external inputs with a register clocked at the same time that the Present State Register is clocked, and then use the outputs of the register as the appropriate variables in the next state logic. This is accomplished in Figure 4.7(a) by using as a buffer the same register which is used as the Present State Register. Thus, DO_MULT is buffered by one bit position of the '175, and the buffered signal, B_DO_MULT, is used in the logic. *External inputs must be made synchronous with the system clock, or the system will not behave properly.*

The generation of the signals to control the data path is shown in Figure 4.7(b). Most of the work is accomplished by a decoder which has its select lines connected to the present state (PR_STATE). Thus, when the system is in State 0, only the Y0 output of the decoder will be asserted. Decoders, such as the '138 shown in Figure 4.7, are not perfect devices, and when transitions occur on the select lines, there is the possibility that glitches will occur on the output lines. Thus, one of the low true enables of the present state decoder in Figure 4.7(b) has been connected to SYS_CLK-H. The effect of this is to disable the decoder during the first half of a clock cycle (assuming that the clock is symmetric), during which transitions will occur on the select lines. In this way, SET_NO_PN is asserted during the last half of State 3, EXCEPT is asserted during the last half of State 1, and PROD_CK is asserted during the last half of States 2, 4, 6, and 7.

Some control signals are levels, rather than clock lines, and require a different solution than the clocks generated by the decoder. Figure 4.7(b) shows three flip-flops consisting of cross-coupled NAND gates, each of which is used to create a control signal which must be a level. EXCEPTION is set in State 1 and used to select the error-generated exponent; NO_PN is set in State 3 and is used to select the proper amount for adjustment of the exponent; and PROD_OUT is set in State 5 and used to enable the output to the data bus, as well as to communicate to the external device that the data is ready. Each of these flip-flops is reset by the signal IDLE_CK, so each will be disabled when the system reaches State 0. The remaining control signals generated in Figure 4.7(b) are PROD_SHIFT and ROUND. PROD_SHFT will be asserted in States 3 and 7, and ROUND will be asserted in State 4. Since these signals are to be asserted throughout the state, the present state lines are decoded directly to generate the signal. But since these gates are not perfect either, occasionally glitches may occur on these control lines. However, since these

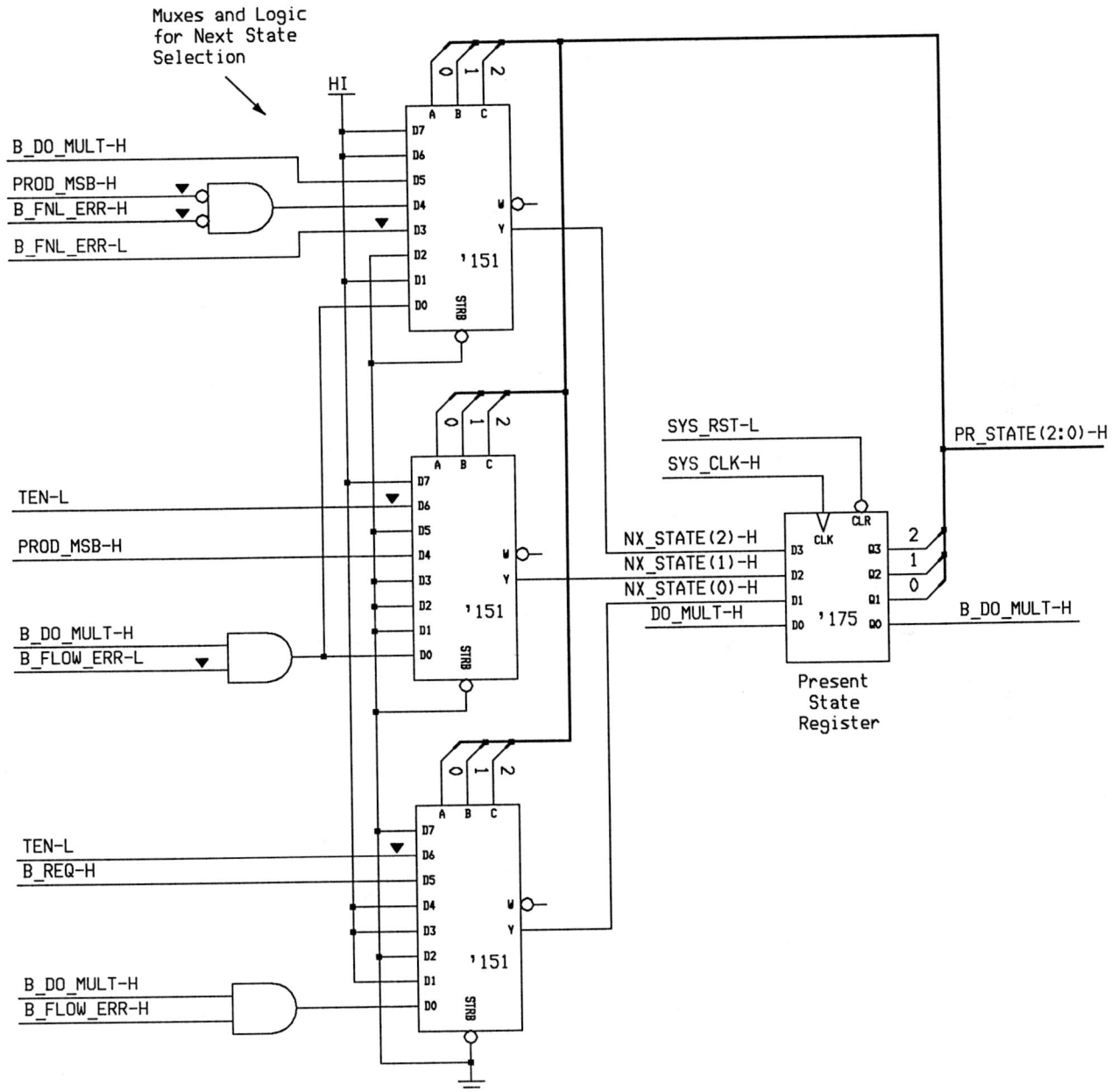

Figure 4.7(a). Detailed Logic Diagram for Control of Floating Point Multiplier: Present State Register and Next State Logic.

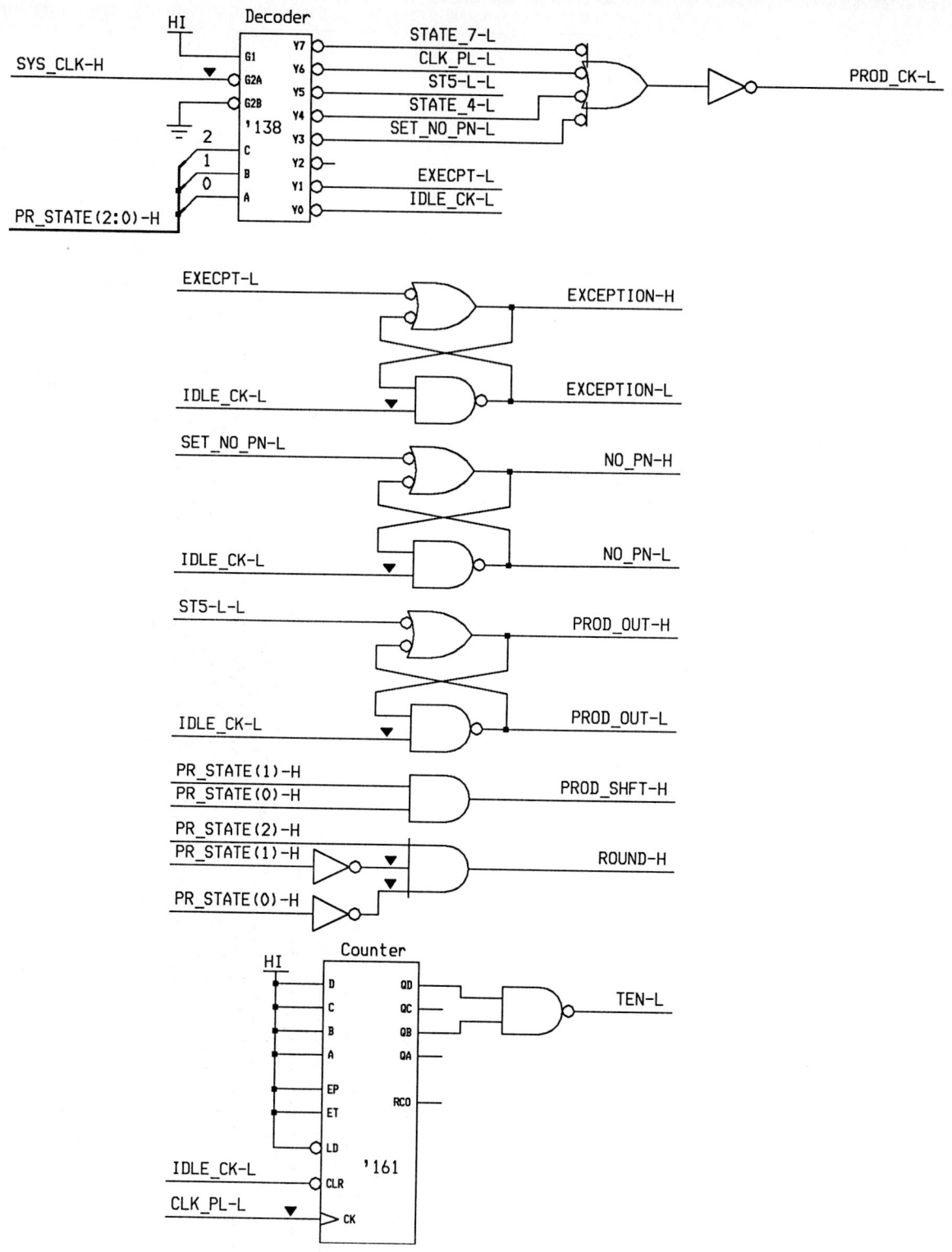

Figure 4.7(b). Detailed Logic Diagram for Control of Floating Point Multiplier: Generation of Control Signals.

control lines are not utilized by edge sensitive functions, glitches on these lines should not cause problems.

The one remaining item shown in Figure 4.7(b) is the iteration counter for partial product addition. This is simply a counter which is cleared by the idle condition, and incremented each time the multiplier shifter is clocked. (Why does the NAND gate on the output of the counter check for nine rather than ten?)

As the details of the system are developed, the detailed logic diagrams for the data path section of the system can also be created. For the Floating Point Multiplier, these details are shown in Figure 4.8. Figure 4.8(a) contains the registers and drivers associated with holding the multiplier and multiplicand. We have been assuming that the external device is responsible for activating the control lines shown (PLIER_LD, PLIER_OUT, PCAND_LD, and PCAND_OUT). The registers are '273s, each of which is capable of holding 8 bits. The tri-state drivers are '541s which each drive 8 bits to the data bus.

The exponent addition and error checking is shown in Figure 4.8(b). The first pair of adders performs the operation of adding the two exponents together. The second pair of adders adds in the appropriate adjustment factor. If NO_PN is asserted, this will be 16; if it is not asserted, then the adjustment value is 15. Check these values with the tables shown earlier in this chapter and verify that the adjustments are correct. Note that the 3 most significant bits of the second pair of adders are tied high, and that the most significant output of the adder is used as an error checker. What condition is checked by this arrangement?

Errors which influence the outcome of the multiplication are also checked by the logic shown in Figure 4.8(b). Gates 1 and 2 check for a zero value on either of the inputs, and gate 3 checks for an underflow condition. Gates 4, 5, and 6 check for the initial overflow condition. (How does this work?) If any of these conditions exists as the multiplication commences, FLOW_ERR will be asserted, and the action of the state diagram will be influenced appropriately. Gates 9, 10, 11, and 12 are used to check underflow/overflow conditions which are detectable at the end of the operation. If such a condition exists, then FNL_ERR will be asserted, and the action of the state machine altered accordingly. If an overflow condition exists, then the output of gate 13 will be asserted, and the exception exponent delivered to the output will be all ones. If an overflow condition does not exist, then the exception exponent will be all zeros. The exponent to be provided, as well as the sign of the result, is selected by the Output Multiplexer and Driver. This device ('257) provides the selection operation of a multiplexer as well as the tri-state capability of the a bus driver. The appropriate output value is selected by the EXCEPTION signal, which will choose the exception exponent only when asserted. The selected exponent (and sign) will be enabled onto the bus by the PROD_OUT signal. Note that the normal sign generation is the exclusive-OR of the signs of the input values.

The Multiplier Shifter is shown in Figure 4.8(c). This pair of shift registers is loaded with the mantissa of the multiplier, and during the partial product addition is made available one bit at a time. The loading of the multiplier mantissa occurs each time that IDLE_CK is asserted, which maintains the most current value of the multiplier. The value loaded consists of the 10 bits of the mantissa supplied by the multiplier register and the single bit which comes from the hidden bit representation. The remaining bits are forced to zero, which is needed for this implementation. (Why?) The bits of the multiplier mantissa are all made available least significant bit first, and appear on the AND line which is then used in Figure 4.8(d).

The portion of the data path logic used for partial product addition, product register, rounding, and post-normalization is shown in Figure 4.8(d). A set of AND

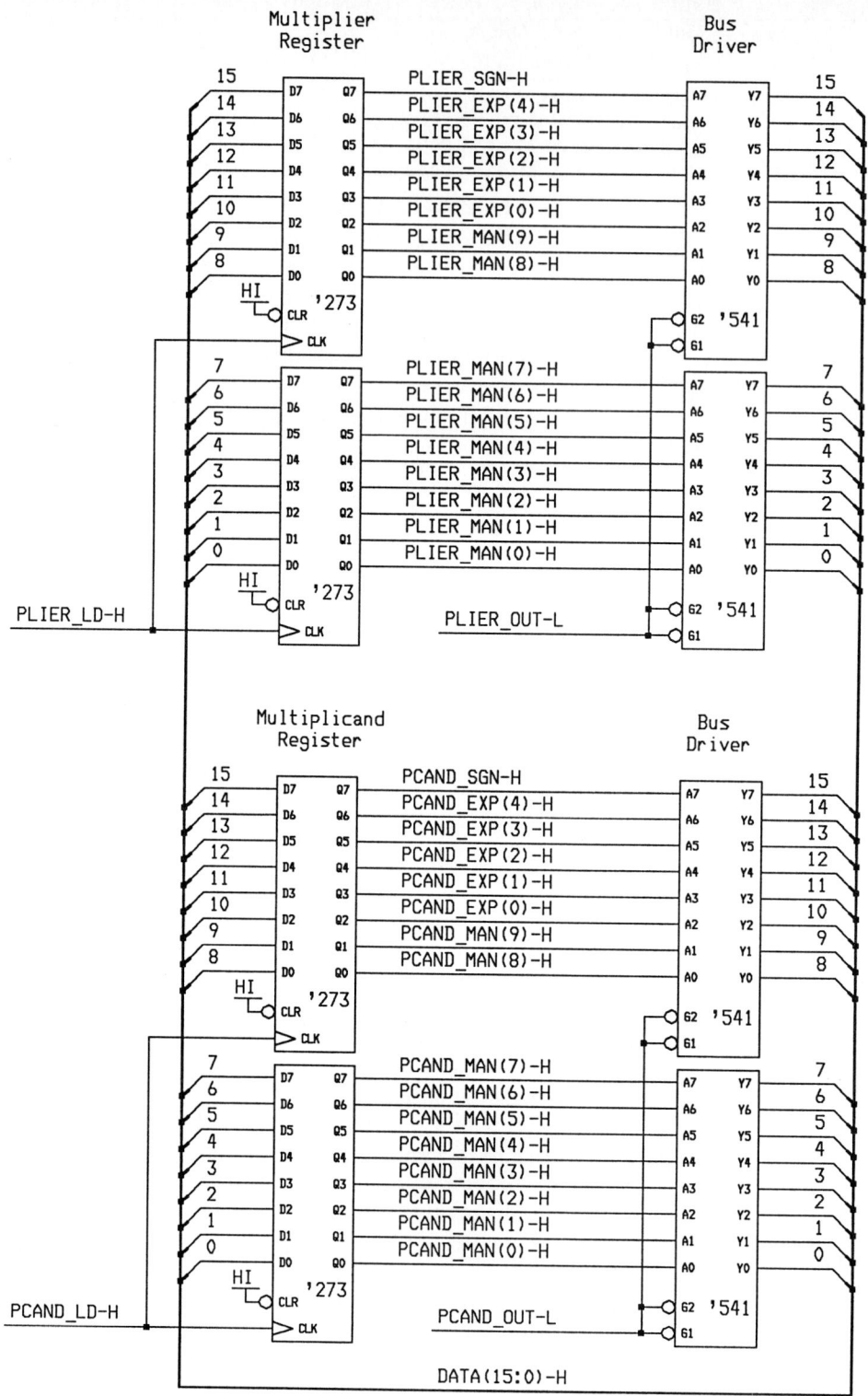

Figure 4.8(a). Detailed Logic Diagram for Data Path of Floating Point Multiplier: Multiplier and Multiplicand Registers.

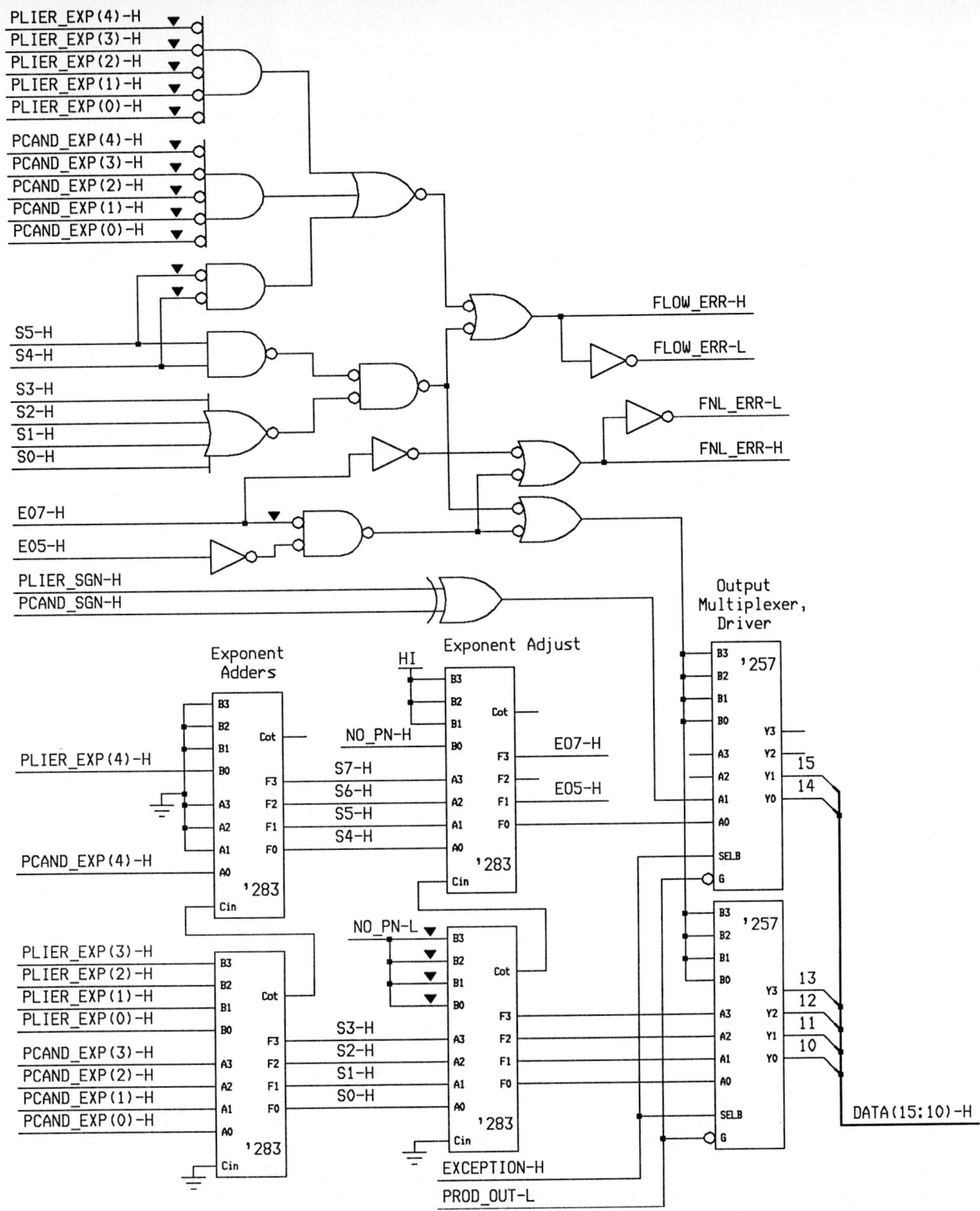

Figure 4.8(b). Detailed Logic Diagram for Data Path of Floating Point Multiplier: Exponent Addition and Error Detection.

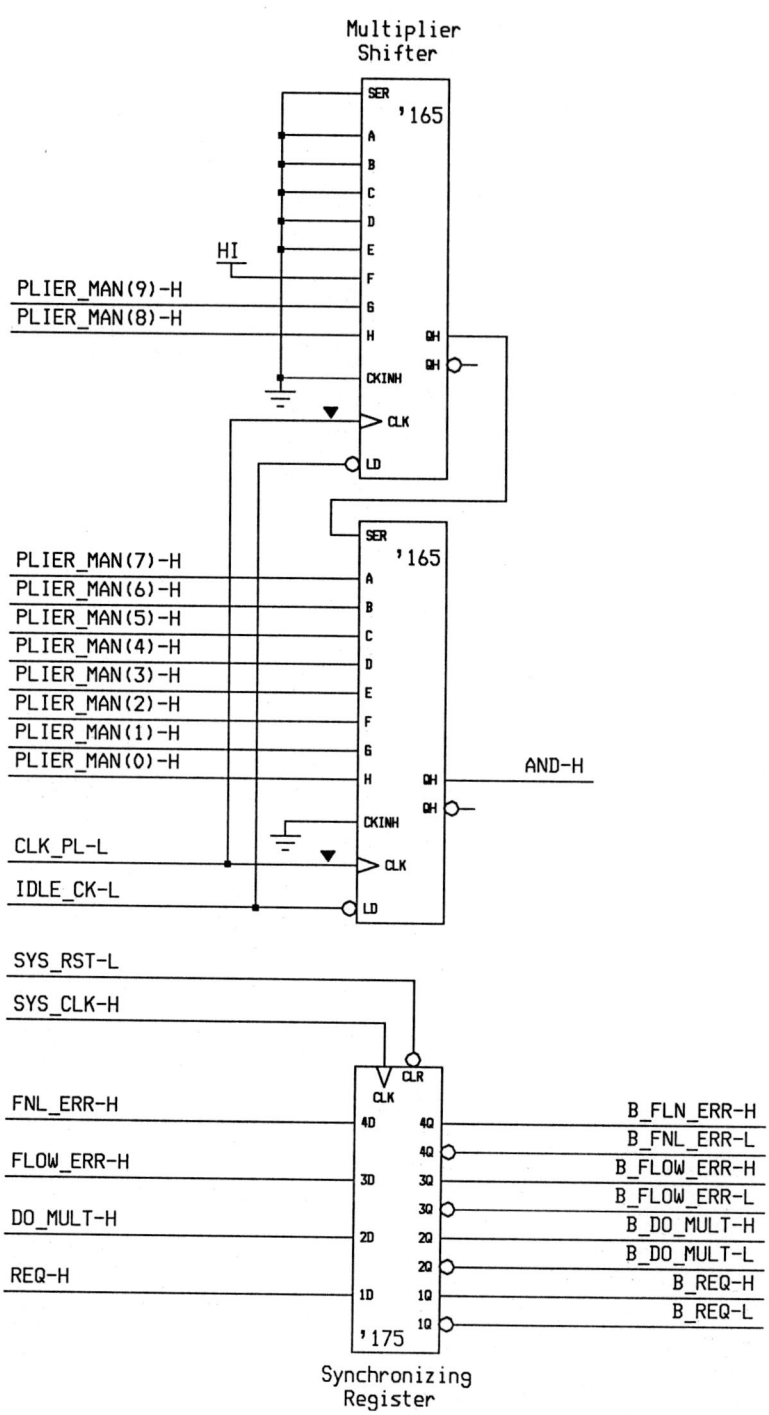

Figure 4.8(c). Detailed Logic Diagram for Data Path of Floating Point Multiplier: Multiplier Shifter Register.

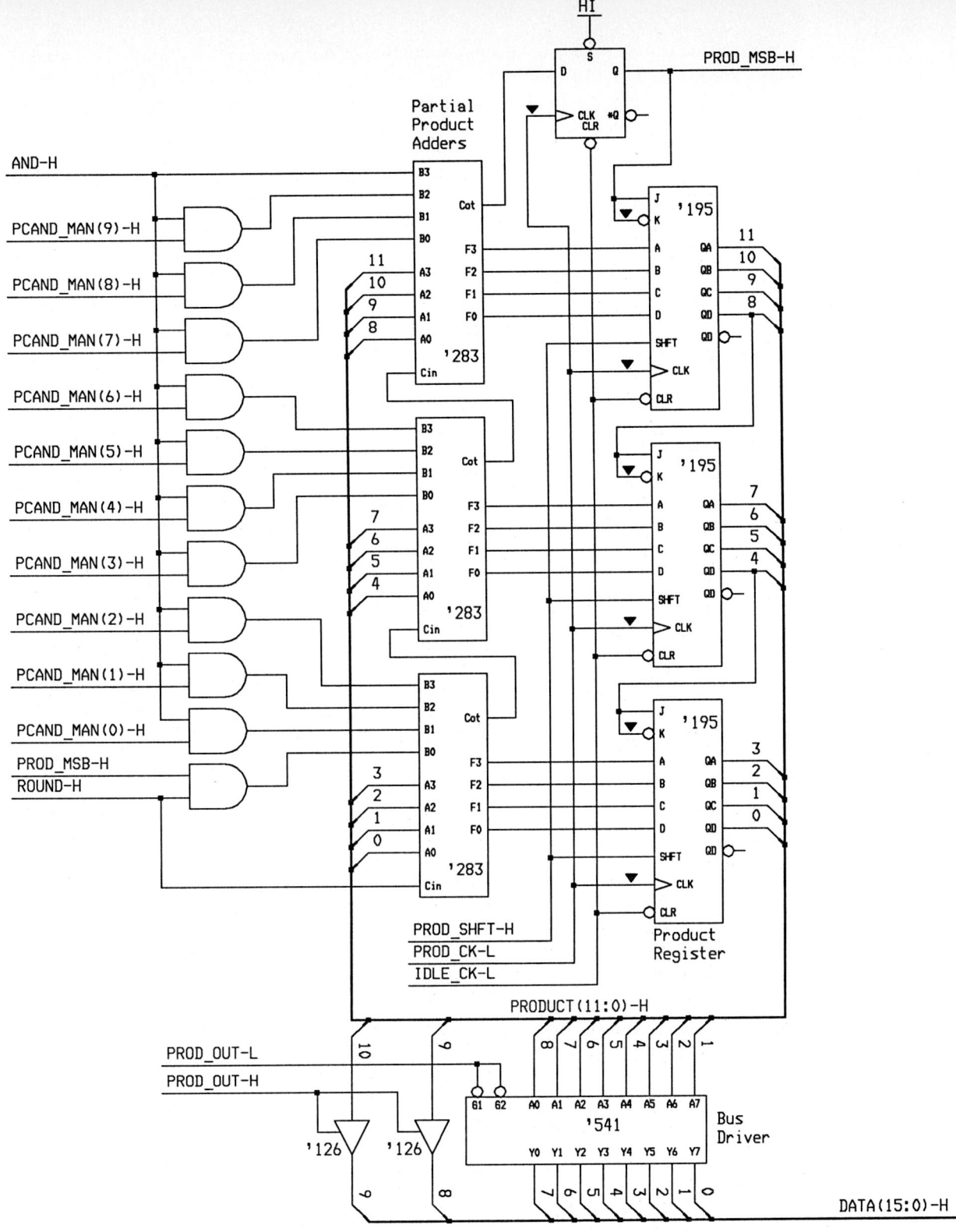

Figure 4.8(d). Detailed Logic Diagram for Data Path of Floating Point Multiplier: Product Register and Partial Product Adders.

gates is used to create the appropriate row of the partial product array. This results from the fact that each row of the partial product array consists of all zeros or a copy of the mantissa of the multiplicand, depending on the bit of the multiplier. In this case, the bit of the multiplier is supplied by the AND signal, and the multiplicand mantissa comes directly from the multiplicand storage register, except for the hidden bit. The hidden bit is handled directly by the AND signal. (How?) The product register consists of '195s, which can be loaded or shifted. The '195s can also be cleared, and this occurs whenever the system is in the idle state. Once the system leaves the idle state, something happens at the product register whenever the PROD_CK line is asserted. If the shift line (PROD_SHFT) is asserted, the action is a shift to a bit of lesser significance. If the shift line is not asserted, then the action is to load the sum of the current value in the Product Register with the other input to the adders. For 10 cycles, this value will be an appropriate row of the partial product array. For the final load, this will be an appropriate round bit. It is left as an exercise for the reader to demonstrate that the rounding works properly when the ROUND signal is asserted. When the entire process is completed, the result is presented to the data bus through tri-state drivers which are enabled when the PROD_OUT signal is asserted.

Alternative control signal generation for the floating point multiplier

The generation of the state machine and the control signals as shown in Figure 4.7 is done in a straightforward and relatively painless fashion. However, it is not the "classical" method. Hence, we will now demonstrate how to construct a state machine which will follow the state diagram of Figure 4.5 using a simple register and the more classical D flip-flop method. The first step in this method, after generating the same information we used to create the state diagram of Figure 4.5, is to generate a next state table. Such a table is shown in Table 4.4. The states shown in Table 4.4 are given in a binary representation rather than a decimal representation. As can be seen from Table 4.4, the creation of the logic necessary for next state generation reduces to a combinational design problem. However, by not using the capabilities of the multiplexers, the number of variables required for each bit of the next state generation is 8, with 3 bits in the present state and 5 other bits. The previous method allowed us to ignore the bits (by using the multiplexers) except when they were needed. Nonetheless, the patterns shown in Table 4.4 are sufficient to generate the logic needed for next state generation. The logic can be generated in a number of ways, from an eight-variable Karnaugh map (very confusing) to a computer program which creates the proper result. The resulting equations are given here, and it is left as an exercise for the reader to verify that they are correct. In the following equations, the 3 present state bits are represented as PS2, PS1, and PS0, and the 3 next state bits are NS2, NS1, and NS0,

$$NS2 = PS2 \cdot \overline{PS0} \cdot \overline{FNL\_ERR} \cdot \overline{PROD\_MSB} +$$

$$\overline{PS2} \cdot \overline{PS1} \cdot DO\_MULT \cdot \overline{FLOW\_ERR} +$$

$$\overline{PS2} \cdot \overline{PS1} \cdot PS0 + \overline{PS2} \cdot PS0 \cdot \overline{FNL\_ERR} +$$

$$\overline{PS1} \cdot PS0 \cdot DO\_MULT + PS2 \cdot PS1$$

$$NS1 = \overline{PS2} \cdot \overline{PS1} \cdot \overline{PS0} \cdot DO\_MULT \cdot \overline{FLOW\_ERR} +$$

$$PS2 \cdot \overline{PS1} \cdot \overline{PS0} \cdot PROD\_MSB + PS2 \cdot PS1 \cdot \overline{TEN} + PS2 \cdot PS1 \cdot PS0$$

Table 4.4. Next State Determination for Floating Point Multiplier.

Present State			Inputs †					Next State		
			DO MULT	FLOW ERR	FNL ERR	PROD MSB	TEN			
0	0	0	0	X	X	X	X	0	0	0
0	0	0	1	0	X	X	X	1	1	0
0	0	0	1	1	X	X	X	0	0	1
0	0	1	X	X	X	X	X	1	0	1
0	1	0	X	X	X	X	X	0	0	0
0	1	1	X	X	0	X	X	1	0	1
0	1	1	X	X	1	X	X	0	0	1
1	0	0	X	X	X	1	X	0	1	1
1	0	0	X	X	0	0	X	1	0	1
1	0	0	X	X	1	0	X	0	0	1
1	0	1	1	X	X	X	X	1	0	1
1	0	1	0	X	X	X	X	0	0	0
1	1	0	X	X	X	X	0	1	1	1
1	1	0	X	X	X	X	1	1	0	0
1	1	1	X	X	X	X	X	1	1	0

† X means don't care

$$NS0 = PS2 \cdot \overline{PS1} \cdot \overline{PS0} \ + \ PS2 \cdot \overline{PS0} \cdot \overline{TEN} \ +$$

$$\overline{PS1} \cdot DO\_MULT \cdot \overline{FLOW\_ERR} \ + \ \overline{PS1} \cdot PS0 \cdot DO\_MULT \ + \ \overline{PS2} \cdot PS0$$

The logic for this system can be implemented directly from the equations, or some relaxation can be allowed, which will permit the combining of some portions of terms. This will result in a system which takes more than two gate delays to determine the result, but which will require fewer overall gates. Similarly, the generation of the output signals in Figure 4.7(b) was accomplished with the use of a decoder to determine the present state, and the use of the decoder allowed the techniques demonstrated in Figure 4.7(b). The generation of the control signals can be done with random logic rather than with ICs such as the decoder. This would necessitate decoding the present state lines with separate gates to generate the various control signals needed. Using these random logic techniques for next state generation and for creation of the control signals results in the system depicted in Figure 4.9.

The next state logic is shown in Figure 4.9(a). Note that the number of levels of logic has been increased to allow the use of gates with fewer inputs, and to allow sharing of common terms. The reader is challenged to compare the gating system involved in the generation of the next state with the equations listed above. Some slight differences will emerge, and the questions are, will the system still function as desired, and do the alterations allow the system to be built with fewer gates?

The random logic approach to the generation of control signals is shown in Figure 4.9(b). An examination of the logic reveals that the same types of signals

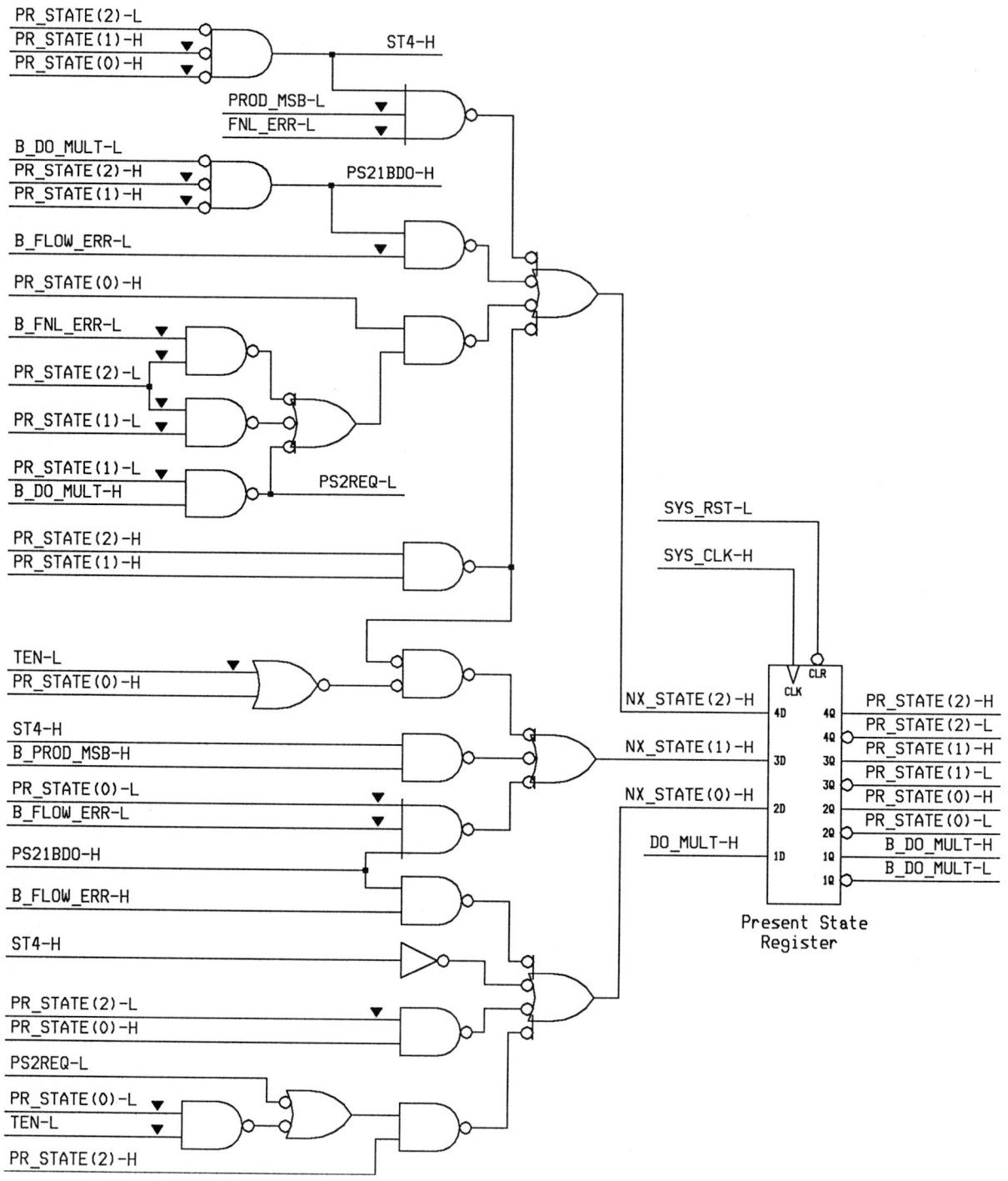

Figure 4.9(a). Random Logic Implementation for Control System
of the Floating Point Multiplier: Next State Generation.

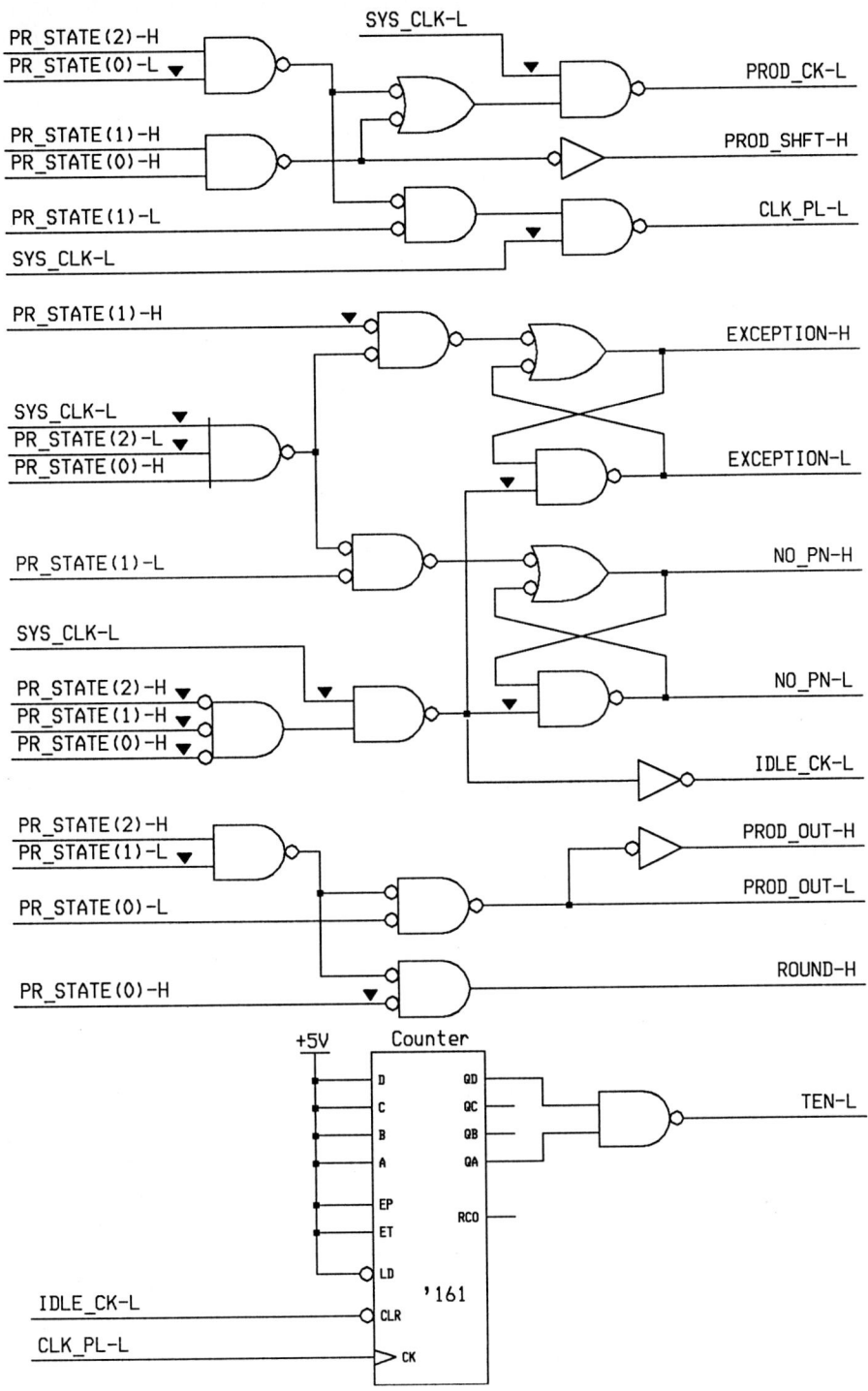

Figure 4.9(b). Random Logic Implementation for Control System
of the Floating Point Multiplier: Creation of Control Signals.

are being generated: signals of the enable type which are asserted throughout the state, and signals of the clocking type which are asserted for the last half of the state only. Enabling the clocking type of signal for only the last half of the state has the effect of disabling the clock signal during the time when glitches could be present due to delays in the inputs to the clocking function.

Another comment which can be made at this time concerns signals such as EXCEPTION, which is set in one state and reset in another. The use of a flip-flop to create this type of signal, as shown in Figure 4.9(b), guarantees that the signal will be asserted, without glitches, until the system arrives at the destination state. Another way in which this type of signal could be generated is to create an enable type of signal which is asserted in all the intermediate states. A signal generated in this fashion has the possibility of glitches as the state changes are made, so it should be used with caution. Both techniques can be useful, and a designer must make the choice of the technique used based on all the appropriate design criteria.

We will now consider still another method for creating logic to follow a state diagram. This is the J-K flip-flop method, which has been used over the years in the design of sequential systems. This method was the preferred method of sequential design because the logic required for next state generation was generally smaller than the logic used in the D flip-flop method demonstrated above, and the amount of logic required was a primary consideration when logic was very expensive. Other criteria, such as total power, board space, ease of implementation, silicon real estate, and so forth, may be more important to logic designers today. As with any design, all the criteria should be evaluated before settling on an approach.

The J-K approach begins in exactly the same fashion as the D method shown above. A table which defines the action of the system is created. The table is exactly like Table 4.4, but with additional information. This information is the excitation requirement for the J-K flip-flop. That is, the next state information shown in Table 4.4 is used to create another field, which represents the values present on the J and K inputs to cause the transition called for in the table. This entry is derived from the excitation table for the J-K flip-flop, which is shown in Table 4.5. The table identifies the state of the system before the occurrence of the system clock (Q_N) and the state of the system after the occurrence of the system clock (Q_M). In order to cause this transition to occur, the J and K inputs of the flip-flop must be as shown in Table 4.5. The presence of the don't care conditions shown in Table 4.5 allows the logic required for the respective inputs to be minimized. To demonstrate how this works, the information in Table 4.4 has been repeated in Table 4.6, with additional fields for the excitation information needed to use J-K flip-flops for the present state register. The generation of the logic for the J-K flip-flops requires six equations, rather than the three for the D implementation above, but the equations are simpler:

Table 4.5. Excitation Table for J-K Flip-Flop.

Q_N	Q_M	J†	K†
0	0	0	X
0	1	1	X
1	0	X	1
1	1	X	0

† X means "don't care."

Table 4.6. Next State Determination for Floating
Point Multiplier Using J-K Flip-Flops.

Present State	Inputs †					Next State	J-K Flip Flop Excitation		
	DO MULT	FLOW ERR	FNL ERR	PROD MSB	TEN		J K	J K	J K
0 0 0	0	X	X	X	X	0 0 0	0 X	0 X	0 X
0 0 0	1	0	X	X	X	1 1 0	1 X	1 X	0 X
0 0 0	1	1	X	X	X	0 0 1	0 X	0 X	1 X
0 0 1	X	X	X	X	X	1 0 1	1 X	0 X	X 0
0 1 0	X	X	X	X	X	0 0 0	0 X	X 1	0 X
0 1 1	X	X	0	X	X	1 0 1	1 X	X 1	X 0
0 1 1	X	X	1	X	X	0 0 1	0 X	X 1	X 0
1 0 0	X	X	X	1	X	0 1 1	X 1	1 X	1 X
1 0 0	X	X	0	0	X	1 0 1	X 0	0 X	1 X
1 0 0	X	X	1	0	X	0 0 1	X 1	0 X	1 X
1 0 1	1	X	X	X	X	1 0 1	X 0	0 X	X 0
1 0 1	0	X	X	X	X	0 0 0	X 1	0 X	X 1
1 1 0	X	X	X	X	0	1 1 1	X 0	X 0	1 X
1 1 0	X	X	X	X	1	1 0 0	X 0	X 1	0 X
1 1 1	X	X	X	X	X	1 1 0	X 0	X 0	X 1

† X means "don't care."

$$J_{PS2} = \overline{PS1} \cdot DO\_MULT \cdot \overline{FLOW\_ERR} + PS0 \cdot \overline{FNL\_ERR} + \overline{PS1} \cdot PS0$$

$$K_{PS2} = \overline{PS1} \cdot \overline{PS0} \cdot PROD\_MSB +$$
$$\overline{PS1} \cdot \overline{PS0} \cdot FNL\_ERR + \overline{PS1} \cdot PS0 \cdot \overline{DO\_MULT}$$

$$J_{PS1} = \overline{PS2} \cdot \overline{PS0} \cdot DO\_MULT \cdot \overline{FLOW\_ERR} + PS2 \cdot \overline{PS0} \cdot PROD\_MSB$$

$$K_{PS1} = \overline{PS0} \cdot TEN + \overline{PS2}$$

$$J_{PS0} = \overline{PS1} \cdot DO\_MULT \cdot FLOW\_ERR + PS2 \cdot \overline{TEN} + PS2 \cdot \overline{PS1}$$

$$K_{PS0} = PS2 \cdot \overline{DO\_MULT} + PS2 \cdot PS1 + \overline{PS0}$$

One implementation which corresponds to the equations shown above is found in Figure 4.10. Note that the complexity and the number of gates involved is smaller than other random logic implementations. In general, this will be the case for J-K implementations, although like most other guidelines it is not always true. With fewer gates, the implementation requires less board space, and can lead to a cost-efficient system. However, if board space is a criteria, there is another method which will provide an even better result.

The final method we will consider is an implementation which uses programmable read-only memory (PROM) for creation of both the next state logic and the generation of the control signals. The concept behind this method is very

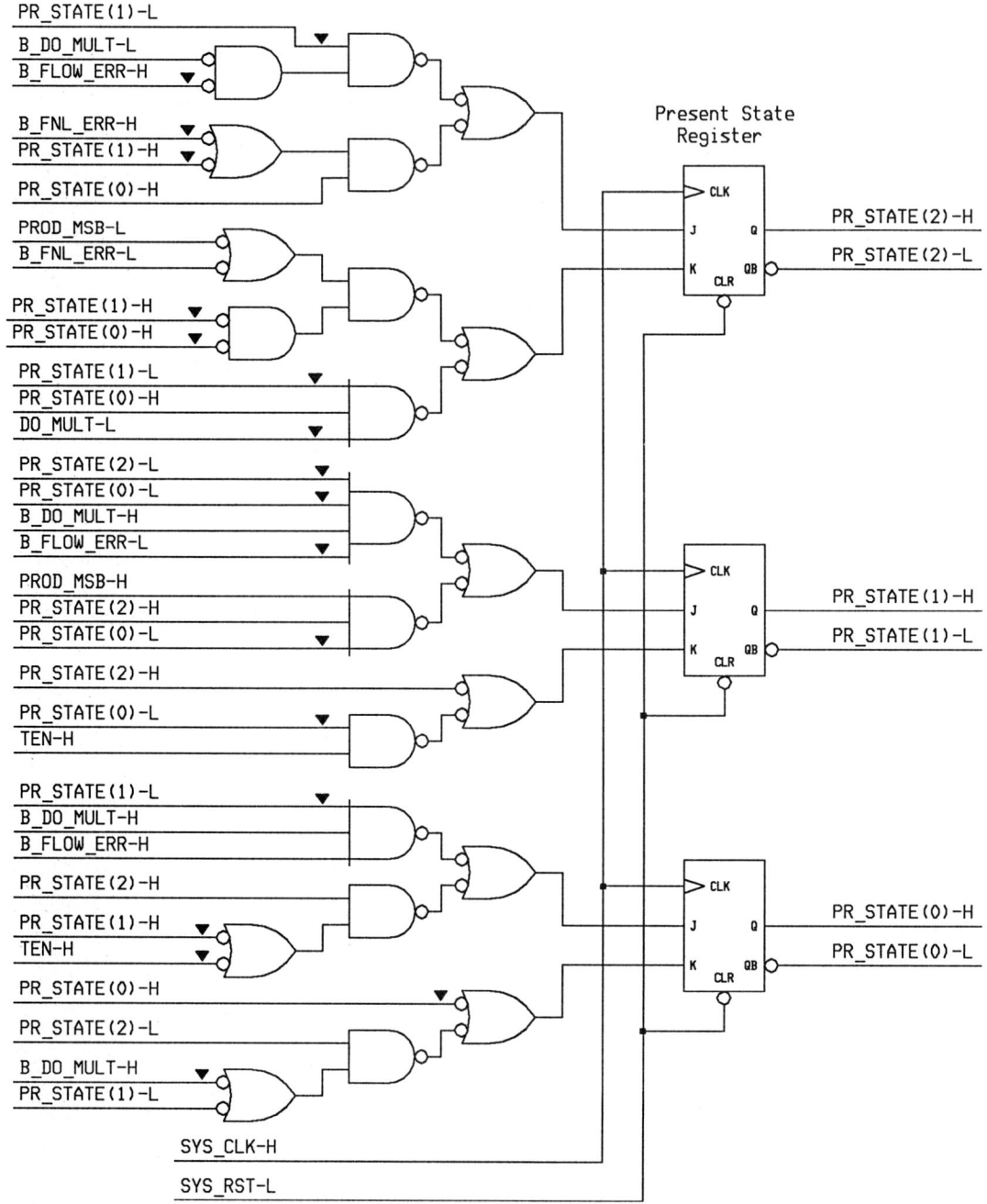

Figure 4.10. Random Logic Implementation for Control System
of the Floating Point Multiplier: Sequencing with J-K Flip-Flops.

simple: directly implement the logic with memory. To see how this might work, consider the block diagram shown in Figure 4.1, and replace the logic blocks with memory. The logic blocks and memory have the same basic function: outputs are asserted according to a set of inputs, with no history function. Thus, in the memory implementation of logic, each input pattern provides an address, and the information stored at that address is the proper assertion level for the outputs, given that input pattern. With memory devices rivaling logic speeds, this can be a very useful mechanism. Also, with a memory implementation, an arbitrarily complex logic system is as easy to implement as a simple system, since each input pattern has its own pattern of output assertions.

We will make one modification to the block diagram of Figure 4.1, which is shown in Figure 4.11. This block diagram has the same basic functional units as shown in Figure 4.1: the Next State Logic block is responsible for determining the correct next state of the system, and the Generation of Control Signals block is responsible for asserting the proper control signals. The difference is one of perception and implementation. The Next State Logic ascertains the correct next state based on the contents of the Present State Register (the current state of the system) and the levels of the External Inputs. However, at the same time that we know what the next state will be, we also know what signals are to be asserted in that state, since that was determined at the time the detailed state diagram was created. Therefore, using the same information as used by the Next State Logic (present state and external inputs) the correct assertion level for signals in the next state can be determined. Thus, the Generation of Control Signals is performed prior to the active edge of the System Clock, so that at the same time a new state is being loaded into the Present State Register, a new set of bits is loaded into the register containing the control signals. Now we implement both logic blocks shown in Figure 4.11 with memory elements rather than random logic, and the entire system can be contained in a very few devices.

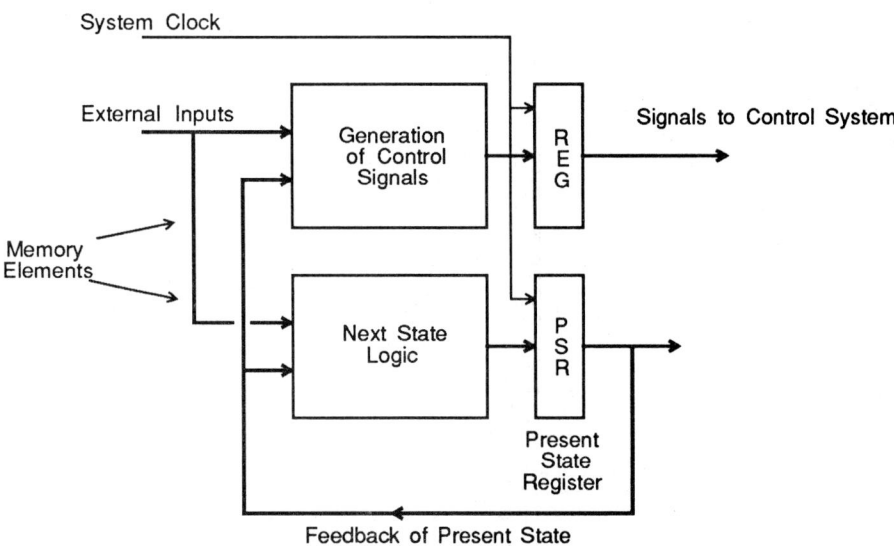

Figure 4.11. Alternative Form of Clocked Sequential System.

The device we will use for this exercise is a registered PROM; that is, it is a Programmable Read-Only Memory which also has in the same package an edge-triggered register. The specific device under consideration here is the Am27S35, which holds 1024 values, each of which is 8 bits wide. This is more than we need, so the action could be expanded if some change is required. A simplified diagram of the Am27S35 is shown in Figure 4.12. As shown in the figure, there are three different features provided by this registered PROM which we will utilize. The most important is the memory array itself, which contains 8192 bits of information. The second is the edge-triggered register, which will pass the value provided by the memory array to the outputs when a low-to-high transition is provided on the clock line. The third element is the initialization feature. This allows any pattern to be forced onto the output lines when the initialize input is asserted, and hence can be used to force the state machine to a known condition with predetermined output levels.

The state machine which uses this device is organized exactly as shown in Figure 4.11. The logic diagram for the system is shown in Figure 4.13 The diagram is deceptively simple, since the 27S35s contain both memory arrays and registers. Nevertheless, the entire system, including generation of outputs, requires only four devices. Note that the memory array is larger than needed in both directions. That is, there are more locations than needed, and there are more outputs than are used. Without any additional devices in the system, up to five additional control lines could be controlled. Also, two more input variables could be used for making decisions. Or the same number of input variables could be maintained, but the complexity of the system expanded to allow for 5 present state bits, which would allow up to 32 states in the state diagram.

The eight control lines generated in Figure 4.13 include both the enable type, which are asserted throughout a state, and the clocking type. The clocked signals are created by ANDing the output of the registered PROM with the inverse of the system clock, in a manner very similar to the earlier implementations. Note that without this modification of the outputs, certain action would not occur when a signal is asserted in two consecutive states. That is, even though PROD_CK is asserted in both State 6 and State 7, no action would occur at the product register without the ANDing function in the control line, since the action occurs on a transition of the control line, not just the asserted level of the signal.

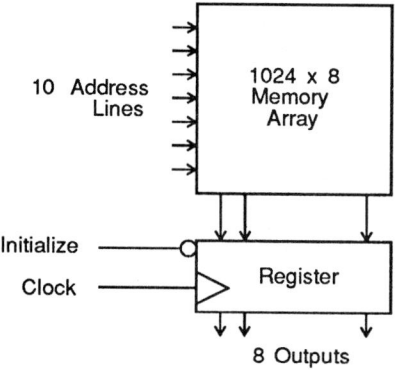

Figure 4.12. Simplified Diagram of an Am27S35.

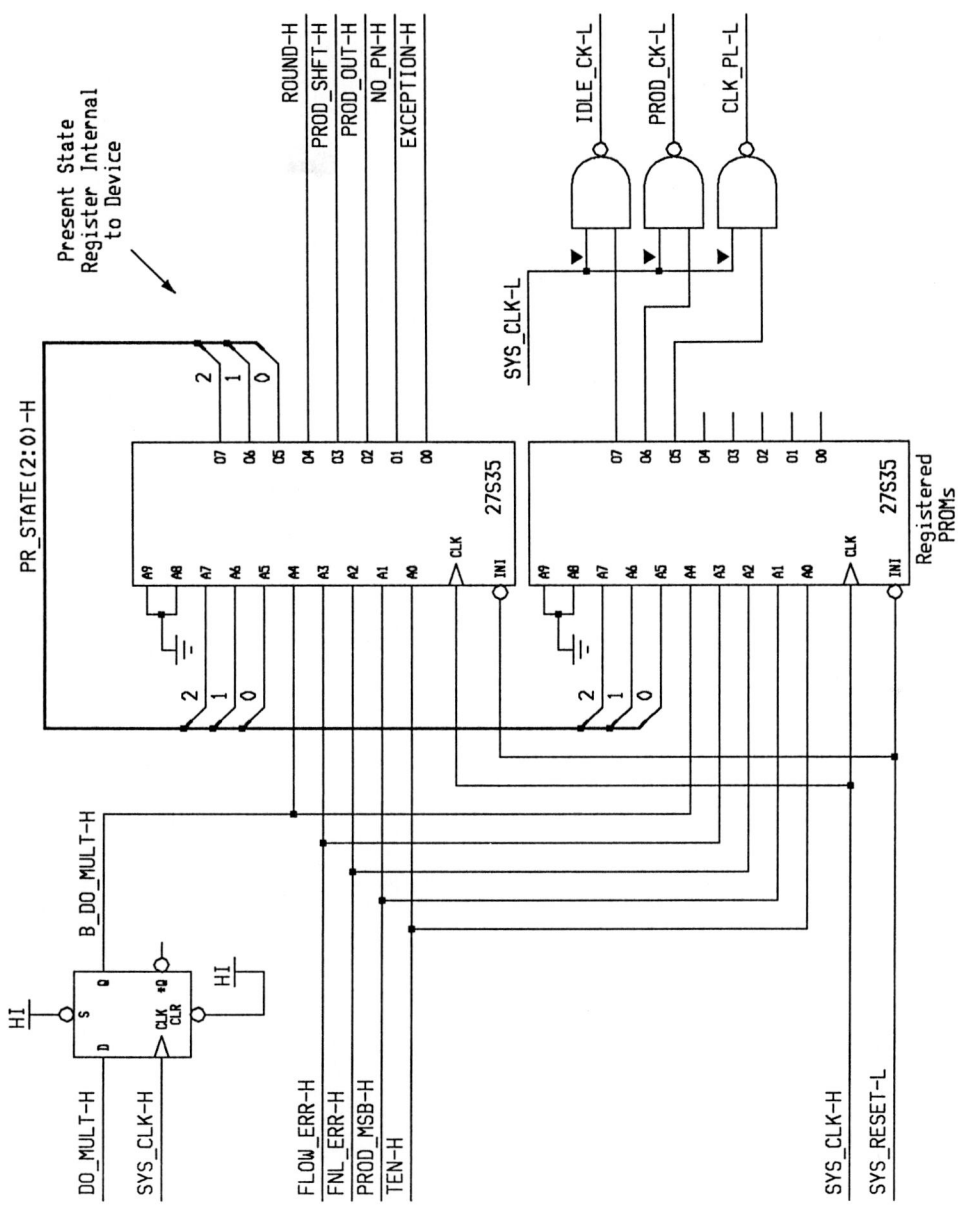

Figure 4.13. Logic Diagram for Memory Based State Machine.

The contents of the memory elements for the system of Figure 4.13 are determined from a combination of the information in Table 4.4 and the state diagram. Table 4.4 determines what the 3 bits used for the next state should be, and using that information (what the next state will be), the assertion levels for the control lines in that state are determined. The contents of the 256 locations needed for the system shown in Figure 4.13, with the state diagram of Figure 4.5, is included in Appendix B. It is left as a challenge for the reader to verify the action of the system by choosing a few cases and comparing them to the table in Appendix B.

A final comment about this method concerns the ease of modification. If, in the process of checking out the system, it is determined that a different behavior of the state machine is required, then no wires need be changed. Rather, new PROMs are created to reflect the new behavior, which is shown in a reorganized state diagram. Thus, this allows for relatively painless modification of system behavior.

Floating point multiplier: Determination of minimum cycle time

As we have seen, the creation of a synchronous sequential system involves both a data path determination and a control path design. When choosing the speed of the system, or the minimum cycle time, both factors must be considered. The process is one of determining the maximum delay through the system, and then selecting an appropriate clock time to match that characteristic. The basic elements of the system cycle time are shown in Figure 4.14. When the active edge of the clock occurs, there is a period of time required (t_{reg}) for the outputs of the register to become stable. First, let us consider the registers involved in the data path, such as the product register. When a low-to-high transition occurs on the PROD_CK line, the output of the register (PRODUCT(11:0)) becomes stable a t_{reg} later. Using advanced low-power schottky (ALS) parts for this design, this time is typically 11 nsec. Not only is the output of the product register stable at this time, but also the signal from the multiplier shifter (AND-H). The next period of time required is shown as t_{logic} in Figure 4.14. In general, this time includes all the time required for the effects of the logic changes to propagate through the random logic which feeds the next register in the system. In the data path for the floating point multiplier, the register involved in the sequential operation is the product register, and the random logic includes the partial product adders and the AND gates which provide one of the values to the partial product adders. For this system, the delay time is on the order of 65 nsec. (Remember that in an adder system all the delays

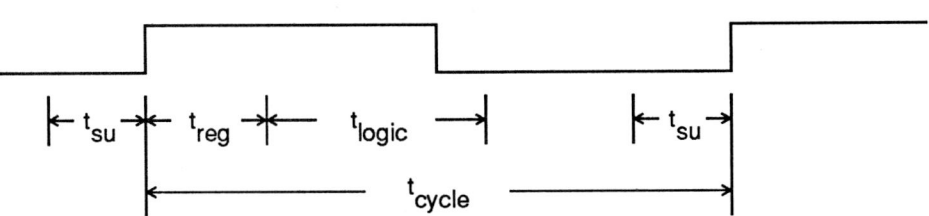

Figure 4.14. Determination of System Cycle Time.

for the carry must be considered, as well as the A and B to F delays.) The final time shown in Figure 4.14 is the amount of time that this information must be stable at the inputs of the register prior to the active edge of the clock. This time is called the setup time (t_{su}), and for the registers used here is 15 nsec. The sum of these three times identifies the absolute minimum allowable time for a clock cycle of the system. Normally, however, the cycle time (t_{cycle}) will be longer to allow for variations in device times and to be a little safe. The times used here have been typical times, and maximum times are also available in data books. If the design is to work no matter what parts are used (no prior testing of parts to weed out slower units, etc.), then the maximum times should be used instead of the typical times.

One time which is not shown in Figure 4.14 is the hold time, the amount of time that data must remain stable at a register input after the active edge of the clock. This is typically a very small time, and usually will not cause any problems. However, a designer must not violate this requirement.

The determination of the minimum cycle time for the control portion of the system is done in the same fashion as the data path. However, the t_{reg} involved is for the Present State Register to become stable (as well as any registers which are buffering input signals. The t_{logic} for the sequential part is the logic time required for next state determination. And the t_{su} is the required setup time at the input of the Present State Register. However, another necessary consideration is the delay time through the decode logic to create any control signals which impact on the timing of the logic for next state determination, and the logic of the data path section, such as the signal ROUND in the above design. Also, the t_{logic} of the control system must allow sufficient time for any signals created *during the current state* which influence the next state determination.

In general, the times required for the control section will be smaller than the times required for the data path of a system. Hence, the determining factor in the cycle time determination for the system will be the amount of time required to do the arithmetic, or carry out other logic requirements, for the data path of a system. However, if a condition is created and tested in the same state, then the effect can be to greatly increase the time required for a clock cycle, as we shall see. Now let us consider the cycle times for the various mechanisms demonstrated in this section.

As shown above, the data path section will require a cycle time longer than 91 nsec to accomplish the work of the system. Thus, any solution must provide at least this much time to allow the arithmetic portion to do its work.

The multiplexer solution to the control section demonstrated in Figure 4.7 seems very straightforward. At first glance the only delays appear to be the delay time through the '175, the logic delays associated with one gate delay plus a delay time through the '151s, and the setup time for the '175. These times total only 46 nsec, much less than the time for the data path. However, closer examination reveals that there are problems with this simple analysis. First, the signal TEN-L is generated one gate delay after an occurrence of CLK_PL, and this time must be included. This adds an additional 12 nsec to the required cycle time. But the real problem with the timing of the system as shown above occurs in State 3, and involves the signal FNL_ERR.

The signal FNL_ERR is generated from the adjusted exponent value to determine if there is an underflow or overflow. This signal will be stable 4 gate delays after the adjusted exponent is stable [see Figure 4.8(b)]. The adjusted exponent will be stable after an adder delay (made up of two '283s) after the assertion of the signal NO_PN. Once the FNL_ERR signal is stable, an additional 35 nsec will be required before the next active edge of the clock. However, the way the system has been designed, NO_PN is not asserted until halfway through State 3 (plus some gate delays), since it is the inverse clock which enables the state decoder and asserts SET_NO_PN. Thus, the following delays are required with the current solution: inverse of SYS_CLK to SET_NO_PN (decoder delay), gate delays to NO_PN-H and NO_PN-L (latch delay — two gate times), assertion of NO_PN to adjusted exponent stable (adder delay), adjusted exponent stable to FNL_ERR stable (four gate delays), FNL_ERR stable to next state stable (one gate delay, one multiplexer delay), plus the setup time for the register. These delays total to a time greater than the time for the arithmetic for a section. Therefore, unless a much slower cycle time (on the order of 200 nsec) is acceptable, then some alternate solution must be considered. The problem with State 3 will exist not only for the multiplexer method of sequential control, but also for other methods as well, since the time required contains the same constituent parts for each method. Thus, solution mechanisms will be proposed, and the reader can determine how to implement each of the mechanisms.

One of the simplest solutions to the problem presented in the previous paragraph is to modify the shape of the system clock so that the duty cycle calls for the assertion of NO_PN earlier in the cycle. With this solution mechanism, the clock signal would remain asserted for, say, 20 percent of the cycle time, and then be released. This is a sufficient amount of time to allow glitches to be removed by the '138 before being enabled. Thus, NO_PN (and other signals asserted by the '138) would occur earlier in the cycle, which would decrease the overall cycle time required.

Another simple solution would be to remove this timing problem completely by changing the state diagram. The state diagram of Figure 4.5 contains seven states, and since eight states are possible with 3 bits in the present state register, there is an unused state number. Inserting a state between State 4 and State 3, and asserting NO_PN in that state, causes the propagation delays to occur before the stable logic value is needed, and the timing problem associated with the current mechanism has been removed.

A third solution mechanism, perhaps the most complicated mentioned here, is to design the system clock in such a way that State 3 has a longer time to do the required work than other states. This complexity is usually not required, but it can be an effective solution method for designs with a sufficiently complex data system to justify the additional clock logic. There are a number of other solution mechanisms for this timing problem, and the reader is invited to generate alternatives to accomplish the same work.

The timing requirements for the "D" and "J-K" implementations are very similar to the timing for the multiplexer method. The "D" mechanism has four gate delays, plus the register times, and the "J-K" mechanism has three gate delays plus the register times. These times are 53 nsec and 45 nsec, respectively. Again, these delays do not include the delays associated with generation of the control signals.

The times involved in the PROM based sequencer system include the propagation delay time from clock to register out, plus the time from register (and inputs) stable until the clock can be activated again, which is a combination of the delay associated with the PROM and the setup time of the register. For the 27S35 shown in the figure, these times total to 65 nsec. For an alternative solution using a low-power, slower PROM which needs an external register, the times total 70 nsec.

Floating point multiplier: Comparison of implementation methods

In the process of creating a design, it is important to determine which system characteristics are important to the system under construction. The resources of the system must be utilized in an efficient manner, whether those resources are silicon real estate, printed circuit board space, power consumption, or processing speed. At this time we pause to consider the results of the designs which have been suggested for this implementation of a floating point multiplication unit. The resources which we will enumerate are the board space required, the power consumption, and minimum cycle time. The cycle time presented in Table 4.7 is the minimum time, not including additional delays required to generate control signals within the system itself. The results are shown in Table 4.7.

The data path itself would require a little over 2 watts of power, and over 13 square inches of board space. Thus, the data path section of the system dominates the resource requirements. Of the control mechanisms, the smallest solution is the first PROM solution, where the PROM and register are included in a single device. This solution requires only 1.76 square inches of board space However, it also has the highest power requirement of all the control mechanisms. The lowest power control solution is the multiplexer method, with a power requirement of only 62.625 milliwatts. By obtaining the information shown in Table 4.7, a designer can make choices of a design method based on the use of resources available in the system. The choices can be further clarified by an enumeration of other system resources, or other requirements. Still, the method remains the same: determine alternative solutions and select one based on valid engineering criteria.

Some of the criteria which are difficult to quantify, and which will have different significance for system designers, are concerned with the more ''intangible'' aspects of a system design. One of the requirements of any system is

Table 4.7. Comparison of Resource Utilization for the Floating Point Multiplier Designs.

Design	Power Dissipation (mW)	Board Space (in^2)	Minimum Cycle Time (nsec)
Data Path	2007.0	13.16	91
Multiplexer Control	62.6	3.44	46
''D'' Flip-Flop Control	192.7	5.52	53
''J-K'' Flip-Flop Control	160.5	5.16	45
PROM Control A	1867.0	1.76	65
PROM Control B	767.0	2.4	70

not only that it works, but also that information concerning the system can be communicated to others. Not only does the designer need to understand how the system works, but also those responsible for layout, fabrication, checkout, field test, and so forth. Thus, one of the "intangibles" is, how easily can this design be understood by others? The reader is now invited to review Figures 4.7, 4.9, 4.10, and 4.13, and determine which of the control mechanisms would be easiest to explain to someone else. Alternatively, given that you do not have the state diagram of the system, which of the systems shown in the figures would be the easiest to analyze? Finally, are these criteria important for your system?

This section has examined solutions to the design of a floating point multiplier system. A detailed analysis of system requirements allowed progress from understanding the system to understanding the algorithms. Understanding the solution requirements and the solution mechanisms allowed creation of a data path block diagram which permitted the proper transfers to be made, and a state diagram which described the action of the system. Using the details made available in the block diagram and state diagram, this section has presented alternative mechanisms for the design of the control system. The mechanism which will be the "best" depends upon criteria determined by the system designer, but alternative solutions should be examined before a final choice is made. Further alternative solutions should be examined not only for the control system, but also for the data path. We will now examine an alternative solution for the data path of the floating point system and consider some of the differences.

4.3.   Floating Point Multiplier: Method 2

The design presented in the previous section was based on a fairly straightforward solution to the floating point problem. The design was created to demonstrate several methods for control system implementation. The work of floating point arithmetic was performed in a data path which almost directly conformed to a "natural" solution to the problem. However, as shown in Table 4.7, this data path required 2 watts of power and over 13 square inches of board space. Another solution is presented in this section, a solution created not for simplicity, but for minimum data path resources. First, let us approach this solution in the same fashion as was utilized in the previous section, and then we will compare the results.

Floating point multiplier: Understanding the problem

The understanding of the problem is the same for this solution as for the last one, since the processing requirements are the same. That is, we will assume that the task at hand is to create a system which will accept inputs from an external source, provide the results to a bus read by the external source, and also provide the contents of the multiplier and multiplicand registers upon demand. In the following explanation, we assume the existence of five signals, which are given in Table 4.8. The format for the information is as presented in the preceding section, so we will not repeat that information. Our objective in this section is to do the work done in the previous section, but with fewer parts, if possible.

Table 4.8. Interface Signals for Floating Point Multiplier.

Signal Name	Function
LOAD_PCAND	External device provides value for multiplicand register
LOAD_PLIER	External device provides value for multiplier register
READ_PCAND	External device asks for value from multiplicand register
READ_PLIER	External device asks for value from multiplier register
DO_MULT	Do multiply and provide result

Floating point multiplier: Understanding the processing algorithms

The understanding of the algorithms involved is very similar in this implementation to that presented in the previous section. All the analysis of the permissible values for the exponents, such as the information presented in Tables 4.1 and 4.2, is still in force. The difference will be in the way that we manipulate the data and handle the various aspects of the system. The effect of this will become apparent as we proceed with the design. As stated above, one of the objects here is to create a system which does not require a large amount of board space or power to implement. Thus, steps will be taken to do things in a way which may not be straightforward, but which will do the same work.

Since much of the activity of this system is based around two different parts, we will take the time to describe them briefly. The first part is a '299, which is an 8-bit shift register with both a parallel load and a parallel read capability, both on the same set of lines. That is, the same lines are used for both input and output. Hence, this device can be directly connected to a bus to store information, then return the information to the bus as needed.

The second part is the one responsible for doing the work. This is the '681, a 4-bit accumulator that contains two registers and an arithmetic unit on the same chip. A block diagram of this device is shown in Figure 4.15. The two registers are labeled "A" and "B," and both receive their information from the data bus. The "B" register is also capable of shifting information in a serial fashion, a capability not present in the "A" register. The arithmetic unit allows the contents of the "A" and "B" registers to be combined in a number of ways, only some of which will be utilized in this implementation. The results of the ALU are enabled to the bus, which can then be loaded back into the "B" register or utilized by some other function connected to the bus. The activity of the registers is controlled by three Register Function Select lines; the operation performed by the ALU (and whether the results are enabled onto the bus) is controlled by three ALU Function Select lines and a mode control.

There is a basic difference between this device and the devices used in the previous section, which must be understood to properly use the system. The difference is in the manner in which the clock is utilized. The registers shown in the diagrams of the previous section are edge-triggered registers, and the activity of the device was controlled by asserting the clock lines of the registers only when a specific action was required. The accumulator shown in Figure 4.15 is also an edge-triggered device. However, this device has been created so that the clock line of the accumulator is connected to the system clock. For whatever activity is identified by the control lines, the Register Function Select lines, and the ALU

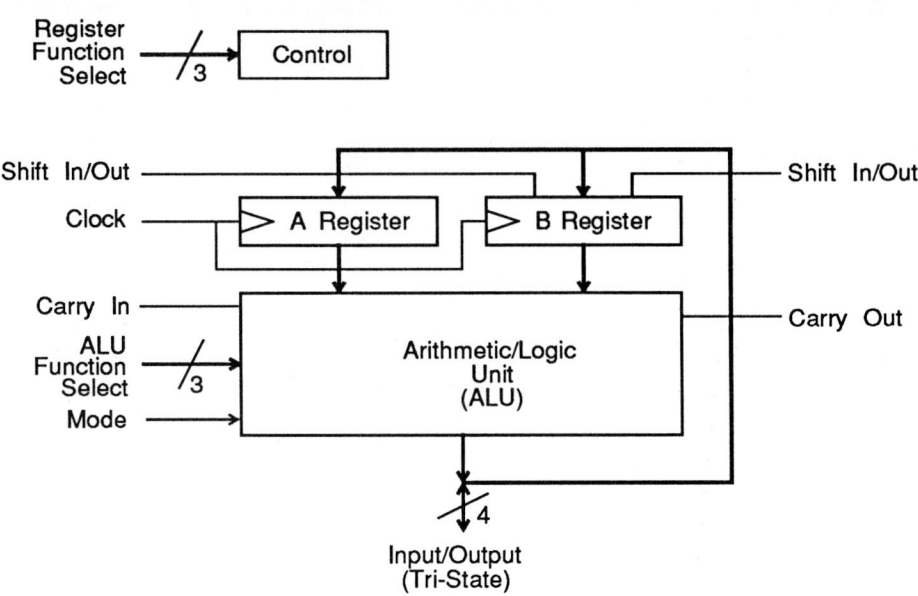

Figure 4.15. Block Diagram of '681 Accumulator.

Function Select lines, transfers occur on the active edge of the clock. If no activity is to occur, then the control lines are configured in a no-operation mode. There are two principal advantages to this type of operation. The first is that all the action in the system is guaranteed to be synchronous, since the action occurs on the active edge of a single clock signal, and a possible skew between different clock lines is avoided. The second advantage is that glitches on control lines do not adversely affect the operation of the system, and the control section need be concerned only with the generation of levels, rather than strobe signals.

The mechanisms used in this implementation of the floating point multiplier require that the multiplicand and multiplier be stored in registers made of '299s, and that the arithmetic involved in the exponent and mantissa manipulation be done in the accumulators ('681s). Constants which will be needed for the process and any other activity will be accomplished by using the capabilities of the accumulators.

Floating point multiplier: Conceptual level data path block diagram

The block diagram for this second approach to the floating point multiplier is included as Figure 4.16. Like the previous data path, which is in Figure 4.2, this diagram shows the major parts of the system needed to perform the work, but not all the details. However, at this point we know which registers we want to use to implement the Multiplier and Multiplicand Registers, as well as the Mantissa Accumulator and Exponent Accumulator. Figure 4.16 indicates that the major data path in the system is an internal 16-bit tri-state bus, which will be used to move all information as needed. There is a Bus Transceiver which isolates the action of the multiplier from the external system; this is done to provide electrical isolation and to allow activity in the multiplier to proceed independently from the activity on the external bus.

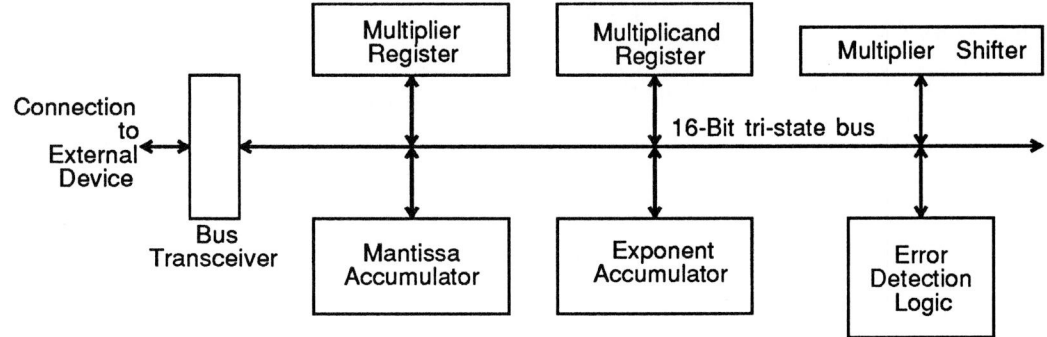

Figure 4.16. Data Path Block Diagram for Floating Point Multiplier.

The capabilities of the previous system are also evident in this system. There are two registers which can be filled from an external source, and the values contained in the registers can be provided to the external device. There is a facility for manipulating exponents, so that the exponent addition required in the floating point multiplication process can be accomplished. There is a facility for the shift-and-add implementation of the multiply operation. Finally, there is an indication that underflow and overflow errors will be detected in the multiplication process, and the results adjusted appropriately. The next step, then, is to add details to the block diagram of Figure 4.16 to indicate just how this work will be accomplished.

Floating point multiplier: Detailed data path block diagram

The block diagram of Figure 4.16 provides a starting point to create a block diagram with a sufficient amount of detail to continue the design process. Each of the blocks shown in Figure 4.16 is expanded to the point that the control lines involved are identified, so that the required manipulation can be accomplished. The detailed block diagram which results from this process is shown in Figure 4.17.

In this system the multiplier and multiplicand registers are formed by '299s. These registers behave in a fashion similar to the accumulator action described above. That is, the clock is tied to the system clock, and the loading of the register occurs only when the appropriate load line is asserted. One of the responsibilities of the control system is to enable the bus transceiver and configure its direction so that the value supplied by the external device is directed to the internal bus. When the contents of the registers are to be read out, this information is made available to the internal bus by asserting the appropriate enable line, and the control section must enable the transceiver and configure its direction out of the multiplier system. Another difference in this implementation is that the control section will handle all the transfers, rather than assume that the external device will move the multipliers in and out. This is to guarantee that only one value is enabled onto the internal data bus at a time. Thus, the control section must include capabilities for reading and writing the multiplier and multiplicand registers.

The exponent manipulation in this implementation is handled by the Exponent Accumulator. Basically, this device works by loading first one exponent and then the other, finally adding them inside the accumulator. Also, a constant is

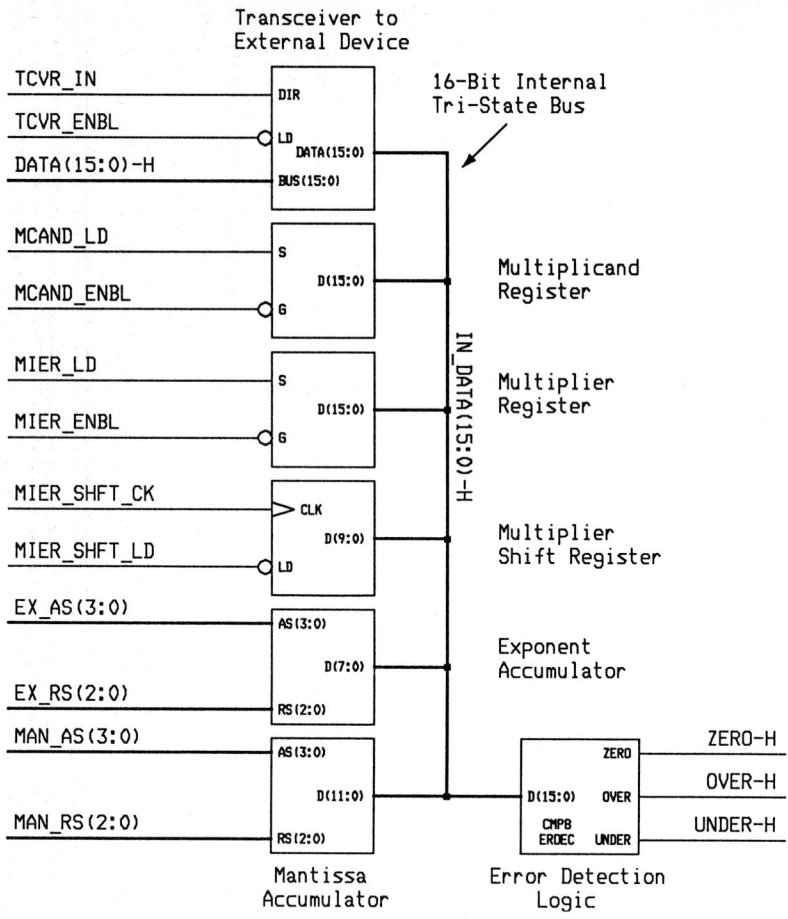

Figure 4.17. Detailed Data Path Block Diagram for Floating Point Multiplier.

utilized to create the final result. As we describe the activity which occurs in the Exponent Accumulator, we will refer to the exponent "A" register as EA, and the exponent "B" register as EB. The sequence of events in the previous section was to add the exponents, then adjust the sum by adding either 16 or 15 to this intermediate result. In this implementation, the constant is loaded first, then the exponents added accordingly. The operation of this mechanism will be explained in conjunction with the state diagram. When the final exponent is needed, it will be enabled onto the internal bus, which will then be sent to the external device through the transceiver. One of the responsibilities of the exponent section is to provide the correct exponent for error conditions as well as for normal operation. This will be accomplished by the ability of the accumulator to provide a result of all zeros or all ones.

The Error Detection Logic is used to identify underflow, overflow, and zero conditions. Since first one operand and then another is enabled onto the internal bus, the same logic is used to check both operands for a zero input value. Thus, only one detection system is needed to find a zero input value. The exponent accumulator makes available on the bus the results of the exponent addition. Therefore, when this process is being performed the Error Detection Logic looks for underflow and overflow conditions. If either is detected, it is the responsibility of the control system to either clear or set all the bits in the exponent.

The mantissa action required in the multiplier is performed in the two registers in Mantissa Accumulator, as well as the Multiplier Shifter. The "A" register internal to the Mantissa Accumulator will be referred to as MA, and the "B" register will be represented as MB. The MA register will be used to store the multiplicand mantissa, while the MB register will be used for the product register. The shifting of the mantissa of the multiplier is handled by the Multiplier Shifter, which is composed of two '165s. In the previous implementation, the rows of the partial product array were individually created, using AND gates to provide either a copy of the multiplicand mantissa or all zeros, depending on the bit of the multiplier shifter. In this implementation, if the corresponding bit of the multiplier is zero, then no addition is performed, and the product register is simply shifted. This accomplishes the same work, but without the external AND gates. Thus, this implementation calls for the storage of the multiplicand mantissa, the storage of the product, and the arithmetic needed in the mantissa manipulation to all be handled by the capabilities of the '681. The post normalization in this system is handled in exactly the same way as in the last implementation. That is, if the result of the multiplication has a "1" in the most significant bit, then a shift is required to make this "1" the hidden bit. If the result of the multiplication has a "0" in the most significant bit, then no shift is required, and there will be a "1" bit in the hidden position. The final result of all this manipulation will be to have the mantissa properly positioned on the bus to output to the external device.

As before, the details which are added to the preliminary block diagram to provide the detailed block diagram provide a basis for the design process. It is imperative that a designer know how these devices operate, and what is required on the control lines to achieve the desired results. It may take many readings of data sheets and hours of detailed study to fully understand a device, but this understanding is required for a good system design.

Floating point multiplier: State diagram

The state diagram enables us to visualize the action which takes place in the process of doing the work of the system. Before we undertake to present a state diagram here, we will identify the transfers which take place as the operations are required of the system. We will break these down into three groups: input operations, output operations, and the multiplication itself.

The process of obtaining the multiplier value from the external device is simply accomplished:

$$\text{External Value} \;\rightarrow\; \text{Multiplier Register}$$

To accomplish this, the bus transceiver must be enabled and the direction configured to send information to the internal bus. Also, the load line of the Multiplier Register must be asserted. Exactly the same operation will be required for loading the Multiplicand Register.

When the external device requires the information which is stored in the Multiplier Register, the reverse of the above operation is implemented:

Multiplier Register $\rightarrow$ External Bus

This occurs when the enable line of the Multipier Register is asserted, and the transceiver is enabled and configured to send information from the internal bus to the External Device. Again, reading the information from the Multiplicand Register is done in exactly the same way.

The real fun begins when the External Device needs a multiplied result. Then a series of operations begins, with a number of possible responses. We now identify the register transfer operations required to implement the multiply. Comments are added to identify the purpose of the transfer, or other important information about the action of the system.

$17 \rightarrow EB$ This occurs when the system is idle, so that when the process begins a known value is stored in the exponent register; the product register will be cleared later.

Multiplicand Mantissa $\rightarrow$ MA
Multiplicand Exponent $\rightarrow$ EA   The first step is to fetch the multiplicand. The mantissa portion is stored in MA, and the exponent portion is stored in EA. Also, at this point the error logic should check for a zero value.

$EA - EB \rightarrow EB$ This is the first step in handling the exponent.

Multiplier Mantissa $\rightarrow$ MS
Multiplier Exponent $\rightarrow$ EA   At this point the multiplier is fetched. The mantissa part is stored in the Multiplier Shifter, and the exponent part is stored in EA. Also, check for zero-valued multiplier.

$EA + EB \rightarrow EB$
Zero $\rightarrow$ MB   This creates a valid exponent. If a final shift is required, then this value will need to be incremented. Also, underflow and overflow can be checked at this point. Finally, the product register is cleared.

At this point, if there is no over/underflow error, then the 11 partial product additions are to be made. This will be accomplished by iterations of the following:

If active bit of Multiplier Shifter is ''1'':

$MA + MB \rightarrow MB$ Do the addition only if the multiplier bit is set.

Shift Multipier Shifter

Shift Product Register   Shifting MS and the product register sets up for the next iteration. However, on the last iteration, don't shift the product register.

Now do the rounding, which will be explained in more detail below:

If most significant bit of product register is ''1'':

$MB + 1 \rightarrow MB$

$MB + 1 \rightarrow MB$ This second increment is done regardless. See below.

If final shift needed:

Shift Product Register

$$EB + 1 \rightarrow EB \qquad \text{Perform final overflow check here.}$$

The final action is to make this information available to the external device:

$$MB \rightarrow bus$$
$$EB \rightarrow bus$$
$$sign \rightarrow bus$$
$$bus \rightarrow external\ device$$

We will create a state diagram to describe the information in the above set of transfers, but before we do, let us identify some of the work to be performed. The transfers to and from the Multiplier Register and the Multiplicand Register are fairly straightforward, but the information about the exponent can be confusing. Initially, the EB register is loaded with 17. The next step is to obtain the exponent of the multiplicand, which will be loaded into EA. While this loading is being performed, the value is checked for zero. If all the exponent bits are zero, then the value of the multiplicand is zero, and the result will be zero. Thus, the rest of the operation can be aborted, and a result of zero established. If the exponent is not zero, then the operation $EA - EB$ is performed, and the result stored in EB. The next operation is to get the exponent of the multiplier. If this exponent is zero, then the process terminates with a zero result. However, if the exponent is nonzero, then the operation $EA + EB$ is performed, with the result being stored in EB. This exponent has the value of $EXP_{PLIER} + EXP_{PCAND} - 17$, which which will be the correct exponent at the end of the process if no final shift is needed. Also, at this point the underflow condition will be true if the result is negative, and the overflow condition will be in force if the sixth bit of the exponent and at least one other bit are "1." It is left as an exercise for the reader to verify that these conditions are correct.

The other condition which needs further explanation is the rounding process. Because of the very nature of normalized floating point numbers, the smallest value which can be represented is 0.10000000000_2, which is just one-half. The largest value which can be represented is 0.11111111111_2, which is almost one. Thus the results of the mantissa multiplications will range from 0.01000000000_2 to 0.11111111110_2. In the calculation process three 4-bit adders (inside the accumulators) are utilized. The alignment of the operands with respect to the adders is such that the final result includes information from the three adders, as well as the carry out of the most significant adder. Let us represent the carry out as C_O, and the adders from most to least significant as ADD_3, ADD_2, and ADD_1. After the final addition step, the result will either have a "1" in the most significant position (actually located on), or a "0" in that position. In either case we are concerned with the most significant "1" bit and the next 10 bits. The most significant "1" bit will become the hidden bit, and the next 10 bits will be made available to the external device. The round should occur by creating a value with a "1" bit in the position one less significant than is to be retained. If C_O is a "1," then the following condition occurs:

C_O	ADD_3	ADD_2	ADD_1	
1	xxxx 0000	xxxx 0000	xxxx 0010	Round value

The round value is created by placing the "1" after the 10 bits of interest. If C_O is "0," then the situation is slightly different:

C_O	ADD_3	ADD_2	ADD_1	
0	1xxx 0000	xxxx 0000	xxxx 0001	Round value

Thus, if the C_O from the final addition is a "1," then the result should be incremented by two for the round process. If the final C_O is a "0," then the result should be incremented by only one. Since one of the capabilities of the ALU is to increment, then this is handled by incrementing the result in one or two steps, rather than incurring the additional resources to provide a constant value of one or two.

Using the register transfers identified above, we are now ready to create a preliminary state diagram for the system. This is shown in Figure 4.18. The idle state in Figure 4.18 causes a constant to be loaded into the B registers of the exponent accumulators. The 17 which is loaded at this point will be used in the formulation of the correct representation of the exponent.

When a signal is obtained to load either the multiplier or the multiplicand, then the system leaves the idle state long enough to cause the appropriate action. This is indicated by the states marked AC and AP. Also, when a signal indicates that the multiplicand or the multiplier is to be made available to the external device, then the system leaves the idle state and makes these values available as indicated in the above register transfer description. The action which causes this is indicated by states OC and OP.

The signal DO_MULT indicates that the external device requires a multiplication to occur. When this condition is detected, the system leaves the idle state and moves to the state marked LDC. This state initiates the multiplication action by moving the multiplicand. The exponent portion of this value is accepted by the Exponent Accumulator, and the mantissa portion is accepted by the Mantissa Accumulator. If a zero condition is detected at this time, then the multiplication process can be terminated by going to the Zero State (ZE). If a zero condition is not detected, then the process moves on to the state marked EBM to perform one of the steps needed for the exponent manipulation.

The EA − EB → EB which is performed in state EBM results in an intermediate value stored in EB to await the arrival of more information. This information comes in the next state (marked LDP), which causes the multiplier to be moved. The exponent portion of the multiplier is accepted by the EA register. The mantissa portion (including hidden bit) is loaded into the Multiplier Shifter register. One of the things which must also be taken care of between the two states which are used to move the multiplier and the multiplicand is a determination of the sign of the result. This sign will be the exclusive-OR of the signs of the two input values. As with the multiplicand value, when the multiplier is available, it is checked for the zero condition. If the zero condition exists, then the Zero State (ZE) terminates the processing, otherwise the action moves on to perform the required exponent addition.

The exponent addition which occurs in the state marked EBA has three effects. The first is to create the value which will be used for the exponent of the result. (This value may be modified if a final shift is required by the mantissa before output.) The second effect is to place onto the outputs of the Exponent Accumulator a value which can be checked for the underflow condition. If this condition exists, then the action moves to the Zero State (ZE). The third effect is to use the value output from the Exponent Accumulator check for an overflow

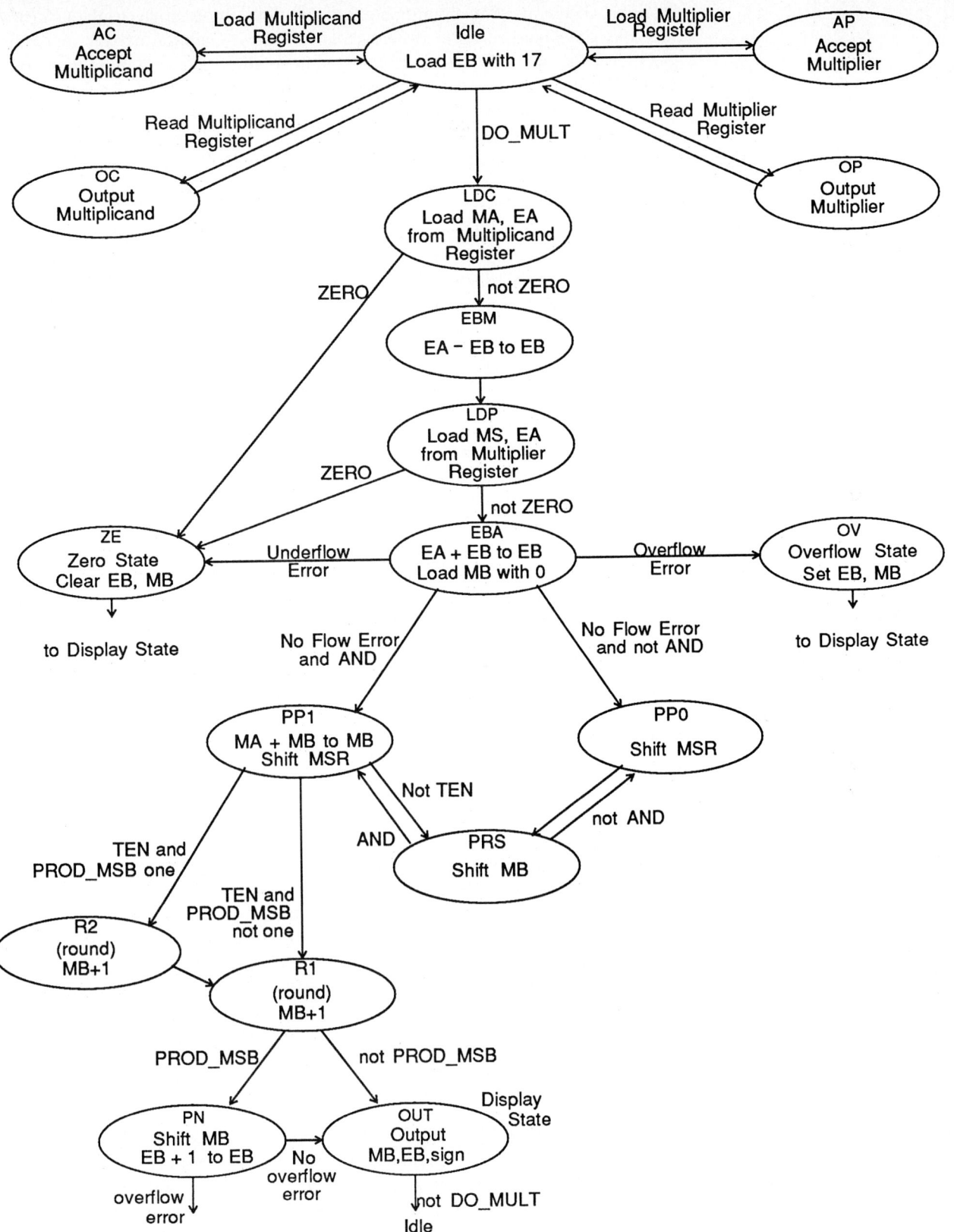

Figure 4.18. Conceptual State Diagram for Floating Point Multiplier.

condition. If the overflow condition exists, then the action moves to the Overflow State (which is marked OV). If neither the overflow condition nor the underflow condition exist, then the actual shift-and-add operation of mantissa multiplication begins. The value of the first partial product will be determined by the serial output of the Multiplier Shifter. If this bit is a "1," then the partial product should be added into the running sum, and the Multiplier Shifter is shifted (which occurs in the state marked PP1). If the bit is a "0," then nothing should be added to the running sum, so a different state (the state marked PP0) is used to perform only the work of shifting the Multiplier Shifter.

If no partial product addition is needed, then the Multiplier Shifter is activated, and the action moves on to shift the product register (MB). However, if the partial product addition occurs, then the next state is one of three. The first choice is that which merely shifts the product register. This choice will be made 10 times; that is, between the state marked PP1 and the state marked PP0, the product register is shifted 10 times (in the state marked PRS). Since the last bit to be provided by the Multiplier Shifter is a "1" (which is the hidden bit), the action of the system will be in the state marked PP1 when TEN becomes asserted, indicating that a sufficient number of iterations have been made. At this time the rounding will occur, as explained above. If the C_O bit is set, then both the state marked R2 and the state marked R1 will be used to provide the rounding needed. If the C_O bit is not set, then only the state marked R1 will be needed.

When the rounding has been completed, then the determination of the proper alignment can be performed. If no further alignment is needed, then the C_O bit will be "0," and the action moves to the output state (OUT). However if the C_O bit is a "1," then the system moves to the state marked PN to cause one final shift. At the same time, the exponent is incremented and checked for an overflow condition. If the overflow condition exists, then the Overflow State (OV) will establish the correct value. Otherwise, the system moves to the output state (OUT) to provide the necessary information to the External Device. This value will remain on the bus until the DO_MULT signal is released.

The state diagram of Figure 4.18 is much more complex than the state diagram in Figure 4.4. One of the reasons for this additional complexity is the simplicity of the data path. Since less is accomplished by the data path in a single step, the control must expand to provide the additional steps needed to accomplish the work. Nevertheless, the principles utilized to create the detailed state diagram are the same: identify the assertion levels needed on the control signals to do the work which was specified in the preliminary state diagram.

The state diagram shown in Figure 4.18 contains 18 states, which would require 5 bits to represent. If there is a way in which we could reduce the number of states required by 2, then the number of states would be 16, which would require only 4 bits for representation. Since the functions of accepting the multiplier and the multiplicand are almost identical, as are the functions of supplying the multiplier and multiplicand, we will combine these functions to produce a system with fewer feedback variables.

The resulting state diagram is shown in Figure 4.19. Note that although the digrams are very similar, they are not exactly the same. We have reduced the number of states from 18 to 16 to be able to implement the system with 4 state bits. As stated before, the state diagram is a tool used to represent the desired action of the system. Proper utilization of the tool will result in a system that operates correctly; therefore, we must be careful in the creation of the state diagram as well as the hardware required to implement the system which implements the state diagram. Since the registers in the '681s will do something every time the system

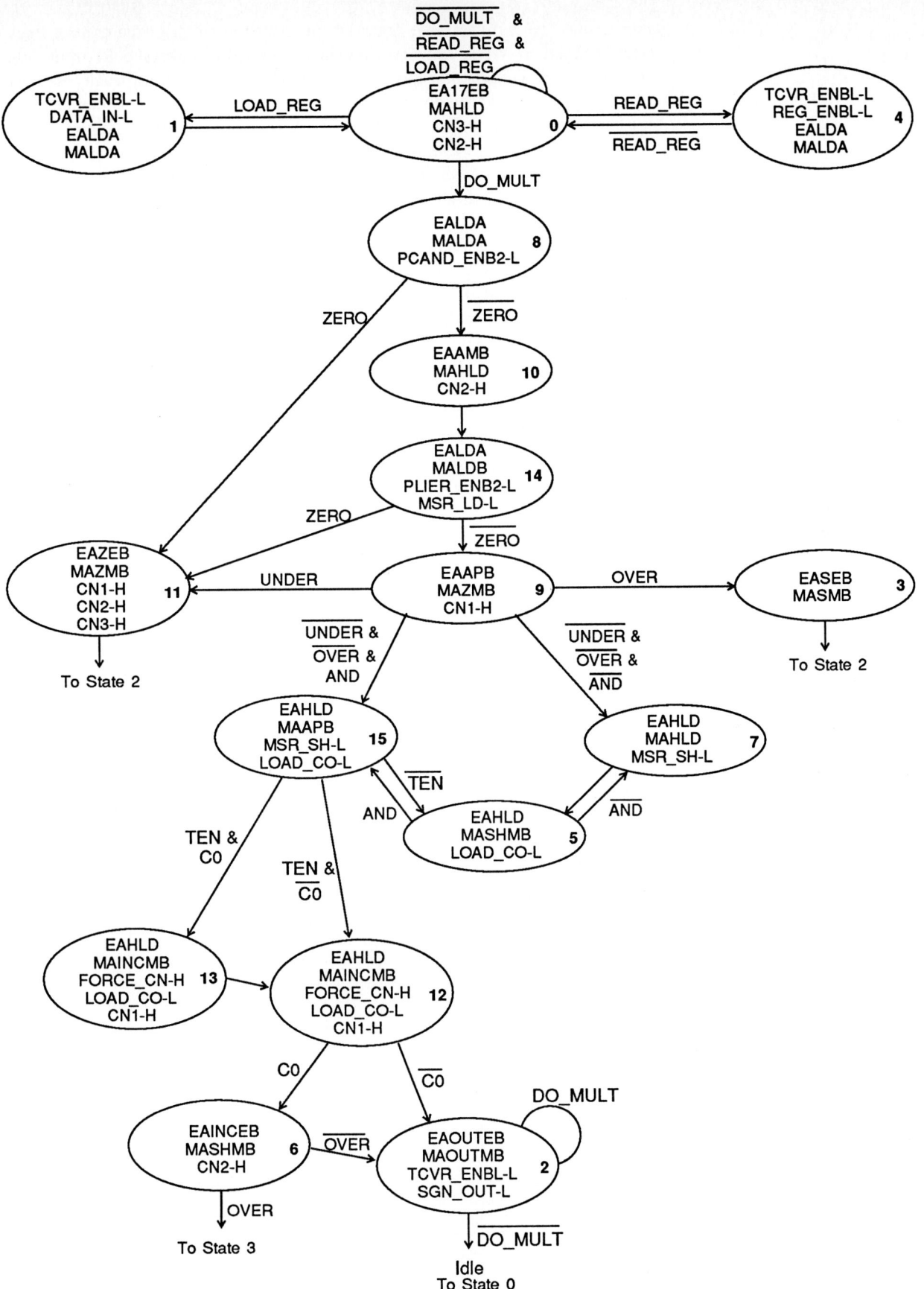

Figure 4.19. Detailed State Diagram for Floating Point Multiplier.

clock occurs, we will carefully consider the assertion levels of the control lines to the accumulators. We have indicated what accumulator action should occur in both the exponent accumulators and the mantissa accumulators for each state. The action of the exponent accumulators is indicated with mnemonics which begin with EA; the action of the mantissa accumulators is indicated with mnemonics which begin with MA. The mnemonic meanings, as well as the bit patterns required on the control lines to accomplish the work, are given in Table 4.9.

The idle state is State 0, since the present state register can be cleared by a system reset to force the system to idle. The action accomplished in this state is to hold the values in the mantissa accumulator and to load 17 into EB in the exponent accumulator. As indicated in the state diagram, the system will remain in this state so long as the DO_MULT signal is not asserted, the register load request is not asserted, and the register read request is not asserted. When any of those conditions is asserted, then the system will leave the idle state to cause the action.

State 1 is used by the system to do the work of loading a register, either the multiplier register or the multiplicand register. The choice of the appropriate register is derived from the input signals. In this way, the two states required for loading registers in the preliminary state diagram have been combined into a single state. Note that the value is also loaded into the A registers of the two accumulators. The reason for this is that the accumulators will either accept or provide a value; that is, there is not a mechanism provided for the accumulator devices used here ('681s) to disable their outputs, except when an input operation is being performed. This is another reason for the transceiver between the internal and external data buses.

State 4 is used to do the work of reading either the multiplier register or the multiplicand register. Again, two states in the preliminary state diagram have been combined, and the selection of the appropriate register to enable onto the internal

Table 4.9. Mnemonics for Exponent and Mantissa Accumulators.

Mnemonic	Bit Pattern F(2:0),M,A(2:0)			Meaning of Mnemonic
EA17EB	000	1	000	Exponent Accumulator, 17 to EB
EALDEA	111	x	xxx	Exponent Accumulator, load EA
EAOUTEB	110	0	100	Exponent Accumulator, output EB
EASEB	000	0	000	Exponent Accumulator, set EB to ones
EAHLD	110	x	xxx	Exponent Accumulator, Hold values
EAINCEB	000	0	100	Exponent Accumulator, increment EB
EAAPB	000	0	011	Exponent Accumulator, EA + EB
EAAMB	000	0	010	Exponent Accumulator, EA - EB
EAZEB	000	0	000	Exponent Accumulator, zero EB
MAHLD	110	x	xxx	Mantissa Accumulator, Hold Values
MALDMA	111	x	xxx	Mantissa Accumulator, Load MA
MAOUTMB	110	0	100	Mantissa Accumulator, output MB
MAZMB	000	0	000	Mantissa Accumulator, zero MB
MASHMB	010	x	xxx	Mantissa Accumulator, shift MB
MAINCMB	000	0	100	Mantissa Accumulator, increment MB
MALDMB	001	x	xxx	Mantissa Accumulator, Load MB
MAAPB	000	0	011	Mantissa Accumulator, MA + MB

bus is determined from the input signals. As in State 1, the accumulators are configured to accept rather than provide information.

The action of the multiply begins in State 8, where the multiplicand is enabled onto the internal bus. The exponent portion is placed in EA, and the mantissa portion, including hidden bit, is placed in MA. During this state the ZERO condition is tested, and if it is determined that the multiplicand value is zero, then State 11 is the next state; otherwise, State 10 is the next state. Note that this choice of state numbers, where two possible next states are configured so that they are logically adjacent, results in a state choice in which the next state is determined by a single bit.

State 10 is used to do the first part of the arithmetic needed for the exponent, and then the action moves to State 14. For the accumulator to perform the $A - B$ operation, the least significant carry must be asserted, which is indicated in the state.

During State 14 the multiplier is moved to the appropriate registers. The exponent portion is loaded into EA, and the mantissa portion, including hidden bit, is loaded into the multiplier shifter. Since the mantissa accumulator cannot output information onto the bus during this operation, the mantissa value is also placed in MB. However, MB will be cleared by the next state. The multiplier is also checked for a zero value during this state, and if a zero value is detected, then the next state is State 11. If not, then the action moves to State 9.

The other addition needed by the exponent is performed in State 9, leaving a valid exponent value in EB. If a final shift is needed, then this value will be incremented. For the remaining states, except for the increment just mentioned, the value in EB must be maintained. The other action which occurs in this state is to clear the product register, MB. At this point the shift-and-add algorithm can begin. However, since the exponent value is available, underflow and overflow conditions are checked at this time. An underflow error will cause the action to proceed to State 11; an overflow error will cause the action to proceed to State 3. State 3 sets values to all "1"s.

If neither an underflow nor an overflow condition exists, then the multiplication process continues. The choice of next state from State 9 is determined by the AND bit, which is the current bit under consideration from the multiplier shifter. If it is a "1," then the partial product should be added, so the next state is State 15. If the bit is a "0," then no partial product addition should be performed, and the next state is State 7.

The only action which occurs in State 7 is to shift the multiplier shifter, then the action moves to State 5 to test the new bit which is shifted by the action of State 7.

State 15 is used to add the partial product and to shift the multiplier shifter. The next state choice depends on the number of iterations which have been performed. If not all the required iterations have been performed, then the signal TEN will not be asserted, and the action moves to State 5. However, on the final iteration TEN will be asserted, and the choice of a next state depends on the value of CO. If CO is asserted, the rounding should be done by incrementing by two, so State 13 is the next state. If CO is not asserted, then rounding will be properly accomplished by a single increment, so the next state is State 12.

The shift of the product register to lower significance is accomplished in State 5. Also, at this time determination is made whether the next row of the partial product array is all zeros or a copy of the multiplicand. The choice is made by looking at the bit from the multiplier shifter. If all zeros is the choice, then the action moves to State 7; if an addition is required, then State 15 is the next state.

Table 4.10. Next State Determination for Floating Point Multiplier.

Present State	Next State Bits 3	2	1	0
0	DO_MULT-H	READ_REG-H	0	LOAD_REG-H
1	0	0	0	0
2	0	0	DO_MULT-H	0
3	0	0	1	0
4	0	READ_REG-H	0	0
5	AND-H	1	1	1
6	0	0	1	OVER-H
7	0	1	0	1
8	1	0	1	ZERO-H
9	$(\text{UNDER or } \overline{\text{FLOW}} \text{ and AND})$-H	OVER-L	UNDER-L	1
10	1	1	1	0
11	0	0	1	0
12	1	0	ZERO-H	1
13	1	1	0	0
14	1	0	0	1
15	TEN-H	1	0	$(\text{TEN and } \overline{\text{CO}})$-H

The rounding is accomplished by States 12 and 13 as described above. If an increment by two is needed, both states are used; if the correct rounded value can be obtained by a single increment, then only State 12 is used. From State 12, the next state is determined by the value of CO. If CO is asserted, then one final shift is required to align the bits properly. This is accomplished in State 6. If CO is not asserted, then the value present in the system is the desired value, so the next state is the output state, State 2.

State 6 is used to perform a final shift of the mantissa to align the bits properly, and also to increment the exponent to match the shifted value. If this increment results in an overflow condition, then the overflow state, State 3, is the next state; otherwise, the next state is State 2 to output the value.

The output state is State 2, where the calculated sign, exponent, and mantissa are enabled onto the internal bus, and the transceiver is enabled to the external device. This condition continues until the external devices no longer need the information, which occurs when the DO_MULT signal goes away. At this time the system returns to the idle state, State 0.

The state diagram shown in Figure 4.19 describes in detail the activity, including the assertion levels of the signals, required of this implementation of the Floating Point Multiplier. As before, we now move on to the implementation of the logic.

Floating point multiplier: Design of control logic and data path

As with the previous implementation of the multiplier, there are several ways in which the control system could be designed. Here we will present only two: the multiplexer method and the memory based method. The implementation of the multiplexer method is easily visualized by creating a table of correct next state possibilities. The table for this system is shown in Table 4.10. Information in Table 4.10 is derived from the state diagram and the input signal information. Using multiplexers allows the system do be directly implemented from the information in the table. In addition to a table which defines the next state requirements, a table which identifies the control signals and the states in which they are generated is very useful. For this system, this information is given in Table 4.11(a) and (b).

The mechanisms involved for generation of the next state logic, as well as for the decoding of the present state to create the levels and strobes needed for operation of the system, are the same as utilized in the previous section. The control logic derived from the information in Tables 4.10 and 4.11 is shown in Figure 4.20. The multiplexers needed for the next state generation need to be 16-line-to-1-line multiplexers. One implementation of this would simply use the '150 1-of-16 multiplexer. However, this is physically a large device, and a smaller board area is required by using two '151s, as shown in Figures 4.20(a) and (b). The inputs to the multiplexers are determined directly from Table 4.10, and this logic is shown in the figure.

The present state register is shown in Figure 4.20(c), along with the logic to decode the present state. Again, the decoder of the output could be created with a single '154 4-line-to-16-line decoder, but an alternative which requires less board space is shown in the figure. Note that in this implementation the system clock is not used to disable the decoders when glitches might be present. Thus, glitches may be present on signals such as REG_LD-L when the state change occurs. However, these signals will be asserted for the duration of a state, rather than for only half of it as is the case with the decoded present state lines in Figure 4.7(b). Those signals which need to be glitch-free are conditioned with the clock, and are shown in Figure 4.20(c).

The logic required to control the accumulators is shown in Figure 4.20(d). The gates are used to OR together the states in which the signals must be asserted. However, note that the logic for MAN_F(1) and EXP_F(1) generates the inverse function, and then inverts the signal for the proper value.

The interface between the external device and the multiplier system is shown in Figure 4.20(e). The five signals from the external device are synchronized with the system clock, and the results are used in the next state logic, as well as generating the necessary enable signals for the multiplicand and multiplier registers. The two signals for requesting the loading of the system are used to set flag flip-flops, and these flags are sent to the synchronizing register. This arrangement allows the action to occur as specified, and the clearing of the flag signal is an indication to the external device that the value has been accepted. Figure 4.20(e) also contains the iteration counter.

The data path logic for this system is quite simple when compared to the data path needed for the previous implementation. Figure 4.17 identifies the required parts, and these are shown in detail in Figure 4.21. The devices used for the Transceiver, the Multiplier Register, and the Multiplicand Register are shown in Figure 4.21(a). As described above, the transceiver provides isolation from the external device, and the two storage registers receive input from and provide output

Table 4.11(a). Output Signal Generation for Floating Point Multiplier.

Present State	Outputs												
	LOAD CO	REG ENBL	REG LD	PCAND ENBL	PLIER ENBL	CN 1	CN 2	CN 3	MSR SH	MSR LD	SGN OUT	TCVR ENBL	DATA IN
0	0	0	0	0	0	1	1	0	0	0	0	0	0
1	0	0	1	0	0	0	0	0	0	0	0	1	1
2	0	0	0	0	0	0	0	0	0	0	1	1	0
3	0	0	0	0	0	0	0	0	0	0	0	0	0
4	0	1	0	0	0	0	0	0	0	0	0	1	0
5	1	0	0	0	0	0	0	0	1	0	0	0	0
6	0	0	0	0	0	0	1	0	0	0	0	0	0
7	0	0	0	0	0	0	0	0	0	0	0	0	0
8	0	0	0	1	0	0	0	0	0	0	0	0	0
9	0	0	0	0	0	0	0	1	0	0	0	0	0
10	0	0	0	0	0	0	1	0	0	0	0	0	0
11	0	0	0	0	0	1	1	1	0	0	0	0	0
12	1	0	0	0	0	0	0	1	0	0	0	0	0
13	1	0	0	0	0	0	0	1	0	0	0	0	0
14	0	0	0	0	1	0	0	0	0	1	0	0	0
15	1	0	0	0	0	0	0	0	1	0	0	0	0

Table 4.11(b). Accumulator Control Signal Generation for Floating Point Multiplier.

Present State	Exponent Control Lines $(F(2:0),M,A(2:0))$	Mantissa Control Lines $(F(2:0),M,A(2:0))$
0	EA17EB	MAHLD
1	EALDEA	MALDEA
2	EAOUTEB	MAOUTEB
3	EASEB	MAZMB
4	EALDEA	MALDEA
5	EAHLD	MASHMB
6	EAINCEB	MASHMB
7	EAHLD	MAHLD
8	EALDEA	MALDEA
9	EAAPB	MAZMB
10	EAAMB	MAHLD
11	EAZEB	MAZMB
12	EAHLD	MAINCMB
13	EAHLD	MAINCMB
14	EALDEA	MALDMB
15	EAHLD	MAAPB

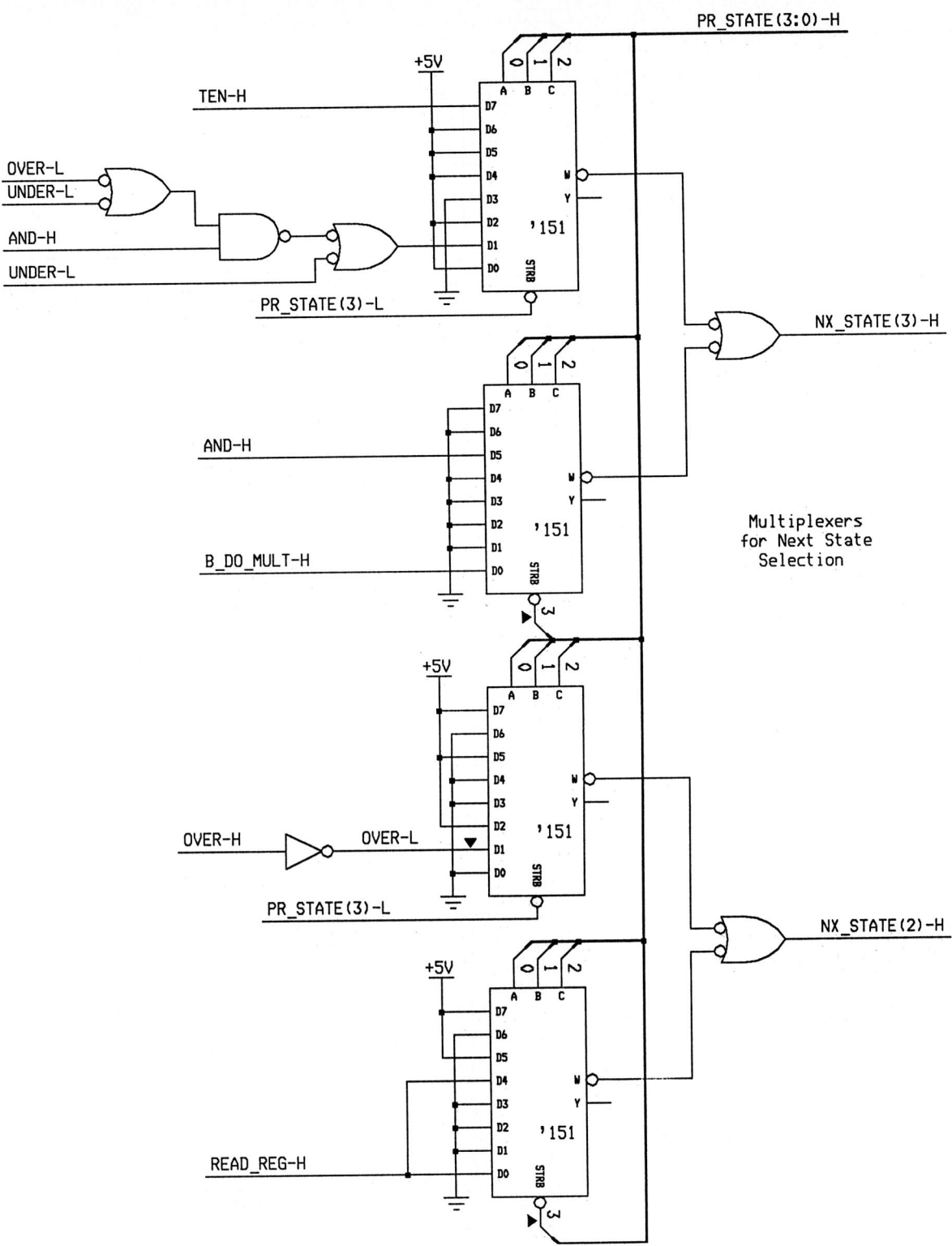

Figure 4.20(a). Detailed Logic Diagram for Control of Floating
Point Multiplier: Next State Logic.

Chap. 4: Design of Synchronous Sequential Systems

135

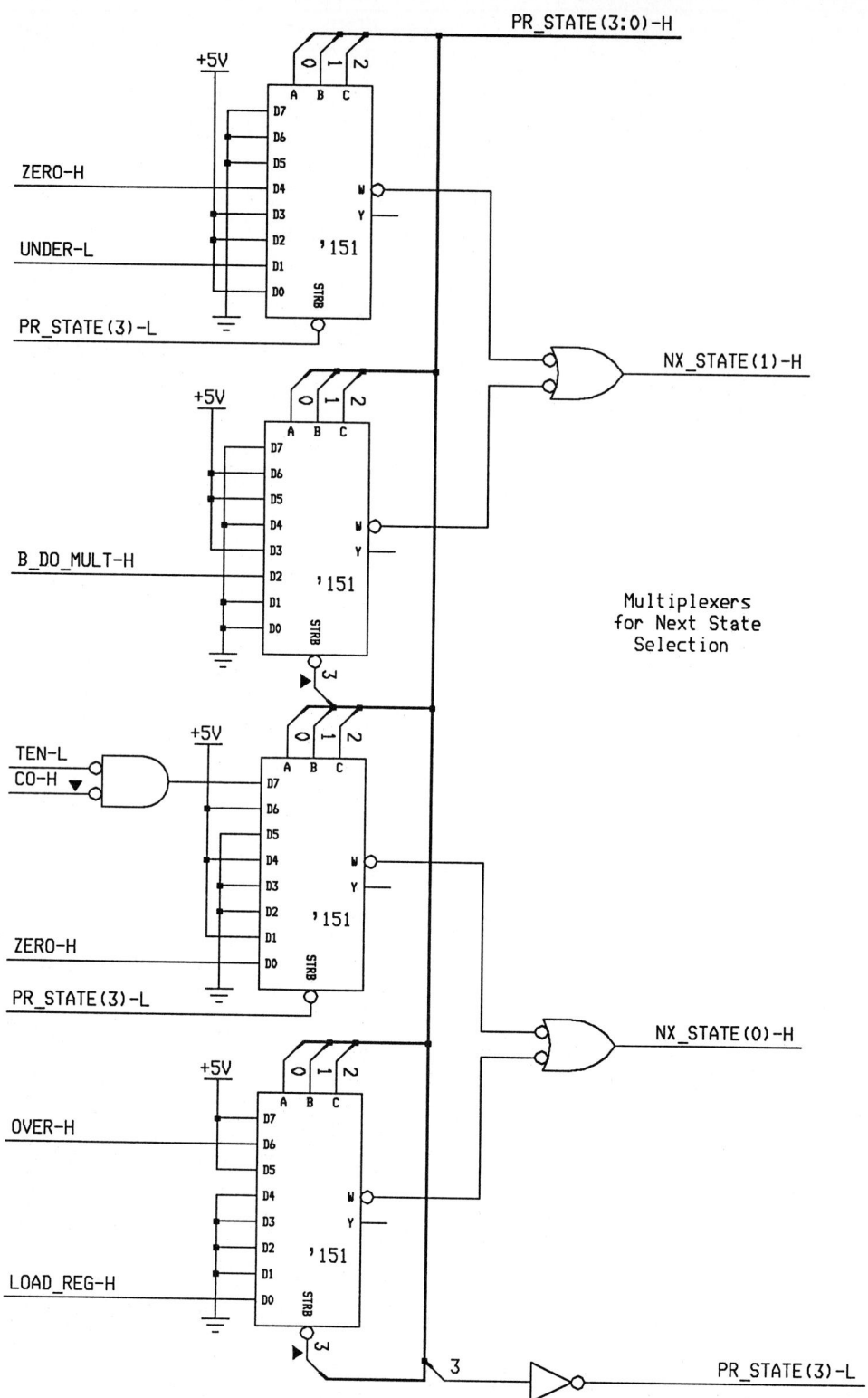

Figure 4.20(b). Detailed Logic Diagram for Control of Floating Point Multiplier: Next State Logic.

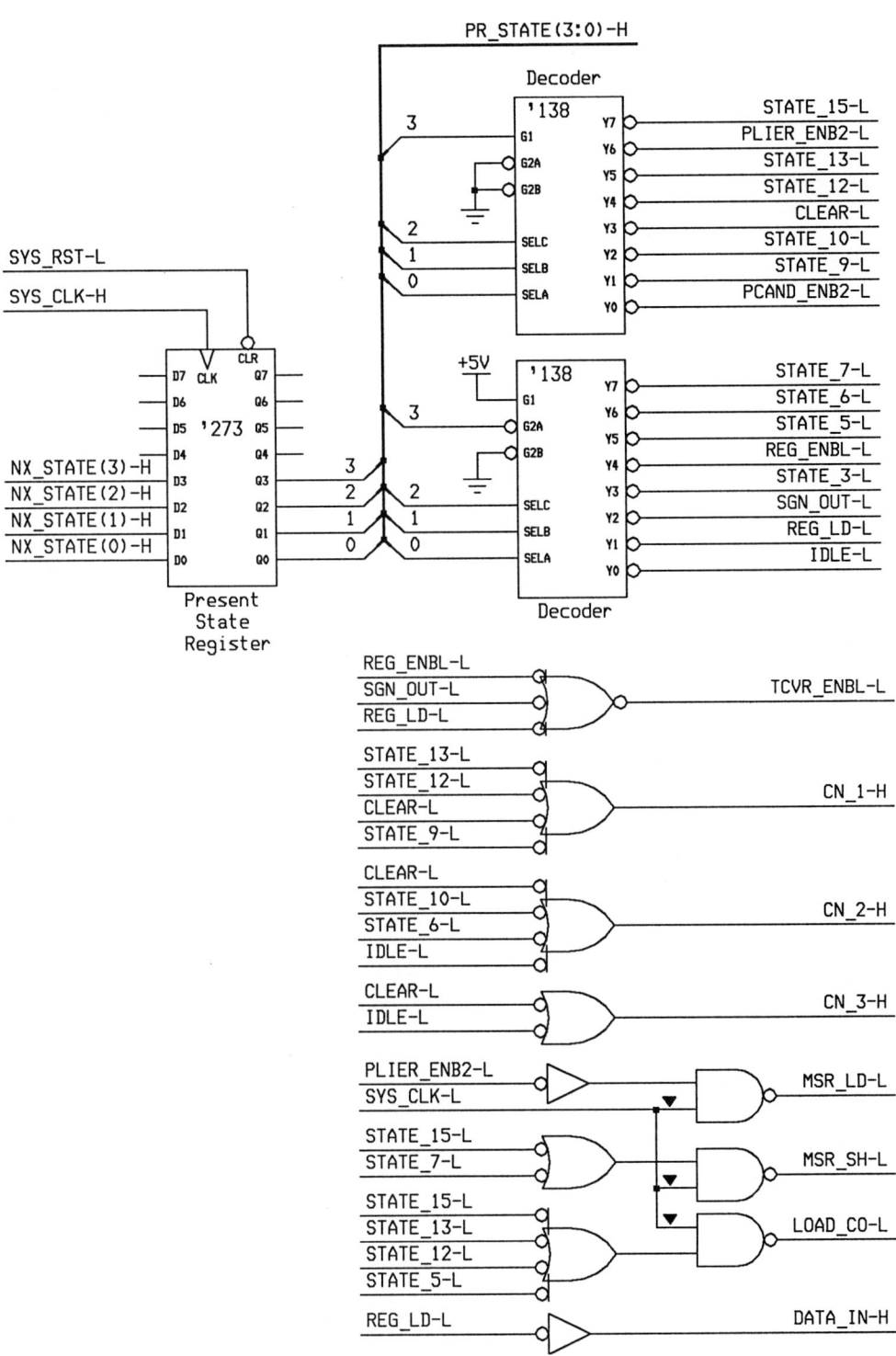

Figure 4.20(c). Detailed Logic Diagram for Control of Floating
Point Multiplier: Present State Register and Decode Logic.

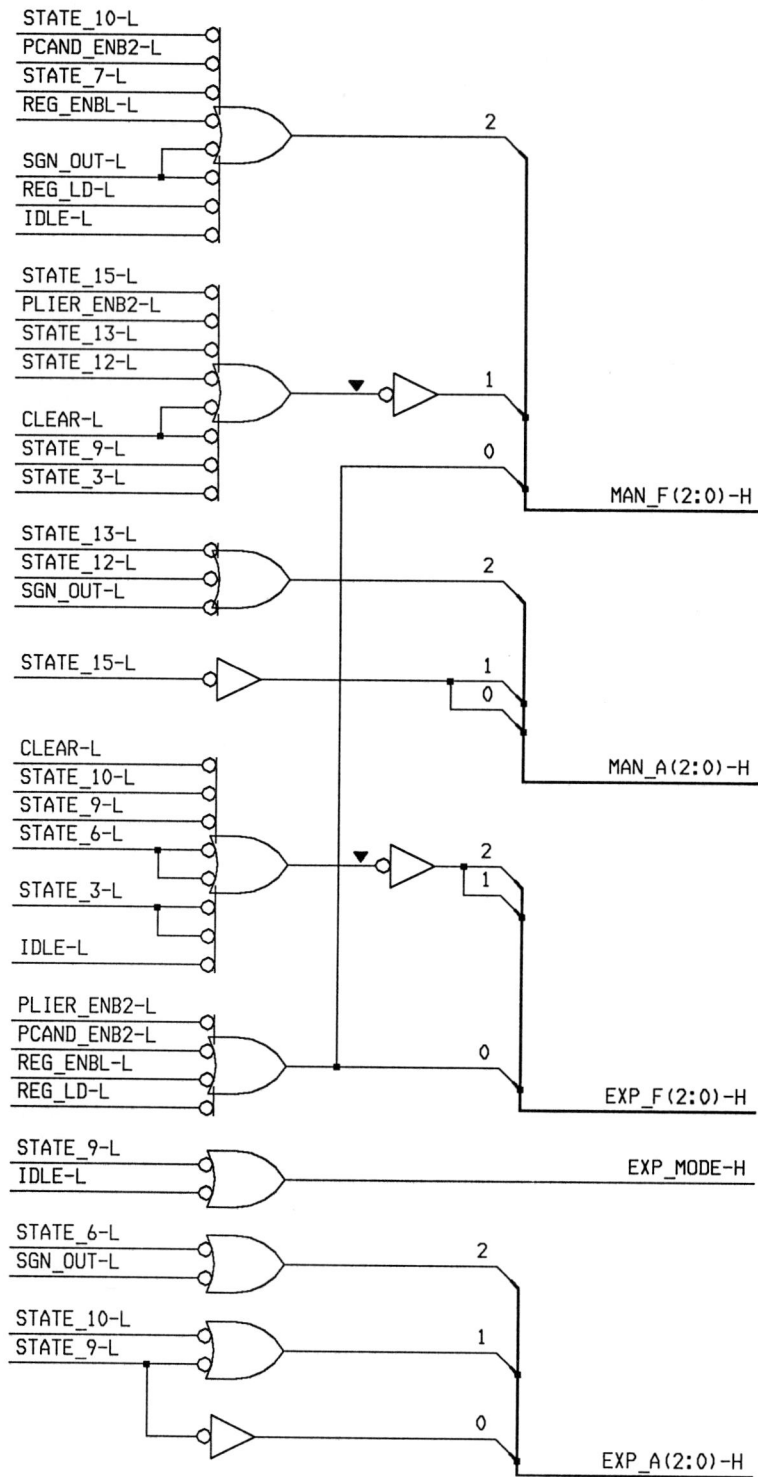

Figure 4.20(d). Detailed Logic Diagram for Control of Floating
Point Multiplier: Decode Logic for Accumulator Control.

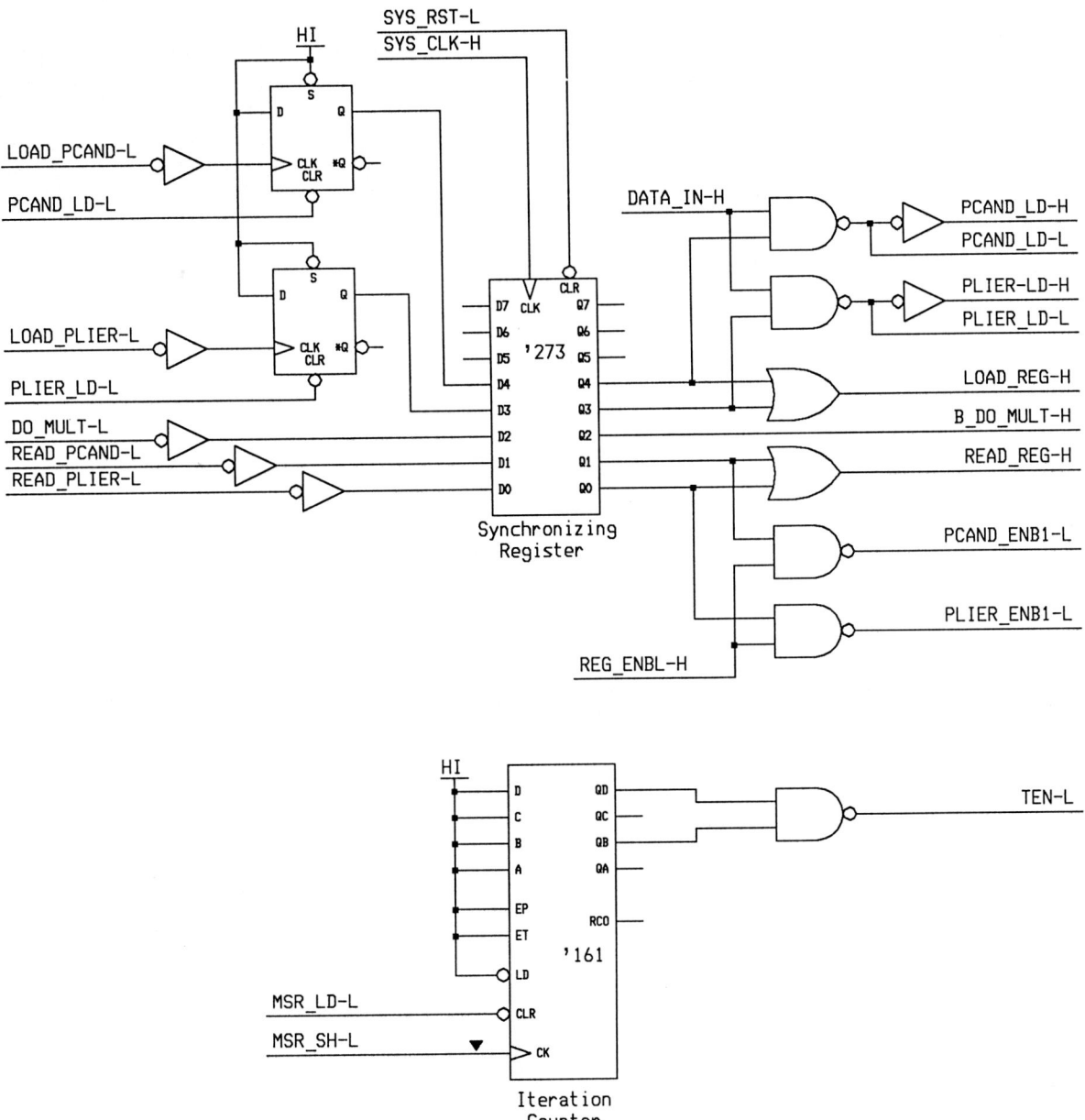

Figure 4.20(e). Detailed Logic Diagram for Control of Floating
Point Multiplier: Synchronizing Register and Control Signal Generation.

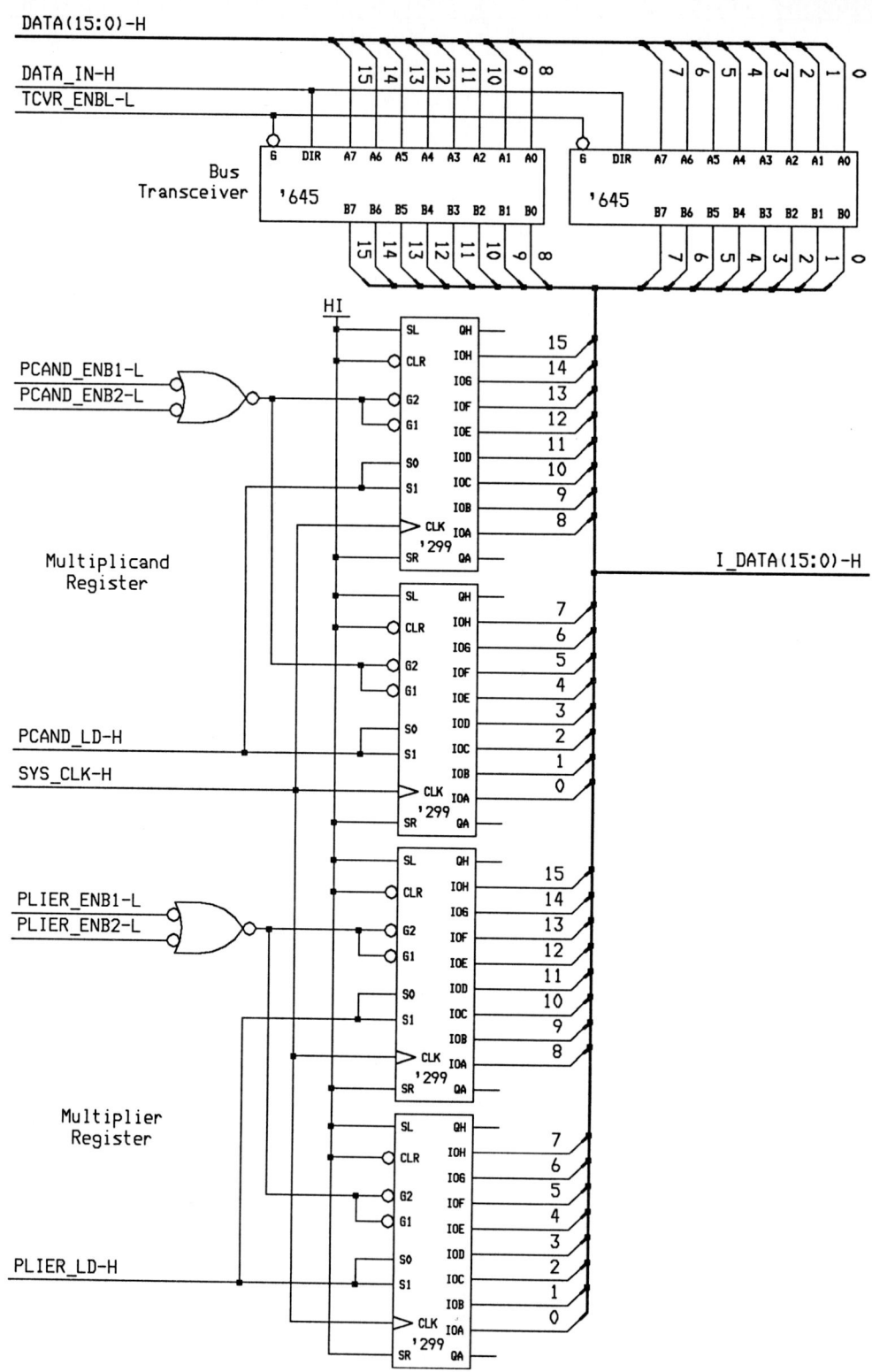

Figure 4.21(a). Detailed Logic Diagram for Data Path of Floating Point Multiplier: Transceiver, Multiplier Register, and Multiplicand Register.

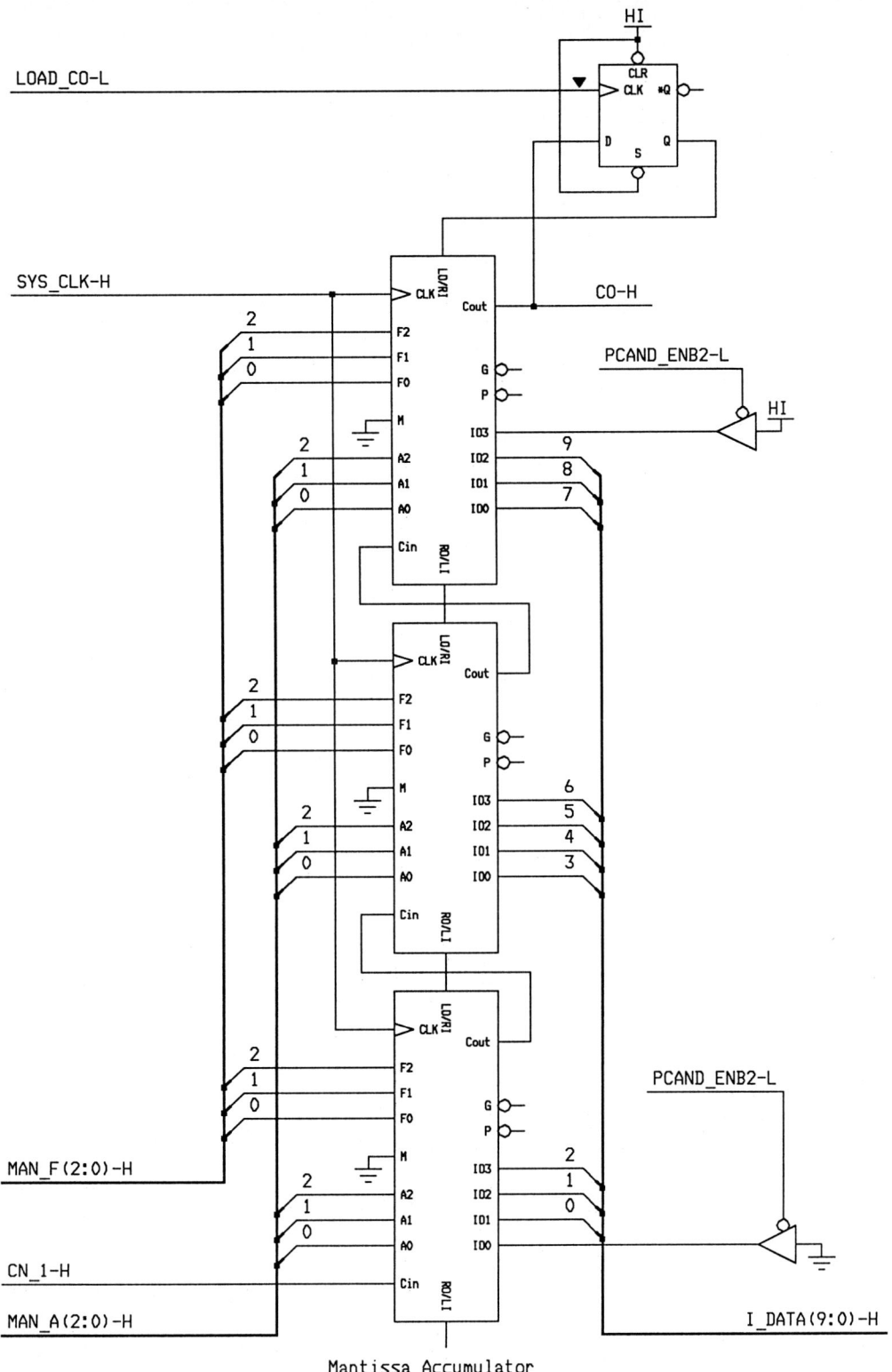

Figure 4.21(b). Detailed Logic Diagram for Data Path of Floating Point Multiplier: Mantissa Accumulator.

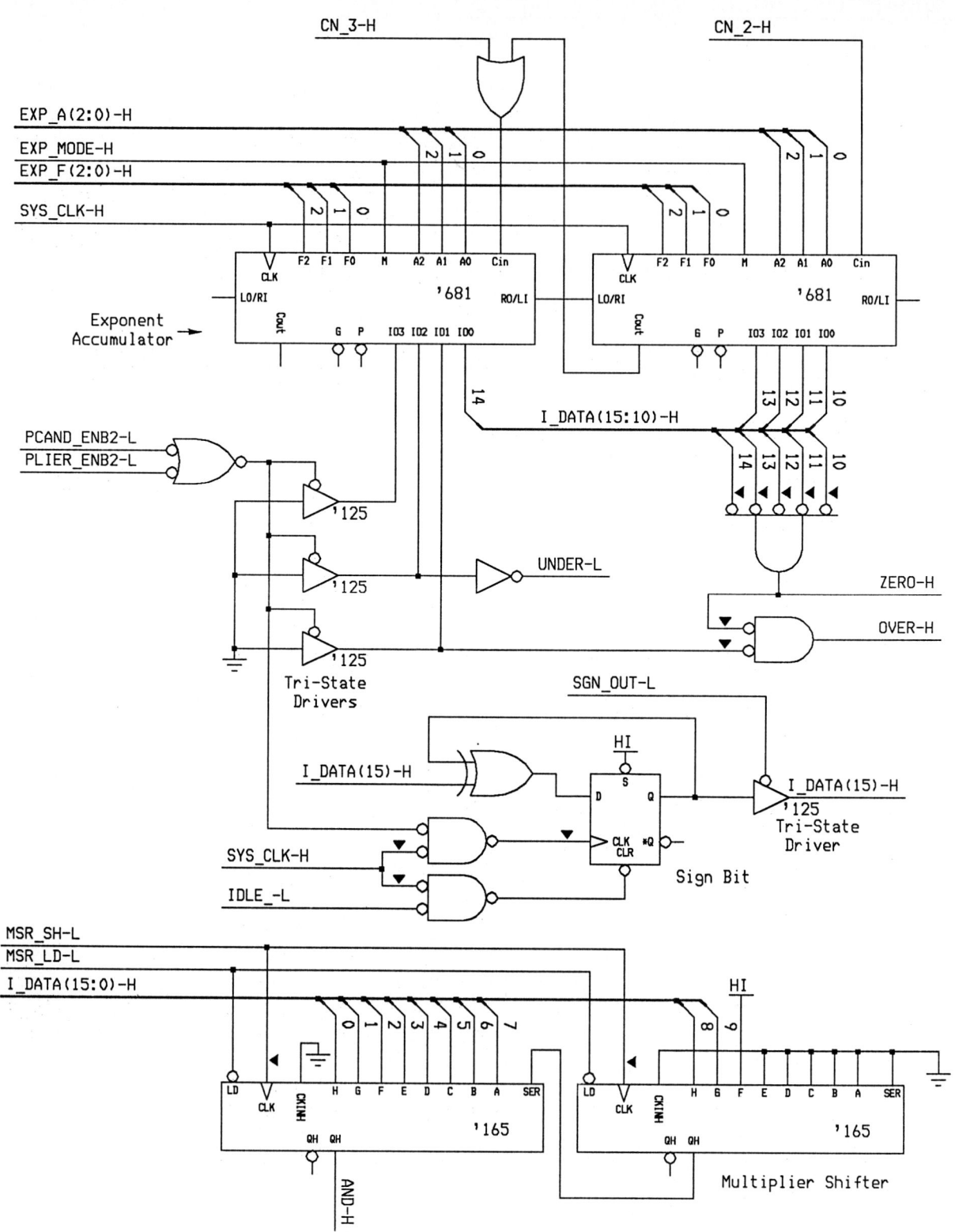

Figure 4.21(c). Detailed Logic Diagram for Data Path of Floating Point
Multiplier: Exponent Accumulator, Sign Bit, Multiplier Shifter, and Error Checking.

to the internal data bus. The activity of the registers and the transceiver is orchestrated by the control section.

The Mantissa Accumulator is shown in Figure 4.21(b). This is composed of three '681s and a single flip-flop. Note that when the PCAND_ENB2 signal is asserted the hidden bit is also made available to the mantissa register internal to the devices. Also note that the internal data bus has been aligned with the accumulator in such a way that the rounding will work and also the alignment as needed.

The final data path drawing is shown in Figure 4.21(c). This drawing includes the Exponent Accumulator, the Multiplier Shifter, and the sign bit, as well as the logic needed to check for errors. Since the derivation of the sign bit must be done as the multiplier and the multiplicand are extracted from their respective registers, this information is stored in a simple flip-flop which is clocked when the values are available. When the system is in the idle state this flip-flop is cleared to provide a correct starting point for the operation. The logic required for overflow, underflow, and zero detection is shown, and all three conditions can be identified with a very few gates in this system. The multiplier shifter provides the single bit (AND) which is used to determine the addition activity of the Mantissa Accumulator.

One immediate observation can be made concerning the two Floating Point Multiplier implementations which have been included here. In the the first system the control was kept to a minimum, which resulted in a very simple sequential system to perform the required wo.k. However, the data path was quite extensive. In the second implementation, the data path was kept, if not simple, at least small. The result was a system which required fewer parts for the data section, but which required a much more complex control system. Before we form a more direct comparison between the two systems, let us look at a memory-based controller for the unit.

Alternative control system for the floating point multiplier

Table 4.10 contains all the necessary information for next state determination, but it is not in the most convenient form for a memory-based system. Therefore, we present the same information in a more classical format in Table 4.12. This type of table could be used to generate solutions like the ''D'' solution or ''J-K'' solution of the previous section. Another use of the table is to generate a PROM-based solution, which is what we do in this implementation. All the logic required for next state determination and control signal generation is performed in a PROM before the active edge of the clock. This reduces the complexity of the system, as far as the number of chips and interconnections involved in the control system. A logic diagram of this solution is included in Figure 4.22. In addition to the logic shown in Figure 4.22, this solution will also need the synchronizing register and other logic shown in Figure 4.20(e). The logic of Figure 4.22 differs in the previous memory-based implementation, shown in Figure 4.13, only in that the registers and memory are not contained in the same device. However, in this implementation three sets of PROM/REGISTERs are required. This will provide for 4 present state bits and 20 control lines. Two additional control lines are needed, and these are derived from the Present State lines by simple gates. An implementation more consistent with the technique would include a fourth PROM/REGISTER pair for these signals. This would leave six unused outputs of the PROM, which may or may not be acceptable.

The number of inputs to the next state logic would require a large PROM, so steps have been taken to reduce the number of locations required. The next state

Table 4.12. Next State Determination Table.

Present State	Inputs									Next State
	LOAD REG	READ REG	DO MULT	AND	ZERO	UNDER	OVER	TEN	CO	
0 0 0 0	0	0	0	X	X	X	X	X	X	0 0 0 0
0 0 0 0	1	0	0	X	X	X	X	X	X	0 0 0 1
0 0 0 0	0	1	0	X	X	X	X	X	X	0 1 0 0
0 0 0 0	0	0	1	X	X	X	X	X	X	1 0 0 0
0 0 0 1	X	X	X	X	X	X	X	X	X	0 0 0 0
0 0 1 0	X	X	0	X	X	X	X	X	X	0 0 0 0
0 0 1 0	X	X	1	X	X	X	X	X	X	0 0 1 0
0 0 1 1	X	X	X	X	X	X	X	X	X	0 0 1 0
0 1 0 0	X	0	X	X	X	X	X	X	X	0 0 0 0
0 1 0 0	X	1	X	X	X	X	X	X	X	0 1 0 0
0 1 0 1	X	X	X	0	X	X	X	X	X	0 1 1 1
0 1 0 1	X	X	X	1	X	X	X	X	X	1 1 1 1
0 1 1 0	X	X	X	X	X	X	0	X	X	0 0 1 0
0 1 1 0	X	X	X	X	X	X	1	X	X	0 0 1 1
0 1 1 1	X	X	X	X	X	X	X	X	X	0 1 0 1
1 0 0 0	X	X	X	X	0	X	X	X	X	1 0 1 0
1 0 0 0	X	X	X	X	1	X	X	X	X	1 0 1 1
1 0 0 1	X	X	X	X	X	1	0	X	X	1 0 1 1
1 0 0 1	X	X	X	X	X	0	1	X	X	0 0 1 1
1 0 0 1	X	X	X	0	X	0	0	X	X	0 1 1 1
1 0 0 1	X	X	X	1	X	0	0	X	X	1 1 1 1
1 0 1 0	X	X	X	X	X	X	X	X	X	1 1 1 0
1 0 1 1	X	X	X	X	X	X	X	X	X	0 0 1 0
1 1 0 0	X	X	X	X	X	X	X	X	0	0 0 1 0
1 1 0 0	X	X	X	X	X	X	X	X	1	0 1 1 0
1 1 0 1	X	X	X	X	X	X	X	X	X	1 1 0 0
1 1 1 0	X	X	X	X	0	X	X	X	X	1 0 0 1
1 1 1 0	X	X	X	X	1	X	X	X	X	1 0 1 1
1 1 1 1	X	X	X	X	X	X	X	0	X	0 1 0 1
1 1 1 1	X	X	X	X	X	X	X	1	0	1 1 0 0
1 1 1 1	X	X	X	X	X	X	X	1	1	1 1 0 1

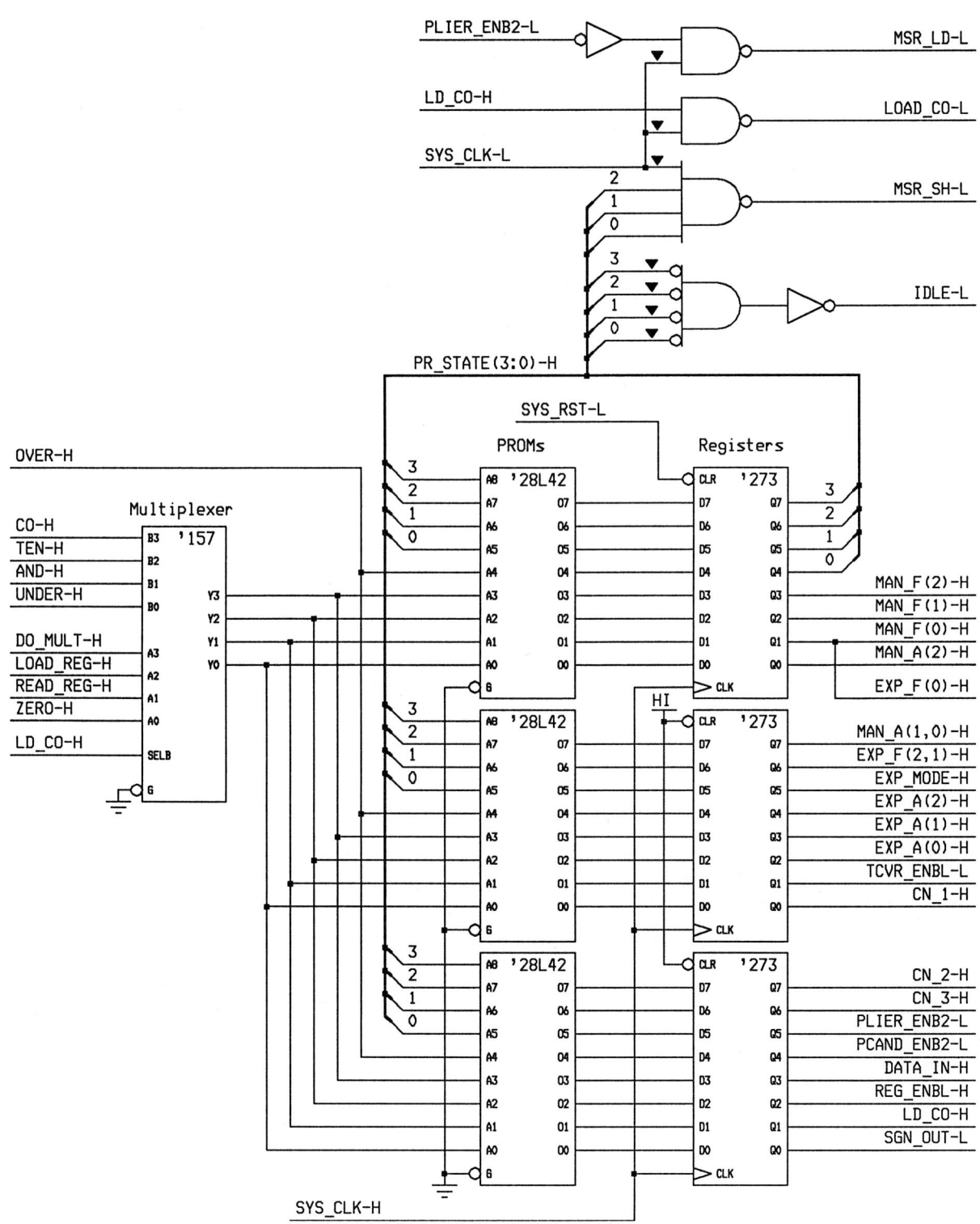

Figure 4.22. Detailed Logic Diagram for PROM-Based
Control of Floating Point Multiplier.

determination is made by considering four Present State lines and nine other logical values. A memory which would be capable of considering all these lines independently would need 2^{13} locations, which is an 8K PROM. Rather than require that large of a PROM, the technique illustrated in Figure 4.22 uses a multiplexer to select a subset of the inputs. Using a 4-line-to-1-line multiplexer as shown in the figure reduces the number of variables from 13 to 9, and the required number of locations in the PROM from 8K to 512. The requirement on the grouping of the signals into the multiplexer is simply that inputs which influence the next state determination from a single state must be present as addresses to the PROM during that state. The addition of the multiplexer expands the possible inputs to the system, but at the same time care must be taken in generating the correct contents of the PROM, since the PROM address lines will be logically connected to different inputs depending on the current state of the system.

The selection line of the multiplexer in Figure 4.22 is derived from a signal used in the system. A better solution would be to have a separate line from the PROM/REGISTERs system to control the multiplexer, but we ran out of lines too soon. Another approach would be to renumber the states in such a fashion that one of the present state lines could be used as the select line on the multiplexer, which is an exercise left to the reader.

Floating point multiplier: Another comparison of implementation methods

Now that we have done another design of the same system, we are prepared to make some comparisons of techniques. The best design choices will be made when many alternatives are available, and the system resources are utilized to their fullest capability. A summary of some of the resource utilization for the designs of the previous section are shown in Table 4.7. To compare the current designs we provide similar information about the designs done in this section. Since the cycle times are basically the same, we will not repeat the cycle time information. The new information is shown in Table 4.13.

Comparing the information in Tables 4.7 and 4.13 reveals some interesting information. The data path in the second implementation requires 40 percent less

Table 4.13. Resource Utilization for the Second Implementation of the Floating Point Multiplier.

Design	Power Dissipation (mW)	Board Space (in²)
Data Path	3665.7	7.8
Multiplexer Control	582.8	9.84
PROM Control A	1197.1	6.0
PROM Control B	2997.1	5.1

board space and 45 percent more power than the first implementation. This is a tradeoff that a designer would have to make in deciding which of the two resources, power or board space, is more important in the current system. Since the devices used in the second implementation are more complex, the control logic required for the second implementation is more complex. The multiplexer method requires more than nine times the power and about three times the board space. The PROM Control (method A) uses only 56 percent more power and over three times the board space. The other PROM method shown is the same implementation, but with 27S35s for the memory and storage elements; this reduces the board space and increases the power requirement.

One final comparison concerns the actual time required to complete the floating point calculations. This time reflects how long it takes to get through the states required to do the work, which is a reflection of the complexity of the state diagram for that system. The longest successful time for the first solution is 23 state times; at 100 nsec per state, this would be 2.3 usec. The longest successful time for the second solution is 29 state times, which would be 2.9 µsec at 100 nsec per state. Since errors are detected and handled earlier in the first system, the times for the zero, underflow, and overflow conditions will be correspondingly shorter for the first implementation.

In this section we have duplicated some of the things done in the previous system, but with a different emphasis. Nevertheless, the concepts and ideas involved are very similar. The design process is composed of a number of decisions, and the manner in which those decisions is made determines the type of design which results. The solutions to the floating point multiplier problem presented in this and the previous section provide two different solutions of the same problem, either of which will provide correct values. Other solutions to the problem are also possible, and those solutions will add still more sets of information to analyze for the "best" solution. If the requirement of a shift-and-add type of algorithm is removed, then direct implementations of the mantissa multiplication could be used, and these change the complexity of the problem and its solution. We hope that the designs shown here demonstrate that a designer must not artificially limit the choices. Different devices and techniques will lead to different solutions, and the designer's task is to intelligently identify the methods and parts which will make most effective use of the available system resources.

4.4. Sine and Cosine Generation with CORDIC Iterations

The floating point multiplication process is just one of a number of systems which can be implemented in a sequential fashion. Another such system is the generation of functions with cordic iterations. In 1956, Jack Volder explored a system of equations for calculating trigonometric and hyperbolic functions; in 1959 he described a machine based on the technique called the CORDIC (COordinate Rotation DIgital Computer) [Vold59]. This system has also been used in hand-held calculators and a number of other applications. A good description of the theory and mathematics involved is found in [Walt71]. The system can be used to do trigonometric functions, square roots, multiplication, division, and hyperbolic functions. The object of this design exercise is to create three different data paths, with corresponding state diagrams, to implement the sine/cosine functions by using the CORDIC iterations. The first task is to explore a minimal power solution which uses serial arithmetic. The second implementation is similar to the first, but uses parallel arithmetic. Finally, design a CORDIC system which could be implemented in a pipelined fashion.

Cordic iteration system: Understanding the problem

A cursory examination of the cordic system indicates that three input values are required, and that up to three output values can be obtained. Thus, the problem we will address is the creation of a system to manipulate data in three registers to create the three output values. We will refer to these values simply as z, x, and y. The system we will be concerned about here is similar to the floating point systems of the previous sections. That is, we will assume that the registers in the system are 16 bits wide, and that the interface is with a tri-state bus. Therefore, in this system we will assume that we have three registers to fill, and six registers to read. The three registers which can be filled and read correspond to the values for z_{in}, x_{in}, and y_{in}; the three registers which are basically read-only registers correspond to the values for z_{fin}, x_{fin}, and y_{fin}. Also, we will assume the existence of a signal, DO_COR, which asks that a cordic iteration be performed on the current contents of the input registers.

This is a sufficient amount of information to begin the design process, which we will do by examining in more detail the iteration system itself.

Cordic iteration system: Understanding the processing algorithms

The cordic iterations for sine/cosine functions are represented in a very simple fashion:

$$z_{i+1} = z_i \pm \alpha_i$$

$$x_{i+1} = x_i \pm \delta_i \times y_i$$

$$y_{i+1} = y_i \mp \delta_i \times x_i$$

This is only one of the many modes in which the cordic system will operate, and it can be used for the required sine and cosine functions. Notice that the operations listed above for the generation of the new x value and the new y value are opposite one another, and that the operation required to generate a new x is the same as the operation for a new z. The obvious questions which arise concern the initial values, final values, and constants. The constants, represented in the above equations as α_i and δ_i, determine the applicability and convergence characteristics of the system of equations. For this sine/cosine generation system, we will choose the δ_i to be 2^{-i}, and we will also choose the α_i in such a way that $\alpha_i = \tan^{-1}(2^{-i})$.

Another piece of information needed for this system concerns the $\pm$ decision. That is, in the equations above, what factor determines whether α_i is added to or subtracted from z_i to generate a new value for z, z_{i+1}. For this system, the decision is performed in such a way that z_{i+1} is closer to zero than z_i. Since all the α_i are positive (getting progressively smaller), the decision is based on the sign of z_i. If z_i is positive, then z_{i+1} is created by subtracting α_i from z_i; if z_i is negative, then z_{i+1} is created by adding α_i. The operation performed to create the new x, x_{i+1}, is the same operation used in the equation for z_{i+1}, and the operation for the x_i is the inverse operation.

The multiplications called for in the above equation can be very simply accomplished, since the δ_i have been chosen as 2^{-i}. Thus, each multiplication is merely a shift of the corresponding operand. The value of $\delta_i \times y_i$ is y_i shifted right by i bit positions. With this selection for the constants involved, the creation of the sine and cosine values reduces to a series of shifts, additions, and subtractions. Identifying the initial values of the variables as z_{in}, x_{in}, and y_{in}, and the final values as z_{fin}, x_{fin}, and y_{fin}, we make the following observations. The value of z_{fin} will be very close to zero, since the addition/subtraction decisions along the way have been made to force that situation. The cordic iterations shown above will create the situation that x_{fin} will be $K \times [x \times \cos(z) - y \times \sin(z)]$, and y_{fin} will be $K \times [y \times \cos(z) + x \times \sin(z)]$. The value of K is determined by the number of iterations in the system and the constants being used. The value of K will be constant for a given system, with a set number of iterations and the appropriate constants. We can seek to create a system which will operate according to the above equations and provide the values which we want. If we choose x_{in} to be one and y_{in} to be zero, then the final values of x and y will be $K \times \cos(z)$ and $K \times \sin(z)$. Thus, the final values are based on an initial angle, z. If we now choose x to be the inverse of K, then the final value will be the desired sine and cosine values. The accuracy of the final value depends on the number of iterations utilized. For a binary system, the value improves approximately one bit position per iteration.

The description given above will lead to a hardware system which produces the desired values. However, often in a system of this nature there is a combination of confusion about the expected results and a lack of confidence that the system will work as advertised. When these conditions exist, it is very beneficial to explore the operations with simulations until confidence is gained in the ability of the system of equations to produce the desired results. This can be done with a very simple computer program which directly implements the equations above, printing out the various values at each iteration. This also produces data which can be used in the checkout process to verify that the correct results are being produced. By creating such a program and experimenting with the various values, we have determined that a good value for x_{in} is 0.607253. Using this initial value, we have created the values used for 16 iterations of the system in calculation of the sine and cosine of 30°. This system has been created to use a radian measure rather than a degree measure, so the angle in question is actually 0.5236 radians. The resulting data is shown in Table 4.14. The reader is invited to reproduce this table with a simple computer program.

Creating the information shown in Table 4.14 will give confidence in the function of the system, but to get to the bit level it is beneficial to take the results one step further and produce the same information in a binary format. Table 4.15 contains the same information as Table 4.14, but represented in a binary format. The data values are represented in a hexadecimal format for 20 bits, where the bits represent a value between ± 1. That is, there is an assumed radix point directly to the right of the most significant bit, which is also the sign bit. All other bits are fractional in value. The information in Table 4.14 is more intuitively appealing to our base 10 background; the information in Table 4.15 is what would be directly implemented in the hardware.

The basic task, therefore, is to provide a system which will accept the values for input, create the appropriate output values, and make them available to an interface.

A conceptual level block diagram for this system can be represented as shown in Figure 4.23. That is, there must be storage for the z_i, x_i, and y_i for each iteration. There must also be a mechanism for providing the constants as needed, doing the multiplications, and performing the additions and subtractions as needed.

Table 4.14. Cordic Iteration Information for 0.5236 Radians.

Iteration i	x_{i+1}	Cosine x_i	Sine y_i	Constant δ_i	Constant α_i
0	0.52361	0.60725	0.00000	1.00000	0.78540
1	−0.26178	0.60725	0.60725	0.50000	0.46365
2	0.20186	0.91088	0.30363	0.25000	0.24498
3	−0.04312	0.83497	0.53135	0.12500	0.12435
4	0.08124	0.90139	0.42697	0.06250	0.06242
5	0.01882	0.87471	0.48331	0.03125	0.03124
6	−0.01242	0.85960	0.51065	0.01562	0.01562
7	0.00320	0.86758	0.49721	0.00781	0.00781
8	−0.00461	0.86370	0.50399	0.00391	0.00391
9	−0.00070	0.86566	0.50062	0.00195	0.00195
10	0.0012	0.86664	0.49893	0.00098	0.00098
11	0.00027	0.86616	0.49977	0.00049	0.00049
12	−0.00021	0.86591	0.50020	0.00024	0.00024
13	0.00003	0.86603	0.49999	0.00012	0.00012
14	−0.00009	0.86597	0.50009	0.00006	0.00006
15	−0.00003	0.86600	0.50004	0.00003	0.00003

Table 4.15. Binary Representation of Cordic Iteration Information for 0.5236 Radians.

Iteration i	x_{i+1}	Cosine x_i	Sine y_i	Constant δ_i	Constant α_i
0	4305C	4DBA7	00000	80000	6487E
1	DE7DD	4DBA7	4DBA7	40000	3B58C
2	19D6A	7497B	26DD3	20000	1F5B7
3	FA7B2	6AE06	44032	10000	0FEAD
4	0A660	7360C	36A71	08000	07FD5
5	0268B	6FF65	3DDD2	04000	03FFA
6	FE690	6E076	415CD	02000	01FFF
7	00690	6F0CE	3FA4B	01000	00FFF
8	FF690	6E8D9	4082D	00800	007FF
9	FFE90	6ECE1	40144	00400	003FF
10	00290	6EEE2	3FDCE	00200	001FF
11	00090	6EDE2	3FF89	00100	000FF
12	FFF90	6ED62	40067	00080	0007F
13	00010	6EDA2	3FFF8	00040	0003F
14	FFFD0	6ED82	40030	00020	0001F
15	FFFF0	6ED92	40014	00010	0000F

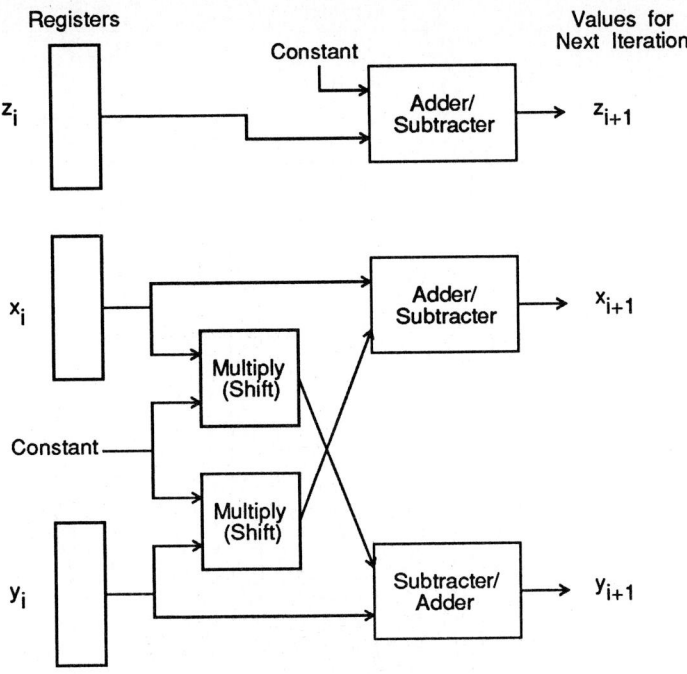

Figure 4.23. Conceptual Level Block Diagram for Cordic System.

There are many ways in which this operation can be performed; we will briefly examine three of these approaches.

Cordic iteration system number 1: Data path block diagram

The first solution is one in which the arithmetic logic is kept to a minimum. A block diagram of this solution is shown in Figure 4.24. This system has some of the characteristics which were utilized by the floating point multipliers represented earlier in this chapter. The block diagram shows a transceiver which provides a path from an external data source [DATA(15:0)-H] to an internal tri-state data path [I_DATA(15:0)-H]. This path can be used to fill and read the initial value registers, Z_IN, X_IN, and Y_IN. The path can also be used to read the other registers which will contain information of interest after the iterations have been performed. These registers are Z_T, X_T, and Y_T.

The arithmetic involved in this system is serial in nature. That is, the values are added one bit at a time, with the carry stored in a flip-flop for the next bit to be added. This trades off logic for time, since the same operations are performed, but only one bit at a time. However, since not much logic is needed for the adder/subtracter, three of them are shown, and the calculations of the values for z_i, x_i, and y_i proceed simultaneously.

The mode of operation for the iteration process begins with the z, x, and y values being copied from the initial values registers to the iteration registers. In addition, the x and y values are also placed in the x and y shift registers, X_SH and Y_SH. Then, depending on the sign of the value stored in the Z_T register, the operations are performed simultaneously in the three adder/subtracter systems. The

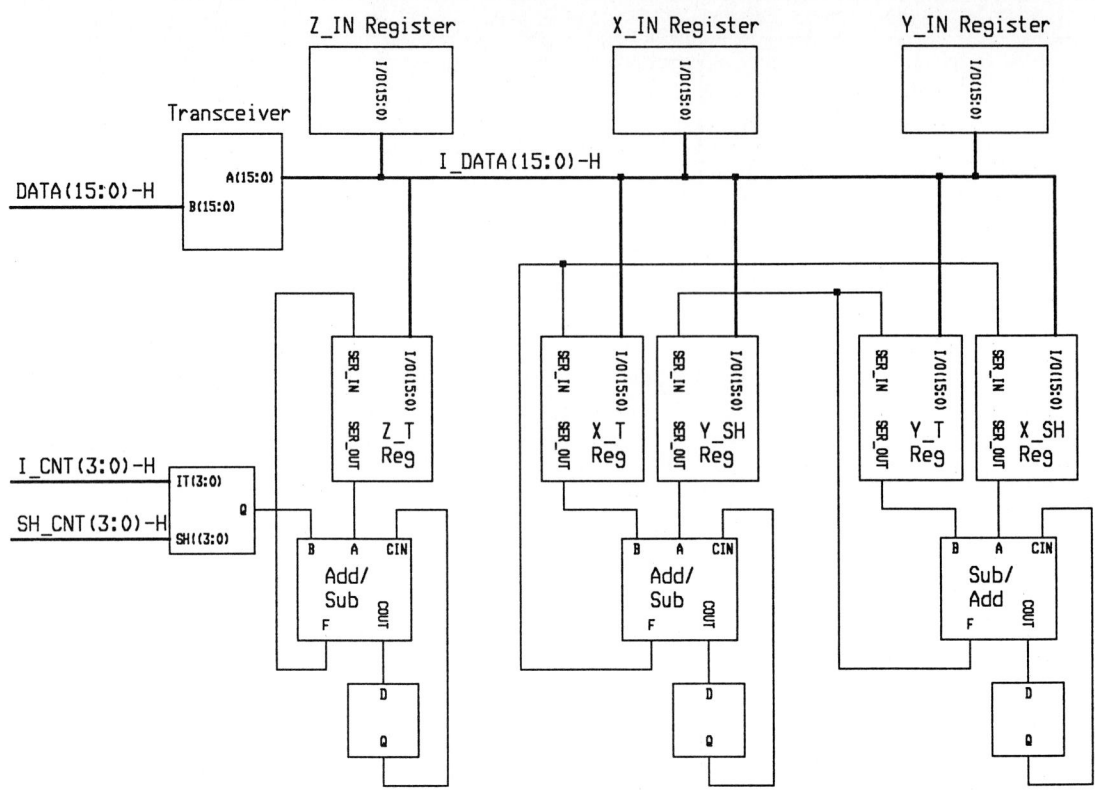

Figure 4.24. Serial Arithmetic Approach to Cordic System.

new value for z is a constant added to or subtracted from the old value for z; the constants are stored in a memory and extracted one bit at a time. The new value for x is a shifted copy of y added to or subtracted from the old value for x. Likewise, the new value for y is a shifted copy of x subtracted from or added to the old value for y. The multiplication is accomplished by shifting, and the shift register is clocked the appropriate number of times before the process starts to perform the actual multiplication. Missing from the block diagram of Figure 4.24 are three flip-flops which are needed in the system. The first retains the sign of the z value, so that the correct operation is performed for all 16 iterations. The second and third flip-flops are to retain the sign of the x and y values which are shifted. These flip-flops are used for sign extension during the shift process so that the proper value results from the multiplication by shifting.

Cordic iteration system number 1: Initial state diagram

A state diagram which shows this process is given in Figure 4.25. The system waits in the idle state for a request for action. If a register read is requested by the external device, then the read state (RD) is entered until the read request goes away. If a load request is made, then the load state (LD) is used to accept this value. The actual cordic process begins when the DO_CORDIC signal is received.

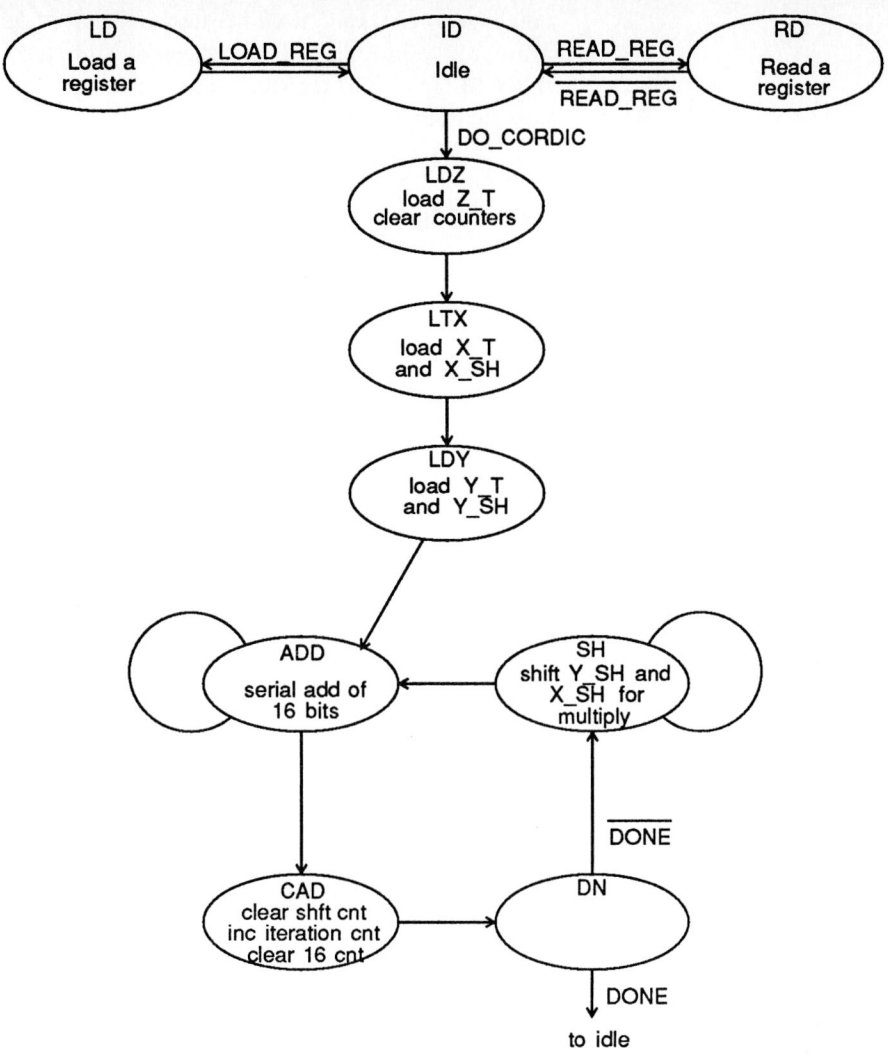

Figure 4.25. State Diagram for Serial Arithmetic Cordic System.

At this time copies of the z, x, and y values are made in states LDZ, LDX, and LDY. Also, three system counters are cleared. These counters are used to control the action of the system. The first is an iteration counter, which identifies which of the 16 iterations is being calculated. The second is a shift counter; this is used to control the number of shifts needed to perform the multiplication. The multiplication is performed by shifting the X_SH and Y_SH registers the appropriate number of bit positions to the right (lesser significance). The third counter is used to count the 16 addition cycles needed to do the actual addition.

When the z, x, and y values have been copied to their respective registers, then the actual cordic process can begin. This is started in the ADD state, where the first iteration is performed. The system will remain in this state for 16 iterations to do the 16-bit addition, then move on to the counter adjustment state (CAD). The counter adjustment state (CAD) is used to clear the addition counter and the shift counter, and increment the iteration counter. Then in the state marked DN the done condition is checked. If all the iterations have been performed, then the system can return to the idle state. If not, then the shift state (SH) is used to shift the values in the X_SH and Y_SH registers to perform the multiplication, and the addition for the next iteration is performed in the ADD state.

Since the addition is serial in nature, the ADD state will be used 16 times for each of the 16 iterations in this implementation. Add to these cycles those required for the shifts and the other functions and over 400 state times are required for the serial form of the function. A faster method is to use parallel adders, which is the solution method we examine next.

Cordic iteration system number 2: Data path block diagram

Each of the adders for the previous system can be implemented with a very small number of gates, hence these were duplicated three times. Another approach to resource utilization is to use a faster, full-word adder, but to use only one adder, and use this adder for all three operations required for each iteration. A block diagram of this type of system is shown in Figure 4.26. This system also has a

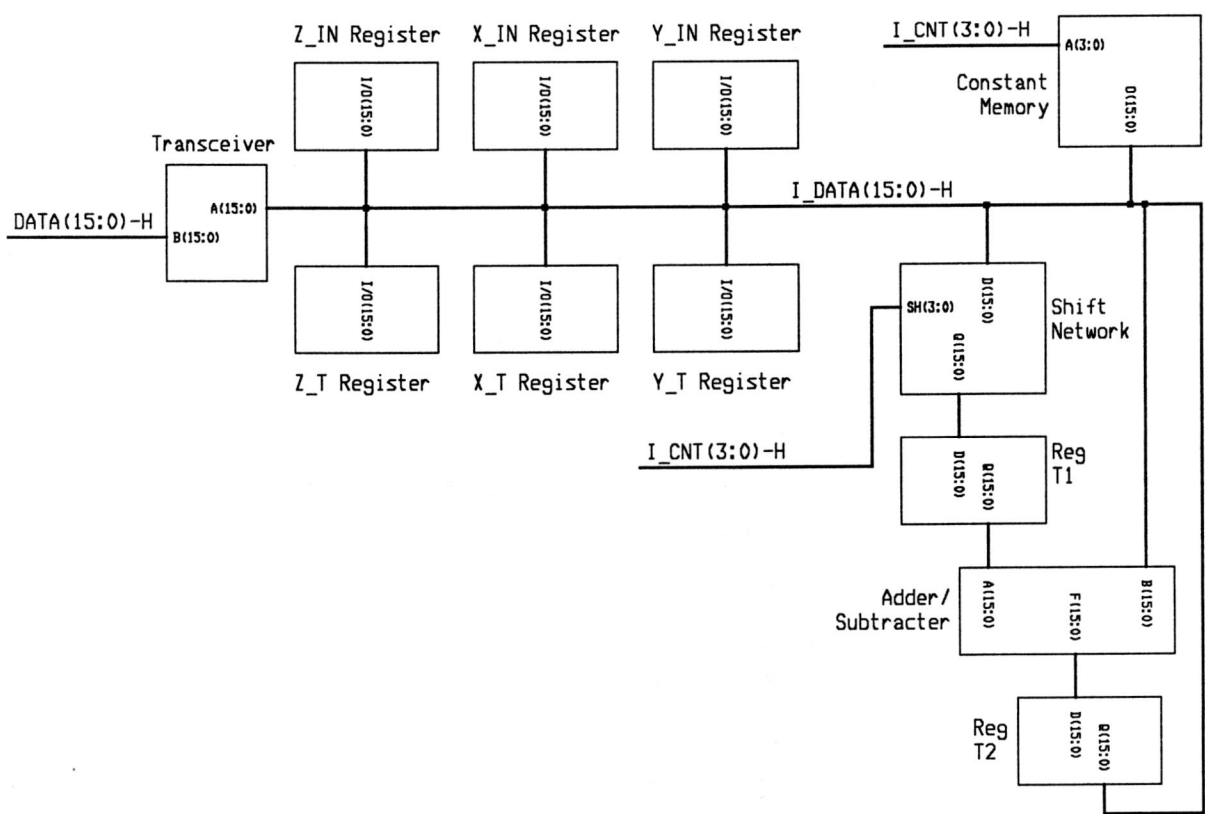

Figure 4.26. Word Adder Approach to Cordic System.

transceiver which is used to logically connect an internal bus [I_DATA(15:0)-H] with the external data path [DATA(15:0)-H]. Again, the three input registers are represented as Z_IN, X_IN, and Y_IN. The registers which are used for the actual cordic iterations are shown as Z_T, X_T, and Y_T. If the system were defined in such a way that the initial values could be destroyed in the process of doing the iterations, then this second set of registers would not be necessary. The arithmetic is performed in an Adder/Subtracter, and the control must identify which of the operations is required for the iteration being performed. The multiplication by shifting is performed by a shift network, which can be purchased as a single integrated circuit, such as the Texas Instruments 'AS8838. The other portions of the system shown in Figure 4.26 are the constant memory, which stores the α_i values, and two temporary registers which are used in the calculations process.

The iterations required in the system can be easily represented in a register transfer fashion as follows:

begin:

$$
\begin{array}{rcl}
Z\_IN & \to & Z\_T \\
X\_IN & \to & X\_T \\
Y\_IN & \to & Y\_T
\end{array}
$$

repeat the following 16 times:

if Z_T is negative, then {

$$
\begin{array}{rcl}
\text{shifted } Y\_T & \to & T1 \\
X\_T + T1 & \to & T2 \\
\text{shifted } X\_T & \to & T1 \\
T2 & \to & X\_T \\
Y\_T - T1 & \to & T2 \\
T2 & \to & Y\_T \\
Z\_T & \to & T1 \\
T1 + \text{constant} & \to & T2 \\
T2 & \to & Z\_T
\end{array}
$$

} else {

$$
\begin{array}{rcl}
\text{shifted } Y\_T & \to & T1 \\
X\_T - T1 & \to & T2 \\
\text{shifted } X\_T & \to & T1 \\
T2 & \to & X\_T \\
Y\_T + T1 & \to & T2 \\
T2 & \to & Y\_T \\
Z\_T & \to & T1 \\
T1 - \text{constant} & \to & T2 \\
T2 & \to & Z\_T
\end{array}
$$

}

The first three transfers load up the iteration registers with the initial values. The remaining transfers perform the cordic iterations as required. This set of operations can be presented in a state diagram as follows.

Cordic iteration system number 2: Initial state diagram

The state diagram for this system is extremely simple. The diagram itself is shown in Figure 4.27. The reading and writing of the registers is handled in exactly the same way as the previous system, using states RD and LD as shown in the figure. The work of the system begins when a DO_CORDIC signal is received, at which point the system leaves the idle state and begins the transfers. The initial values are moved to the iteration registers in states LDZ, LDX, and LDY. Once the transfers have been performed, then the actual work of the system can begin.

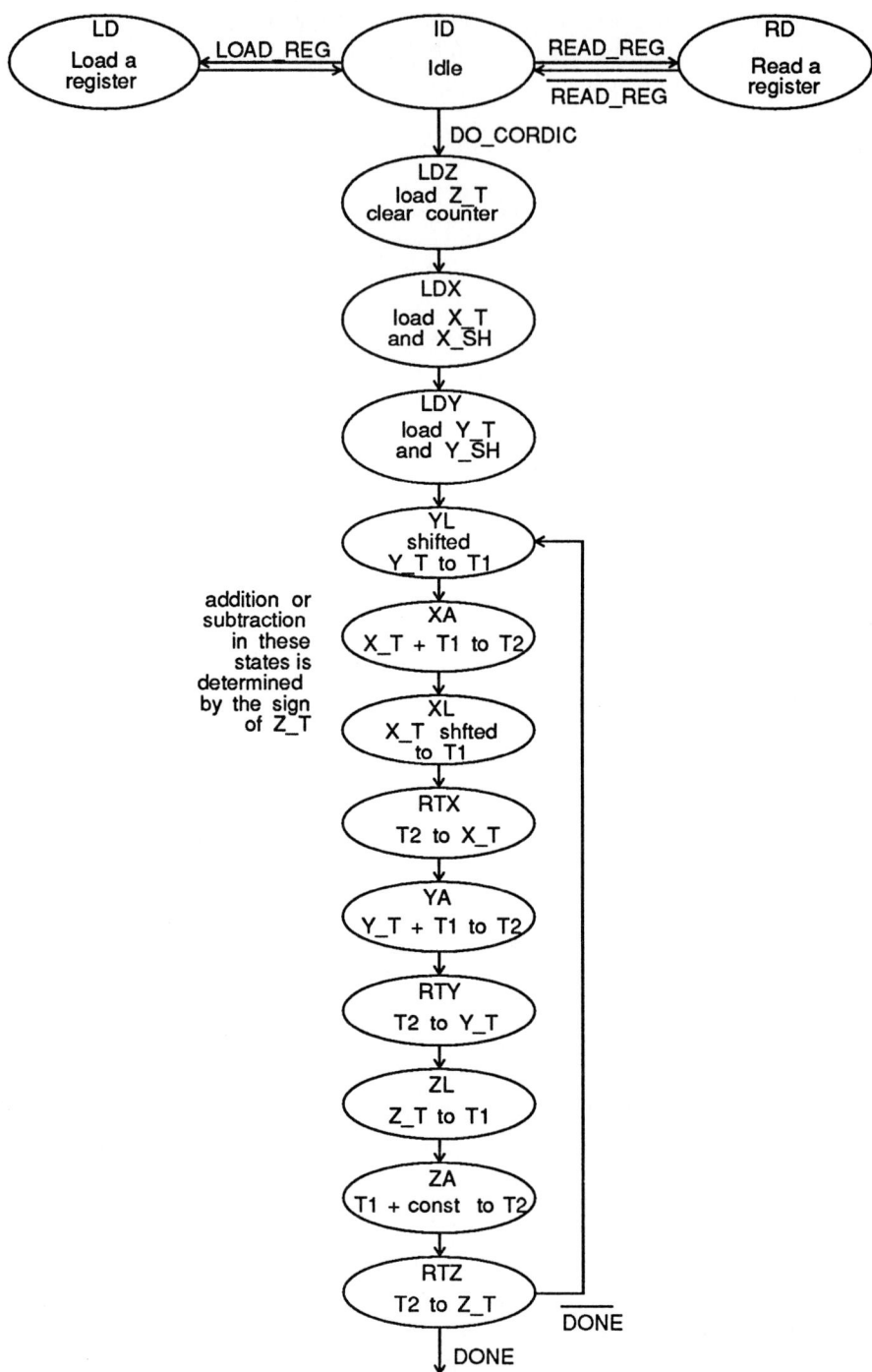

Figure 4.27. State Diagram for Word Arithmetic Cordic System.

The register transfers listed above identify two different sets of action. The choice between the sets is determined by the sign of the Z_T register. Yet the state diagram shown in Figure 4.27 shows only one train of action. The reason for this is that the assumption is being made that the Adder/Subtracter shown in Figure 4.26 will select the proper operation (either addition or subtraction) based on a single bit. Thus, with minimal gating the proper operation can be selected in the three states in which the adder must perform the required work, which are states XA, YA, and ZA. The logic gates to make this decision utilize the sign of Z_T and the current state to determine the necessary operation. Note that the action of the system has been specified in such a way that the sign of the Z_T register does not change until the end of all the transfers required for the iteration. For this reason the state diagram does not show two different paths, as might be inferred from the register transfer system described above. However, a state diagram with two distinct paths could also be utilized.

The determination of the ''DONE'' conditions is handled by the system examining the number of iterations accomplished. Thus, an iteration counter is cleared to begin the process, and each time through the loop the counter is incremented. When 16 iterations have been completed, the process is over, and the system returns to the idle state.

The system described by the block diagram of Figure 4.26 and the state diagram of Figure 4.27 requires 9 states per iteration to perform its work. Thus, the 16 iterations required to produce the answers will be completed in less than 150 clock cycles. This is in contrast to the previous method, which needed over 400 clock cycles for the same action.

Each of the two systems presented so far approaches the implementation of the cordic system with a different view of the importance of various system resources. We now present a third method for doing the cordic operations which has a completely different approach to the utilization of system resources.

Cordic iteration system number 3: Data path block diagram

The third implementation scheme we will examine is an implementation of the cordic system which relies on the concepts of pipelining. For a description of the concepts of pipelining and some of the problems associated with it, see [Poll90]. The basic idea in pipelining is that data will flow through the system, proceeding to a new section of the pipe with each clock pulse. The cordic iterations can easily be pipelined by assigning a different iteration to each section of the pipeline. Thus, once we understand what must be in each section of the pipe, we will be able to identify the action which will take place throughout the system.

A block diagram of one section of a cordic pipeline is shown in Figure 4.28. As shown in the figure, the system contains the same basic parts as the other implementations of the cordic system presented in this section. There are registers for storage of the z, x, and y values, as well as arithmetic elements to calculate the values for the next iteration. Since each section of the pipeline is configured to do only one operation, some of the hardware becomes simpler. The constant can be hardwired into place, since it will never change for this section. Also, the add/subtract decision for the arithmetic units is determined solely on the current sign bit of the Z register. But perhaps the most striking savings in time and silicon is the ability to do the multiply by wiring up the correct shift between the registers and the arithmetic units.

The system represented by the block diagram of Figure 4.28 creates three new values, z_{i+1}, x_{i+1}, and y_{i+1}, one add time after the data in the registers becomes

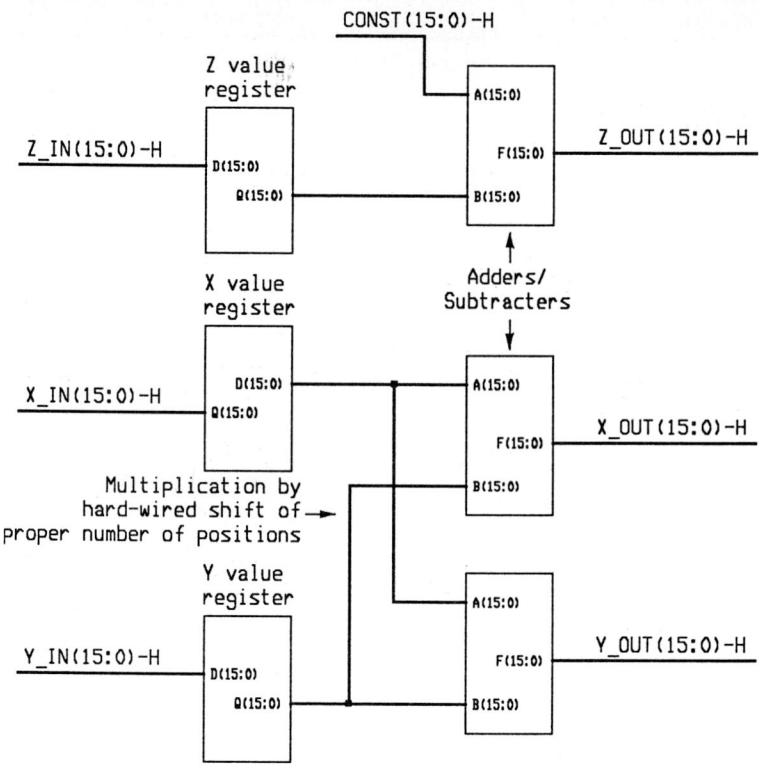

Figure 4.28. Block Diagram for a Cordic Pipeline Section.

stable. At that time, these values can be accepted by the next section of the pipeline, and the current section can accept a new set of values. Thus, a new set of sine/cosine values can be created every clock time, in steady state. However, for this to occur, it is necessary to have a data delivery system which can supply the initial values at a sufficiently fast rate to keep the pipeline full. Also, the results must be utilized at the same rate. This places strict requirements on the data communication system.

Note that no state diagram is given to accompany the pipeline mechanisms for the cordic, since the operations are performed by streaming data through the system, rather than by controlling the action of registers which are utilized many times in the generation of the results.

Comparison of cordic systems

The cordic systems presented in this section have a vastly different view of the important system resource which should be utilized. We have presented the initial data path block diagrams and the preliminary state diagrams for the systems, as well as a description of their operation. It is left as an exercise for the reader to provide the required details to finish these designs to the degree shown in the first two sections of this chapter. However, we would like to compare some of the

Table 4.16. System Resource Utilization of Cordic Systems.

Cordic System	Power Dissipation (mW)	Board Space (in^2)	Data Rate (words/sec)
Serial Arithmetic	2295	10.8	24,000
Word Arithmetic	3585	14.24	68,000
Pipeline Section	2550	7.92	
Pipeline System	40800	126.72	10,000,000

resource utilization of the data paths presented. This information is shown in Table 4.16.

The systems shown in Table 4.16 demonstrate the fact that different designs will produce different results, even for the same problem. If board space and power are to be conserved, then the choice of the designs presented in this section would be the serial arithmetic method. Further refinement to conserve more power or board space is also possible. The word arithmetic system uses over 50 percent more power than the serial method, and about 30 percent more board space, but produces results at a rate almost triple that of the serial method. The pipeline method indicates that, with a sufficient investment in power and area, extremely high data rates can be achieved. This rate would be sustainable only if the other elements of the system were created in such a fashion that data could be delivered, and results utilized, at the specified speed.

The details of the previous sections are not shown for the cordic implementations. The reader is invited to create detailed block diagrams and state diagrams for the designs, then implement them with an appropriate control section. In this way, the concepts of sequential control which are exemplified by the first two sections can be practiced, to enhance the understanding of design methods.

4.5. Summary

In this chapter we have demonstrated the process of sequential design by following different designs through the steps of the design process. These steps include:

Understand the processing requirements
It is imperative that the designer know what the system does, and what interfaces are required.

Understand the processing algorithms
The designer must also know the mechanisms involved in the processing which the unit performs.

Create a conceptual level data path block diagram
This step allows the designer to explore alternative hardware methods of doing the processing of the system.

Create a detailed data path block diagram
>When the data path is ready, the designer identifies the signals which will control data flow in the system.

Create a state diagram which describes the sequential nature of the system
>The state diagram provides a description of the action to be performed, and specifies the sequence of events.

Design the logic of the control section
>The detailed logic of the control section is generated by using information contained in the state diagram, coupled with knowledge of the control signals in the system.

Create the logic design for the entire system, implement and debug
>When a system is properly implemented the processing will match all the specifications and do the work required.

Following all the steps listed above results in a system which will accomplish the desired objectives. Alternative methods of answering the questions which arise in the design process will result in designs which utilize system resources in a different way, conserving time, power, or some other resource. The resources pointed out in this chapter include the power required to drive the system, the board space required for an ALS implementation of the system, and some indication of the speeds involved. Other resources can be dealt with in a similar fashion. These might include the area of a silicon implementation, the purchase price of components, or any of a number of other criteria. The decisions of the design must be based on a comparison of reasonable engineering criteria.

A final comment reiterates the point of the information presented in this chapter. Different mechanisms have been presented, and it is up to a user or designer to decide which of the mechanisms are appropriate for a specific system. The concepts demonstrated by the designs in this chapter can also be applied to the newer control devices being provided by manufacturers. These devices contain, in a single device, most of the elements of the control circuitry which have been described in this chapter. Among these devices are registered PALs, such as a 22V10, which contain register elements and programmable AND-OR circuitry. Another device is a field-programmable logic sequencer, such as the 82S105, which provides for 6 feedback variables and 16 inputs. Still another is the fuse-programmable controller (Am29PL141), which is a memory-based system and provides 64 words of programmable memory, 16 outputs, and 6 test inputs. All these controllers implement sequential systems in a manner similar to the mechanisms described in this chapter.

4.6. Exercises

4.1 Design a system for controlling a traffic light. That is, create a specification of the device behavior, identifying the lights that it is to control (left turn, yellow, red, green, etc.), the sources of input (car present in left turn lane, etc.), and the desired behavior of the system. Then design a system for controlling all of the lights in the system, using random logic and D-type flip-flops.

4.2 Repeat Exercise 4.1 using J-K flip-flops as the storage element of the present state register.

4.3 Design a divide circuit. That is, assuming that you have a divisor and a dividend (16 bits each, two's complement integer format), design a system which will create the correct quotient and remainder. Use '273s for the registers and '283s for the arithmetic unit. Develop the data requirements, then create a data path block diagram, and complete the design to the gate level

4.4 Use the multiplexer method of state machine design to create a system which accepts a single data input which is synchronous with a clock signal. The system will generate an output for a single clock period whenever the sequence 101100 is detected on the input line. Do not use a shift register in the solution.

4.5 Design a digital clock with alarm and chimes. Assume that the sounds which you must make are created by a speaker. Also, find a 7-segment display for the readout, and use as many digits as necessary. Determine a mechanism for setting the time, and for displaying and setting the alarm time. The system should sound the correct number of chimes on the hour, and sound a single chime (of a different tone) on the half hour as well as the hour.

4.6 Design a counter system which provides a count of the number of "1" bits in a data stream. The system has four inputs: CLOCK-H, DATA-H, ENBL-H, START-H. When the START signal is asserted, the counter is cleared. Then, the system waits for the next assertion of ENBL; when that happens, the counter is incremented each time DATA is asserted when the rising edge of CLOCK occurs. This value is available at an output, and will remain stable until the next occurrance of START. Illustrate all of the steps required to complete the design, from system definition to logic diagrams.

4.7 Create a single-bus system which implements a simple 16-bit computer system with a minimal instruction set. The instruction set contains only the following instructions: jump if carry set, add, invert, right shift, store accumulator, clear accumulator, and NAND. The system is a single-address computer, with an arithmetic unit, accumulator, program counter, instruction register, and memory address register. Use '181s for the ALU, '273s for the accumulator and memory address register, and '574s for the other registers. Propose a block diagram for the system, create an instruction format, and design the system. Include direct and indirect addressing capabilities.

4.8 Assume that a system has two inputs, a data line and a clock line. Create a sequential design that will transform a binary input sequence according to the following rules:
a. any input bit that has a one before and after it becomes a one,
b. any input bit that has a zero before and after it becomes a zero,

c. any input bit that has a one(zero) before it and a zero(one) after it is toggled.

Do not worry about startup and termination, just assume steady state. For example, the following portion of a sequence,

will be converted to

Note that the result cannot come out in "real time", i.e. bit n cannot be output until time $n+1$ because of causality considerations.

4.9 Create a dynamic RAM controller. The inputs are an 18 bit address, a 9 bit refresh address, a signal (READ) which is asserted when a valid address is available on the address lines and a normal read cycle is desired, and a signal (REF) which is asserted when a valid refresh address is available on the refresh address lines, and a refresh cycle is needed. The outputs are a 9 bit memory address, the column address strobe (CAS), the row address strobe (RAS), and a signal indicating the data is ready (READY). The system clock is 40 MHz (25 nsec cycles), and assume that all the logic you have will work fast enough to meet that time.

a. Give a block diagram to show the data paths between the inputs and the outputs. Assume that you have 74F157's to work with, which are four bit 2-1 multiplexers. On this diagram identify the control lines which are to be activated by your sequential machine.

b. Design a sequencer which will control the lines identified in Part A as well as the control outputs identified in the problem statement. When a refresh is requested, the refresh address is to be presented, and then 25 nsec later the column address strobe is asserted for 100 nsec; finally, the CAS line is released, and 25 nsec later the address released. When a read is requested, the 9 MSBs of the address are presented, and after 25 nsec the RAS line is asserted; 25 nsec later the address is changed to present the 9 LSBs of the address; 25 nsec later the CAS line is asserted. The ready line is asserted 75 nsec later, and this condition is maintained until the READ line is released. At that point the RAS, CAS, and READY lines are all released, and 25 nsec later, the address is released. If a read and a refresh request are made simultaneously, the refresh wins. The requests are not synchronous with the system clock.

4.10 Design a sequential circuit that will monitor two data lines X-H and Y-H. Your circuit is to give an indication if either the 01 or 10 sample was received once and only once in a four sample sequence. Assume that the data inputs are updated on the falling-edge of the system clock. Hint: use an 8 state sequencer and no counter.

4.7. Additional References

[AMD85] Advanced Micro Devices, *Bipolar Microprocessor Logic and Interface Data Book.* Sunnyvale, CA: Advanced Micro Devices, 1985.

[AMD88] Advanced Micro Devices, *PAL Device Handbook.* Sunnyvale, CA: Advanced Micro Devices, 1988.

[Andr80] Andrews, M., *Principles of Firmware Engineering in Microporgram Control.* Potomac, MD: Computer Science Press, 1980.

[Bart85] Bartee, T. C., *Digital Computer Fundamentals* (6th ed.). New York: McGraw Hill Book Company, 1985.

[Boot84] Booth, T. L., *Introduction to Computer Engineering: Hardware and Software Design.* New York: John Wiley & Sons, 1984.

[Bree89] Breeding, K. J., *Digital Design Fundamentals.* Englewood Cliffs, NJ: Prentice Hall, 1989.

[Come84] Comer, D. J., *Digital Logic and State Machine Design.* New York: CBS College Publishing, 1984.

[Flet80] Fletcher, W. I., *An Engineering Approach to Digital Design.* Englewood Cliffs, NJ: Prentice Hall, 1980.

[HaVr78] Hamacher, V. C., Z. G. Vranesic, and S. G. Zaky, *Computer Organization.* New York: McGraw-Hill Book Company, 1984.

[Haye88] Hayes, J. P., *Computer Architecture and Organization* (2nd ed.). New York: McGraw-Hill Book Company, 1988.

[Lang82] Langdon, G. G., Jr., *Computer Design.* San Jose, CA: Computeach Press Inc, 1982.

[Lee87] Lee, F., "Designing a State Machine with a Programmable Sequencer," *Electronic Products Magazine.* Vol. 29, No. 17, February 1, 1987, pp. 29-35.

[Mano79] Mano, M. M., *Digital Logic and Computer Design.* Englewood Cliffs, NJ: Prentice Hall, 1979.

[Mano88] Mano, M. M., *Computer Engineering: Hardware Design.* Englewood Cliffs, NJ: Prentice Hall, 1988.

[McCl86] McCluskey, E. J., *Logic Design Principles, with Emphasis on Testable Semicustom Circuits.* Englewood Cliffs, NJ: Prentice Hall, 1986.

[Poll90] Pollard, L. H., *Computer Design and Architecure.* Englewood Cliffs, NJ: Prentice Hall, 1990.

[Schm87] Schmitz, N., "Prose Devices Simplify State Machine Design," *Computer Design.* Vol. 26, No. 2, April 1, 1987, pp. 97-102.

[TI85] Texas Instruments, *The TTL Data Book,* Volume 2. Dallas, TX: Texas Instruments, 1985.

[Walt71] Walther, J. S., "A Unified Algorithm for Elementary Functions," *Proceedings Spring Joint Computer Conference,* 1971, pp. 379-385.

[Vold59] Volder, J. E., "The Cordic Trigonometric Computing Technique," *IRE Transactions on Electronic Computers,* Vol. EC-8, No. 3, Septermber 1959, pp. 330-334.

5

Design of
I/O Interface Systems

Computers can be extremely useful devices, but this becomes apparent only when the computer is able to interface to other devices. In this chapter we will examine some of the techniques used to interface computer systems. In particular, we are concerned here with asynchronous buses and mechanisms in which transfers can be made across those buses. The asynchronous bus has been used for transfers in computer systems for many years. One of the most prolific is the UNIBUS, created by Digital Equipment Corporation for their PDP/11 series of computers in 1970 [Bell70]. Another protocol used by Digital Equipment Corporation at a later date is the QBUS protocol, which is a time-multiplexed asynchronous protocol. That is, at one time the information on a set of bus lines is address, and at a later time data will be present. Other asynchronous buses include the MULTIBUS, the S-100 bus, and the VME bus. This chapter presents designs based around the VME bus, although the concepts and methods apply to a number of other asynchronous systems.

5.1. Asynchronous Bus Protocols

Transfers on asynchronous buses take place between master modules and slave modules, as shown in Figure 5.1. The master module is responsible for asserting an address onto the address lines of the bus, and the slave module will decode those address lines to determine if the requested address is in the address space covered by the slave. The master then initiates the actual operation by activating the appropriate control lines. The slave responds to the control lines in a predefined manner, and the data is transferred between the two modules. The set of rules governing the transfer is called the bus communication protocol, and it defines the sequence of signal assertions required to transfer information on the bus, as well as the timing of the signals required.

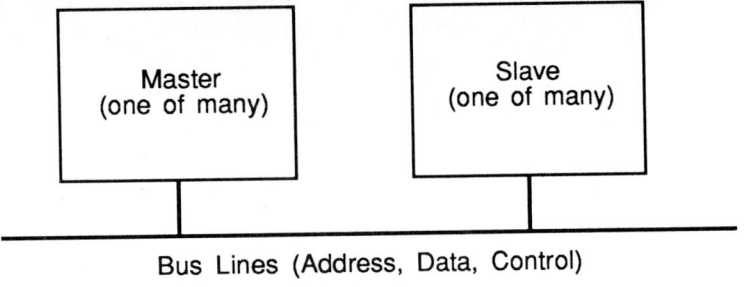

Figure 5.1. Asynchronous Bus Transfer System.

There are three basic types of signals involved in an asynchronous transfer protocol. The first type is the address. For UNIBUS systems the address consists simply of 18 address lines; for a VME bus the address group contains 32 address lines and 6 address modifier lines. The 32 address lines are capable of uniquely identifying over 4.29×10^9 locations; for byte-addressable systems we say that 32 bits of address is capable of addressing 4 gigabytes of memory. The address modifier lines increase this addressing capability further, and are used to specify types of data transfers, such as supervisor or nonsupervisor transfers. In general, address lines will be of a tri-state type, so that different master modules can assert the lines. The process of assuring that only one master module has control of the bus at any one time is called bus arbitration.

The second type of signal involved in the transfer is the data bus itself. The number of lines involved in this transfer depends upon the bus and the system. The UNIBUS is a 16-bit bus. The IEEE 488 bus is an 8-bit bus. The VME bus protocol is so defined that both 16- and 32-bit configurations are permissible. The data lines are bidirectional, allowing information transfer from the master to the slave for a write, or information transfer from the slave to the master for a read.

The third type of signal involved in the transfer is the set of control lines which actually controls the transfer from one module to another. The number of lines involved in the control of the transaction will vary with the protocol, but they will have the same basic functions. Two of the lines form a handshake pair, one asserted by the master module, and the other asserted by the slave module. We will refer to the line asserted by the master as the request line (REQ), and the line asserted by the slave as the acknowledge line (ACK). Another function is to identify the type of transfer being performed. This function can be accomplished with a single line which specifies whether the transfer is a read or write. Or, a number of control lines can be involved. First, let us examine a generic example, and then we will look at a specific protocol

For our purposes of a simple example we will assume that the transfer involves an address bus, a data bus, and three control lines. The address bus is a tri-state mechanism with 32 lines, and these lines are asserted high [ADDR(31:0)-H]. The data bus is also a tri-state transfer path, capable of transferring 32 bits at a time [DATA(31:0)-H]. The control line which identifies the current cycle type is READ-H, which will identify a read transfer when it is high, and a write transfer when it is low. The handshake line asserted by the master module in the transfer is called REQ-L. When this line is asserted the master is requesting a transfer of some kind. The handshake line which is asserted by the slave module is called ACK-L; this line is asserted in response to the action of the master, and indicates that the slave module has done some predetermined function.

The relationship of these lines in a typical transfer is shown in Figure 5.2. The write cycle is shown in Figure 5.2(a), and is used to send information from the master to the slave. The action shown in Figure 5.2 occurs after the ownership of the bus has been given to the master by the arbitration system. The cycle begins at T_A when the master asserts an address on the address lines. Prior to that time the value on the address bus is meaningless, since no address is asserted and the lines on the bus float to a level characteristic of that line. Some systems will terminate the bus in such a way that these lines float to a specific value; other systems do not provide this facility. At the same time that the master asserts the address lines, it

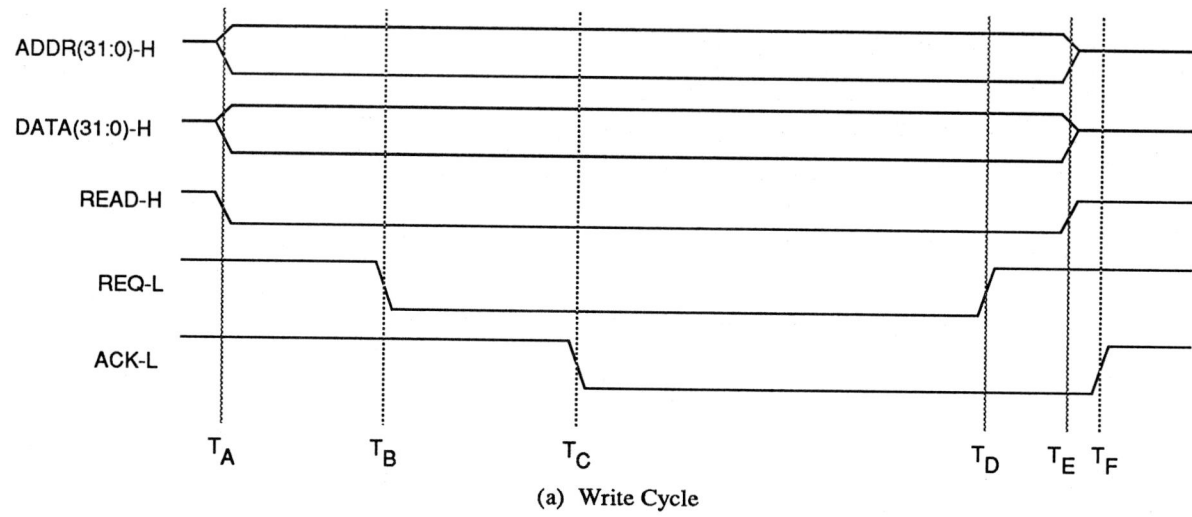

(a) Write Cycle

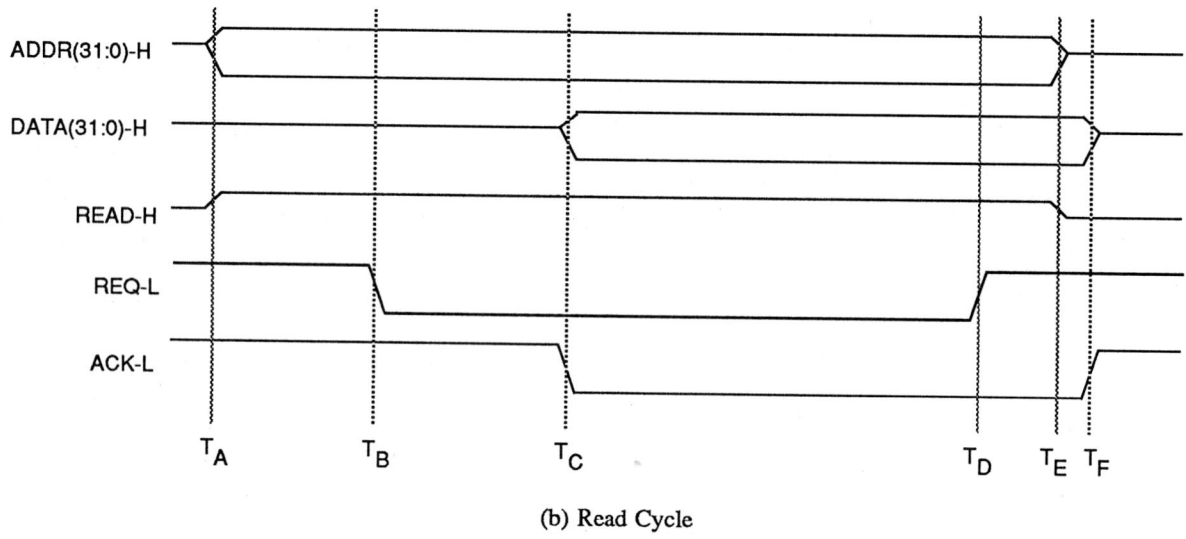

(b) Read Cycle

Figure 5.2. Control Lines for Read and Write on Asynchronous Bus.

also identifies the type of transfer by the level on the READ line. For a write, this line is not asserted, so it is forced to a low level. Since this is a write, the master is providing the data, so it will assert the data at the same time as the address. This is not a strict requirement, but write transfers are usually done this way.

After the master asserts the address and data lines, and unasserts the READ line, it waits for a period of time before asserting the request signal, REQ-L. This period of time is to allow for propagation delay and skew. Propagation delay refers to the amount of time that is required for a signal to propagate from the source to the destination. This includes delays through logic gates as well as delays associated with the physical travel of the signal through wires, printed circuit traces, and connectors. Skew is a term which refers to the time difference between the arrival of the first signal at the destination and the last signal at the destination. The arrival times will differ because of different logic delays and physical delays in the various signal paths. The amount of time allowed for propagation delay and skew is defined by the protocol, and differs from system to system. When this time requirement has been met, the master asserts REQ at time T_B.

The assertion of REQ is a signal from the master to all slave modules indicating that the address is valid, and that the slaves should determine if the address is one that they recognize. The address will identify only one of the slave modules, and the remaining action occurs between the master and that slave module. After the assertion of the request line, the next action is determined by the slave module. The selected slave module ascertains that the requested transaction is a write, and then takes whatever steps are appropriate to accept the data on the data lines. This may involve initiating a dynamic memory cycle if the addressed slave is a memory, or it may merely require placing the information into a register. In any case, when the necessary steps have been completed, the slave signals this fact to the master by asserting ACK at time T_C.

When the master identifies the fact that the data has been accepted, which is signaled by the assertion of ACK, it will respond by deasserting the request line at time T_D. The slave module detects the deassertion of the request line, and it deasserts the acknowledge line at time T_F. These four events, the assertion of the request line, the assertion of the acknowledge line, the deassertion of the request line, and the deassertion of the acknowledge line, are sometime called a four-event handshake protocol. This technique is used in almost all asynchronous bus systems in one form or another.

Since the address and data lines have been asserted by the master, they must also be released by the master to allow a new cycle to proceed. Note that the release of these lines is shown to occur at time T_E, which is after the release of REQ at time T_D. This planned delay is to prevent delays inherent in the system from causing a situation where the address changes (because of being released) in such a way that another slave module could be activated before the deassertion of the request line was detected.

The read cycle proceeds in a manner very similar to that described above. The address and READ line are asserted at time T_A, but the data is not asserted, since that information will come from the slave. At T_B the master asserts the request line, which action activates one of the slave modules. The slave module does what is required to obtain the necessary information, which may be a long time or a short time. Whenever this information is ready, is is placed on the data bus, and the acknowledge line is asserted. This is shown in Figure 5.2 at time T_C. When the master module detects the assertion of the acknowledge line, it waits for a period of time defined by the protocol to allow for skew, and then it accepts the data which has been placed on the data lines. The master module then signals that the transfer has been completed by deasserting the request line. The slave responds to the deassertion of the request line by deasserting the acknowledge line and releasing the data bus; this is shown at time T_F in Figure 5.2(b).

As mentioned above, a number of protocols use this technique to transfer information. The UNIBUS has 18 address lines and 16 data lines. The request line for the UNIBUS is called M SYN, and the acknowledge line is called S SYN. The function of the read line is performed by two lines which select the type of transaction from read word, write word, write byte, and read-modify-write. The MULTIBUS has 20 address lines and 16 data lines. However, the address lines are used in two different spaces, a memory space and an I/O space. Instead of having a single read line, the MULTIBUS identifies the type of transfer by using one of four different request lines. Different lines are used for memory read, memory write, I/O read, and I/O write requests. The acknowledge line is called a "Transfer Acknowledge" signal. The VME bus uses still a different mechanism, as we shall see in the next section.

5.2. Simple VMEbus Interface

One of the more popular buses used in recent years is the VME bus. This chapter presents a number of interfaces for this bus, each created to demonstrate a technique which can be used with the VME bus or with any other asynchronous bus. The basic idea utilized for this interface is an interface board for analog signals. This board contains two D/A converters for creating analog signals from a digital value, and two A/D converters to create a digital equivalent of an analog signal. The basic data path block diagram for the system is shown in Figure 5.3

The diagram in Figure 5.3 indicates that the interface will accept values from two different analog-to-digital converters (A/D), provide values for two different digital-to-analog converters (D/A), and also read and write to a Status and Control

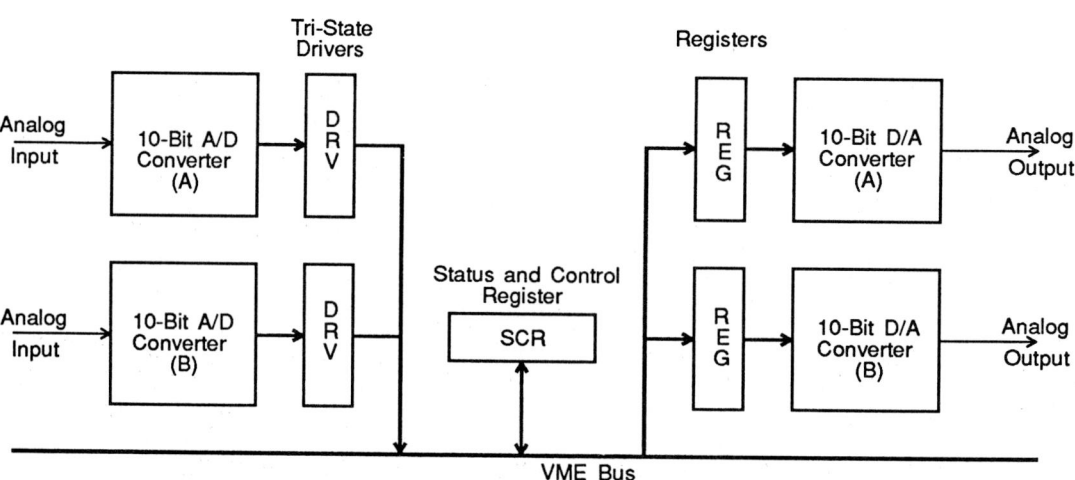

Figure 5.3. Data Flow Diagram for Simple VME Interface System.

Register (SCR). The interface is to be designed in such a way that all these values are available as part of the memory space. That is, the information is transferred by a technique known as programmed I/O, where the I/O transfers occur because of direct instructions in the machine which cause the transfer. The fact that these transfers are caused by accessing locations in the memory space is sometimes referred to as memory mapped I/O. We will design the interface in such a way that the addresses for the locations are placed on a 64-byte boundary, and are as shown in Table 5.1.

As Table 5.1 shows, there are five addresses of interest in this system. The first is used to communicate with a Status and Control Register. For this simple implementation, we will assume that this register contains two status bits for the A/D converters. That is, by reading the register the status of the flags can be determined; the status flags indicate when data is available to be read from the A/D converter. These will be the two least significant bits of a 16-bit quantity. In later sections of this chapter we will add more information and capabilities in the SCR register.

We are making some assumptions about the parts and information transfers involved with this sample interface. We are making the assumption that the A/D converters do not have tri-state drivers built into the devices. Many A/D converters do have this capability, but we are assuming that the units involved here do not. We will position the 10 bits from the A/D converters in the 10 most significant bit positions of the 16 bits involved in the transfer, and we will force the other bits to be zero. This is the proper alignment if values are normalized, and an A/D converter with more precision could be added with no change in the control hardware or software involved with the interface. Also, we will assume that the 10 bits supplied to the registers which feed the D/A converters are obtained in the 10 most significant bits of the bus. Finally, we will assume that the conversion action of the A/D converters is initiated by writing to the addresses indicated in Table 5.1. For some applications this is a sufficient mechanism for the sampling process; alternative methods are discussed later in the chapter.

The logic of the data path for this interface is shown in Figures 5.4(a) and (b). The A/D converters are shown in Figure 5.4(a), along with the flags which indicate that data is ready. The tri-state drivers are simply shown as '244s, but care must be taken to assure that the current carrying requirements of the VME bus are satisfied by the system. The conversion process begins when a start signal (AD_A_STRT or AD_A_STRT) is asserted. The process is complete when the end of conversion (EOC) line on the converter is asserted. We are assuming that the mode of operation of EOC is as follows: when the conversion process is initiated, the EOC line goes high; when the conversion is complete, the EOC line again goes low. This high-to-low transition is converted to a low-to-high transition by the inverter, and

Table 5.1. Addresses for the Simple VME Interface System.

Address	Read Action	Write Action
xxxxxxx0	Read SCR Register	Write SCR Register
xxxxxxx2	Read value from A/D A	Initiate sample process for A/D A
xxxxxxx4	Read value from A/D B	Initiate sample process for A/D B
xxxxxxx6	No action	Write value to D/A A
xxxxxxx8	No action	Write value to D/A B

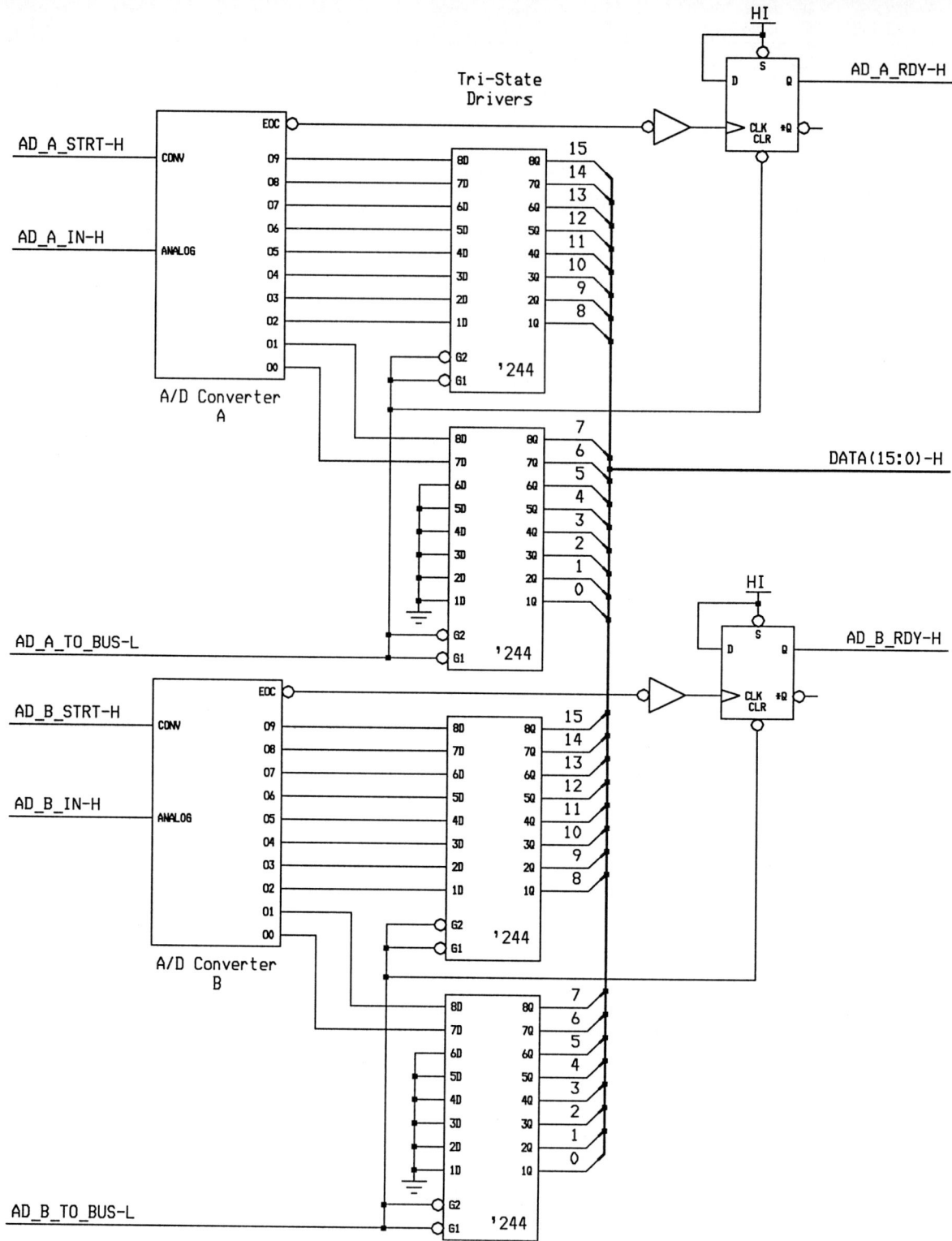

Figure 5.4(a). Logic of Simple Interface System: A/D Converters and Flags.

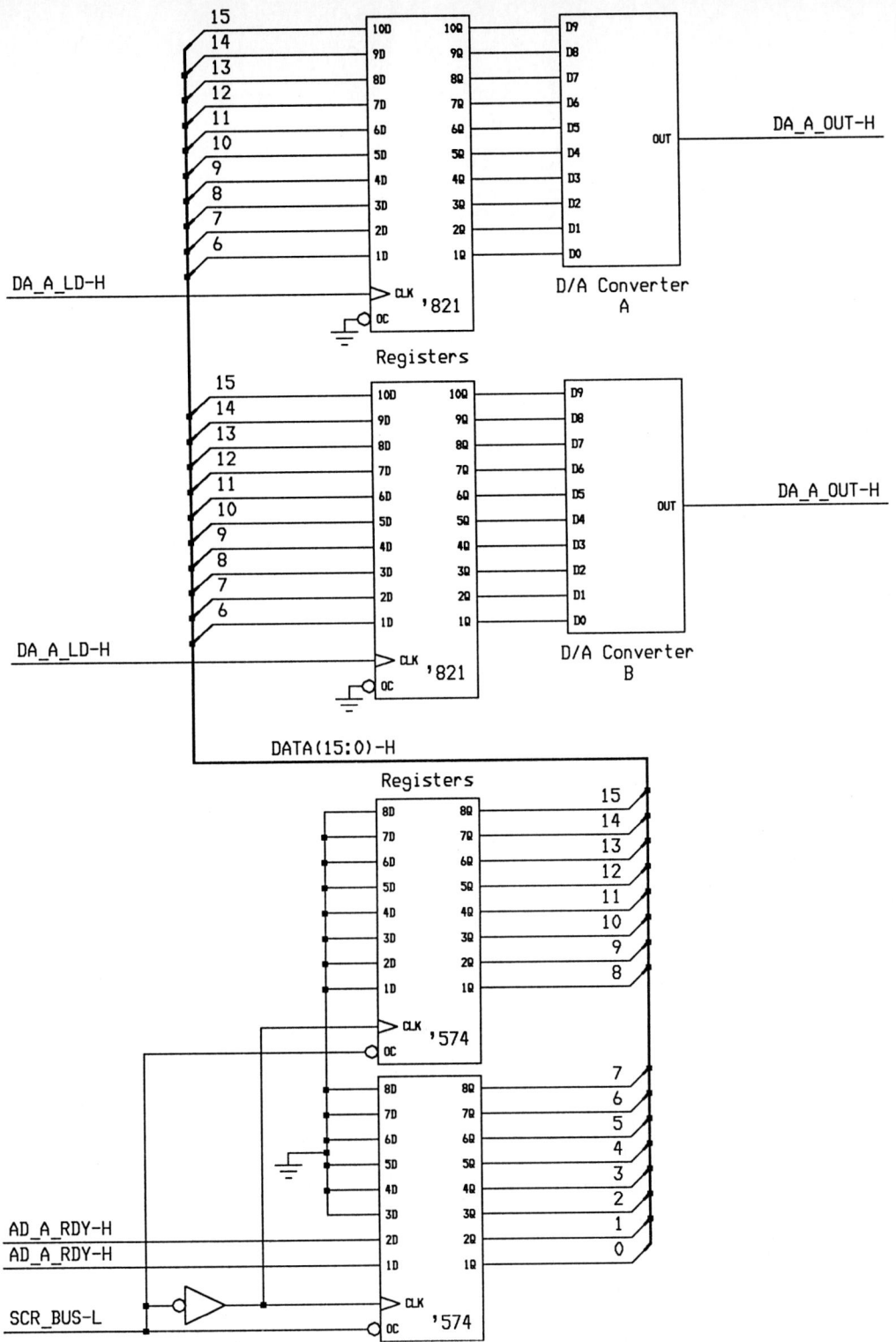

Figure 5.4(b). Logic of Simple Interface System: D/A Converters and Status Register.

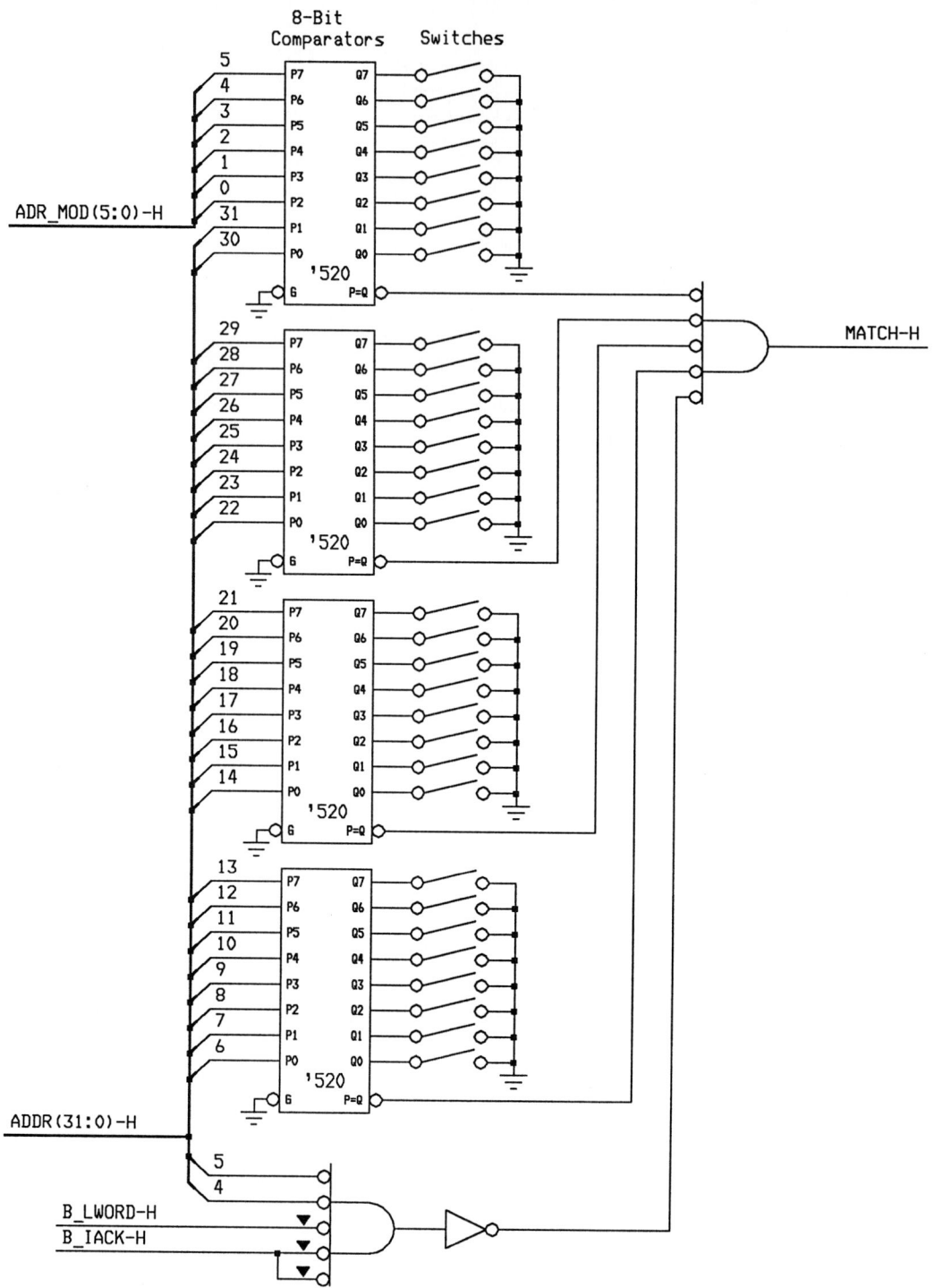

Figure 5.4(c). Logic of Simple Interface System: Address Decoder and Address Modifier Detection.

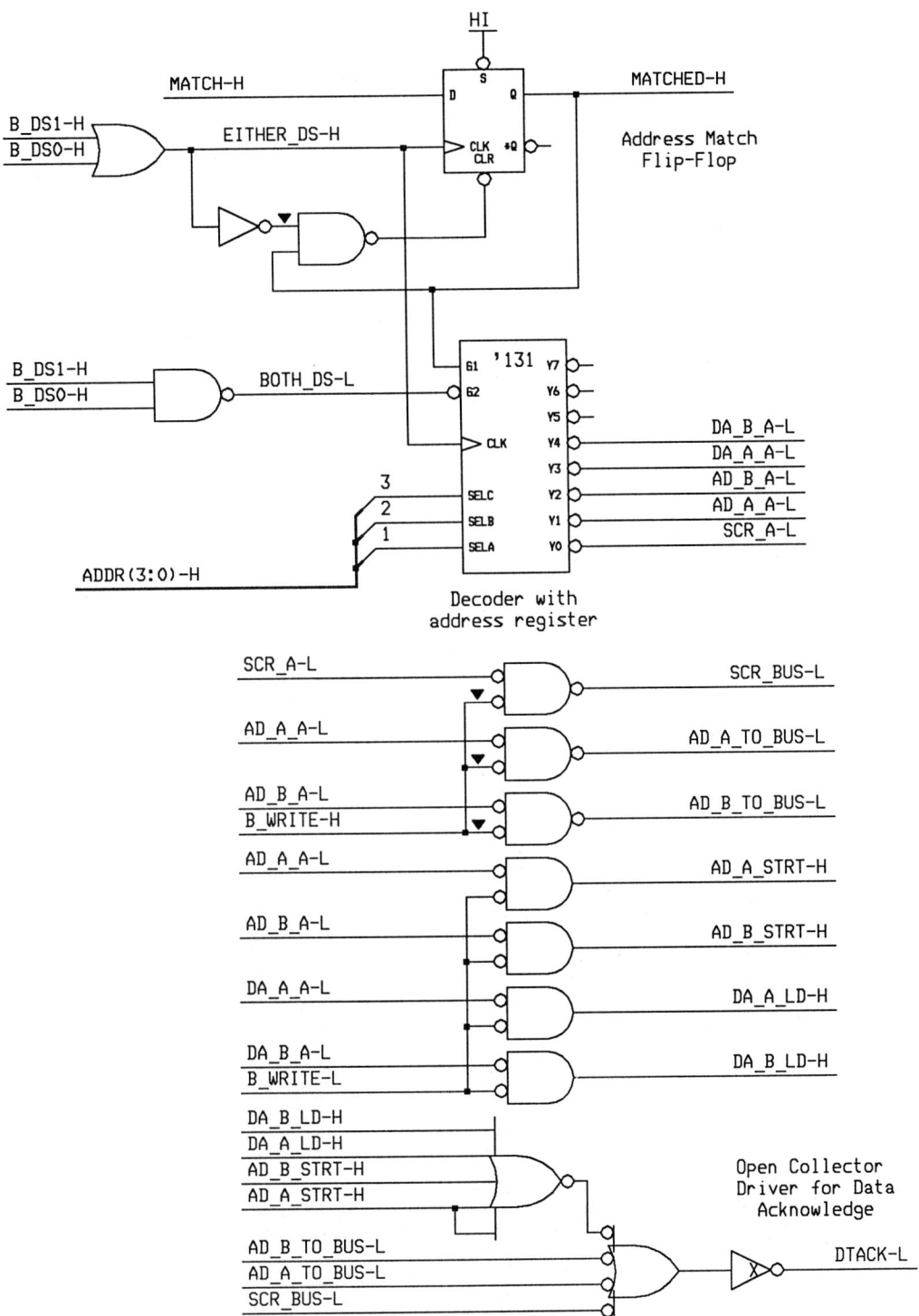

Figure 5.4(d). Logic of Simple Interface System: Control Logic.

when a conversion is complete the flag is set. This flag can be read by a processor, and when a set flag condition is detected (AD_A_RDY is set), the processor can then read the value. The data from the converter is placed onto the bus by asserting the AD_A_TO_BUS signal, which also clears the flag.

The D/A converters and their registers, as well as the A/D ready flags which comprise the status register which we have thus far, are shown in Figure 5.4(b). The registers shown are '821s, capable of storing all 10 bits required by a D/A converter in this system. The value from the bus will be accepted by activation of the appropriate load signal (DA_A_LD or DA_A_LD). Finally, the two registers ('574s) used for the status register in this implementation are shown in Figure 5.4(b). The flags are connected in the manner shown so that when the value is placed onto the bus, it will not change during the bus cycle. Note that the assertion of the SCR_BUS signal will cause the current levels of the flags to be placed into the registers, and then onto the bus.

The matching of an appropriate address is handled by the logic shown in Figure 5.4(c). As mentioned above, the VME bus has 32 address lines. The bus also has 6 lines called address modifiers, which are represented as ADR_MOD in the figure. These 6 lines have the same basic timing characteristics as the address, and are used to identify the type of transaction in progress. One of the functions this performs is to separate program requests from data requests, or supervisor requests from user requests. These lines also identify how many of the address lines are active. That is, the bus can be configured to use 16, 24, or 32 address lines, according to the patterns on the modifier bits. Therefore, the matching hardware must examine both the address lines and the address modifier bits.

The devices used to detect the address are '520s. One of the assumptions we will make here is that the devices connected to the bus are fast enough to do the job, and that they also meet the loading characteristics required by the system. See [Moto85] for detailed specifications. What this means in practice is that an appropriate family of devices is used to connect to the bus. This can be FAST logic (74FXXX), schottky logic (74SXXX), advanced schottky logic (74ASXXX), advanced low power schottky logic (74ALSXXX), high power logic (74HXXX), or other family as appropriate. The '520s are configured so that the user can, with the use of DIP switches, specify the address modifier code to which the system responds, and also specify the matching pattern for bits 6 to 31 of the address bus (ADDR). Hence the reason for the above specification of a 64-byte boundary. What changes would allow specification to a 16-byte boundary?

One of the reasons for using the '520s for comparison is that the devices contain internal pullup resistors, so external devices for that purpose are not necessary. The equal signals from all four comparators are fed to a NOR gate used in an ANDing function, as is one other line. This last line comes from another ANDing function, which looks at two of the address lines and two other lines. The two address lines specify that address bits 4 and 5 must be unasserted (low) for the function to be asserted. The other two lines are electrically buffered (hence the prefix B_ in the signal name) versions of the LWORD signal and the IACK signal. These lines are asserted low on the bus, and have been run through inverting receivers (not shown) to electrically buffer them. The bus protocol specifies that when IACK is asserted, a slave must not respond. Hence, IACK must be unasserted for a match to be made by the slave. Also, for 2-byte transfers, which is the designed mode of operation for this system, LWORD must also not be asserted. When all of these conditions have been satisfied, then the match signal (MATCH-H) will be asserted.

The control logic for this system is shown in Figure 5.4(d). One of the interesting differences between the VME bus protocol and the "generic" protocol described above is the behavior of the address lines. The specification indicates that the address lines can change as soon as the master detects the assertion of the acknowledge line. Hence, the logic of Figure 5.4(d) stores the level of the match signal in a flip-flop, so that the signals need not rely on the correct address for the remainder of the bus cycle. The role of the request signal in the "generic" protocol is taken by two signals in the VME protocol. These are two data strobes, DS0 and DS1, and they are used to indicate transfer of information on different sections of the data bus. The DS0 signal is used for transferring a byte on the 8 LSBs of the bus; DS1 is used to transfer a byte on data lines 8–15. Both lines are used for a 16-bit transfer on data lines 0–15, and when LWORD is asserted a 32-bit transfer can be accomplished using all the data lines. Because of the normal differences in physical and electrical configurations from system to system, the time of arrival of these two signals at a slave may not be the same. Hence, the bus protocol states that the address must be stable 10 nsec prior to the arrival of the first of the signals, whichever that may be. Therefore, these two signals are ORed together and then used to strobe the match detection (MATCH-H). This signal is also used to clock a decoder. The decoder shown in the figure is a '131, which has a 3-bit register on the select lines. Hence, when a rising edge is detected at the clock line of the '131, three of the address lines are placed in a register. The least significant address line is not used, since this interface is being designed for 16-bit transfers. The address match flip-flop will be cleared at the end of a successful cycle, when the flip-flop is set and the data strobes are both deasserted.

Outputs from the '131 will be asserted when the address match is successful AND the data strobes are both asserted. This condition will assert one of the eight lines of the decoder; it is assumed the user will not attempt to access one of the three unused addresses. When the outputs of the decoder are asserted, the WRITE line will determine which of the control signals is activated. If the WRITE line is not asserted, then something will be directed to the bus. The status value is asserted onto the bus by SCR_BUS-L; the value from A/D A is directed to the bus by AD_A_TO_BUS-L; and the value from A/D B is asserted onto the bus by AD_B_TO_BUS-L. If the WRITE line is asserted, then the write action specified in Table 5.1 occurs. The A/D conversion process is started by one of the A/D start signals (AD_A_STRT-H, AD_A_STRT-H), or a bus value is placed into one of the D/A registers (by DA_A_LD-H or DA_A_LD-H). If any of these actions occurs, then the acknowledge line (DaTa ACKnowledge, DTACK-L) is asserted. As shown in the figure, this must be an open collector driver to meet the specifications in the bus protocol.

The design presented in this section demonstrates some of the characteristics of interface systems which are common to asynchronous buses. One module, called the master, is responsible for initiating the action. The master asserts the address and configuration lines, then it asserts the request lines to alert the slaves that a transfer is underway. Each slave checks its own match logic when the request is made, and the selected slave does the work required. The slave responds by asserting the acknowledge when the work has been completed. The master deasserts the request, and the slave deasserts the acknowledge, and the bus can begin a new cycle. The interface shown in Figure 5.4 is sequential in nature, but the sequentiality is that which is caused by the protocol and the handshake mechanism, not sequential in the sense demonstrated by the designs in Chapter 4.

The mechanisms demonstrated in this section can be modified and applied in a number of different application areas. Programmable logic can be used to reduce the logic required for the match determination. Different decoding and gating mechanisms can be used as the situation requires to do the work required by an interface. But the same criteria should be used in determining the applicability of techniques for interface: In what way are system resources most effectively utilized in the design?

5.3. Adding Interrupt Capability

The basic programmed I/O interface described in the previous section can be used to do all the transfers necessary to digitize analog values, and to provide an analog value out. However, there is no mechanism available in that interface to let the computer know when a value is available from an A/D converter. Also, any timing used in conjunction with the D/A converters must be derived from the computer itself, not from the interface. We now want to expand the function of the interface to include an interrupt capability, which will allow the interface to signal the processor that some action is necessary.

The new interface system will contain the constituent parts of the original interface, along with some additional gates and functions to accomplish the new tasks. The definition of the interface will change to include an interrupt capability. The Status and Control Register will now include some bits which govern the activity of the interface, and a counter and an ID capability will be added. Let us consider each of these differences to determine how the original interface must be changed.

The first difference has to do with the SCR. Previously, there was no control portion to this function, but rather only the ability to examine the A/D data available flags. We now expand the function of the system to include 3 other bits. The first 2 bits will be located in the 16-bit status/control word at bits 8 and 9. These bits are A/D interrupt enable bits for the A and B A/D converters. If these bits are zero, then the A/D function is exactly as described in the previous section: writing to an address (xxxxx2 or xxxxx4) initiates the conversion process, and reading the addresses obtains the values. The bits indicating that the conversion process is complete can be obtained by reading the status value from address xxxxx0. However, if bit 8 of the SCR is set (via programmed I/O instructions), then when the conversion process of A/D converter A is complete, the unit should interrupt the action of the processor. Similarly, if bit 9 of the SCR is set, then when A/D converter B has data available (the AD_B_RDY flag is set), the processor will be interrupted. The process of reading the value will clear the ready flag, which in turn will remove the interrupt request.

When an interrupt condition exists, the interface will alert the processor, and the processor will go through a predetermined action to respond to the interrupt. For the VME bus, an interrupt condition is signaled by asserting one of seven interrupt request lines. These lines are labeled IRQ1 through IRQ7, and are prioritized in such a way that IRQ7 has the highest priority, and IRQ1 has the lowest priority. If the processor is capable at that time of servicing the interrupt, it will respond with an interrupt acknowledge cycle; otherwise the processor will continue executing the current program until it is in a situation where it can service the interrupt.

When the processor which services the interrupt is ready, it will perform an interrupt acknowledge cycle. For exact details of the interaction, see [Moto85]. Basically, this cycle is very similar to a read cycle. The fact that it is an interrupt cycle is identified by the assertion of the interrupt acknowledge line, IACK-L.

When this signal is asserted, "normal" slave interaction is suspended, and interrupting modules prepare to respond. A device which has issued an interrupt request will respond to an interrupt cycle if the level of the interrupt it asserted (IRQ1-IRQ7) matches the 3-bit value on address lines ADDR(3:1). It is also necessary that the interrupt handler, which we have assumed is the processor, request a transfer of data which is large enough. That is, if the device is designed to transfer 16 bits on an interrupt cycle, it can do so when activated by an interrupt handler which asks for a 32-bit transfer or a 16-bit transfer, but not one which requests an 8-bit transfer.

When all the above conditions are met, an interrupting device can respond to an interrupt cycle. That is, when the the interface has requested an interrupt, the level of interrupt matches address lines ADDR(3:1), and the width of the requested transfer is equal to or greater than the information to be transferred by the interrupting interface, then an interrupt cycle can proceed. The cycle proceeds when the interrupting interface detects a falling edge on a daisy-chain signal called IACKIN-L. When this signal arrives, the interrupting interface responds by asserting onto the data bus a code which lets the interrupt handler know how to respond. This code is referred to as the "Status/ID" bits, and will be interpreted by the interrupt handler in an appropriate way. That is, the piece of software known as an interrupt service routine will receive this information, and respond in an predetermined fashion. This requires agreement between the software and the hardware as to the nature of the Status/ID bits.

If an interrupting interface receives a falling edge on IACKIN and the conditions listed above are not satisfied, then the interface passes this signal on to the next interface in the system by asserting the IACKOUT-L signal. In this way both parallel and serial arbitration schemes are used. Parallel choices are available because of the different IRQ lines; serial choices are made with the daisy-chain function of the IACKIN-IACKOUT signals.

The final change to be made in the interface is the addition of a timer/counter for the D/A converters. The assumption made here is that the system wants to put out a D/A value periodically. (The same could also apply to asking for values from the A/D converter.) One way this could be done is to create a timer system consisting of a counter which will count some predetermined amount, set a flag, and start over. When the flag is set, the processor should be interrupted to indicate that some action is needed.

The schematic diagrams which show the changes are included as Figure 5.5. The modified SCR and the drivers for the Status/ID information are shown in Figure 5.5(a). The main difference in the SCR is that now there is a register to be filled as well as status bits to be read. Also note that the control bits in the register are also available when the status register is read, so that if the processor needs to ascertain which of the functions is enabled, that can be accomplished. The 3 bits shown are interrupt enable bits; when set, the corresponding function will cause an interrupt. That is, if the AD_A_INT_ENBL bit (bit 8) is set, then when a new value is available from A/D converter A, an interrupt request will be made. The other bits cause action in a similar fashion. The Status/ID section is shown simply as a set of tri-state drivers which place information onto the data bus. The information placed there is determined by the user, and is done in such a way that the interrupt service routine will know what to do when that information is received. This could simply be a bit pattern which identifies the type of interface of the specific interface, or it could incorporate information concerning which of the three interrupting devices actually needs service.

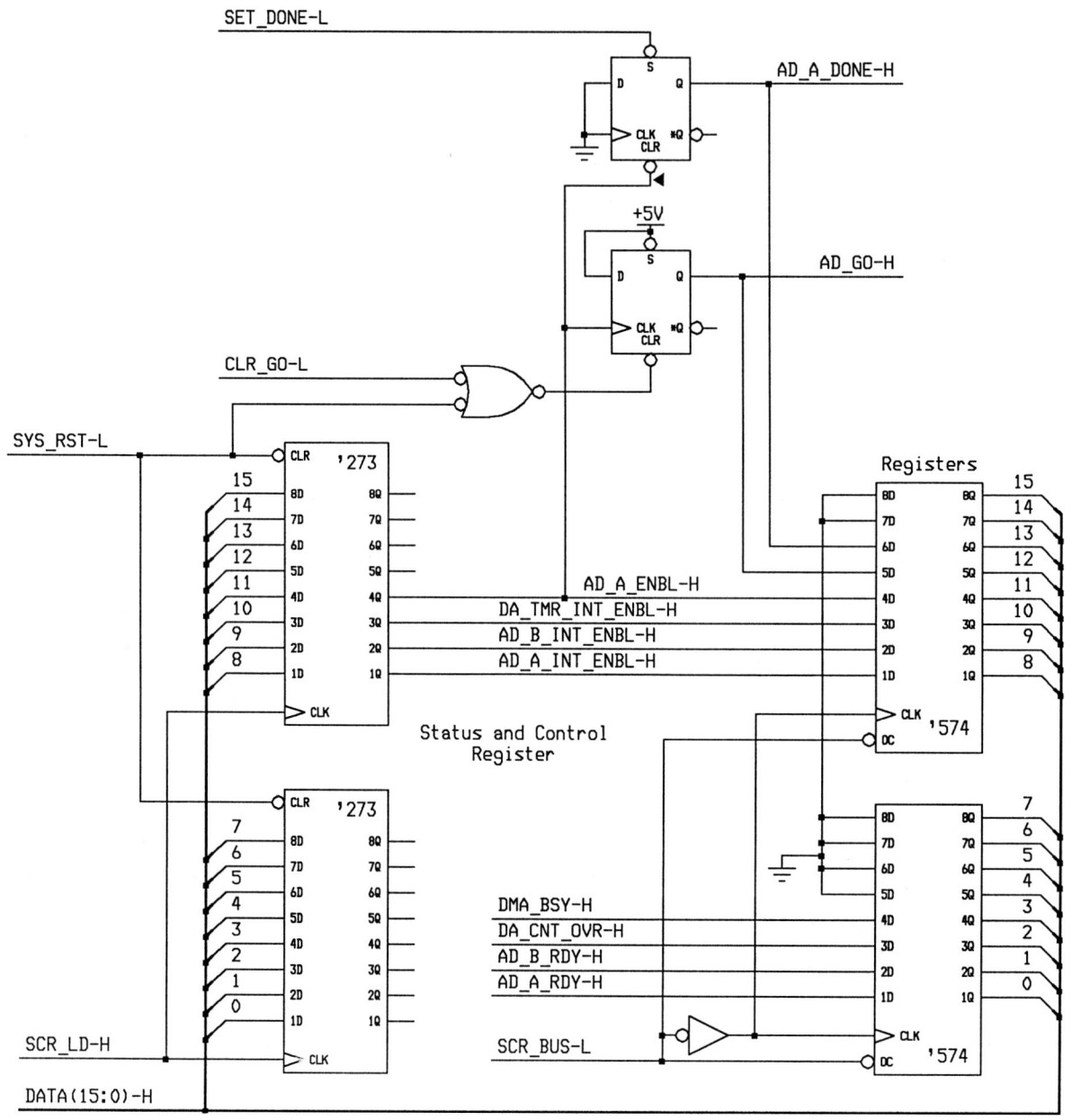

Figure 5.5(a). Additions to Logic of Simple Interface System: Status and Control Register, and Status/ID Drivers.

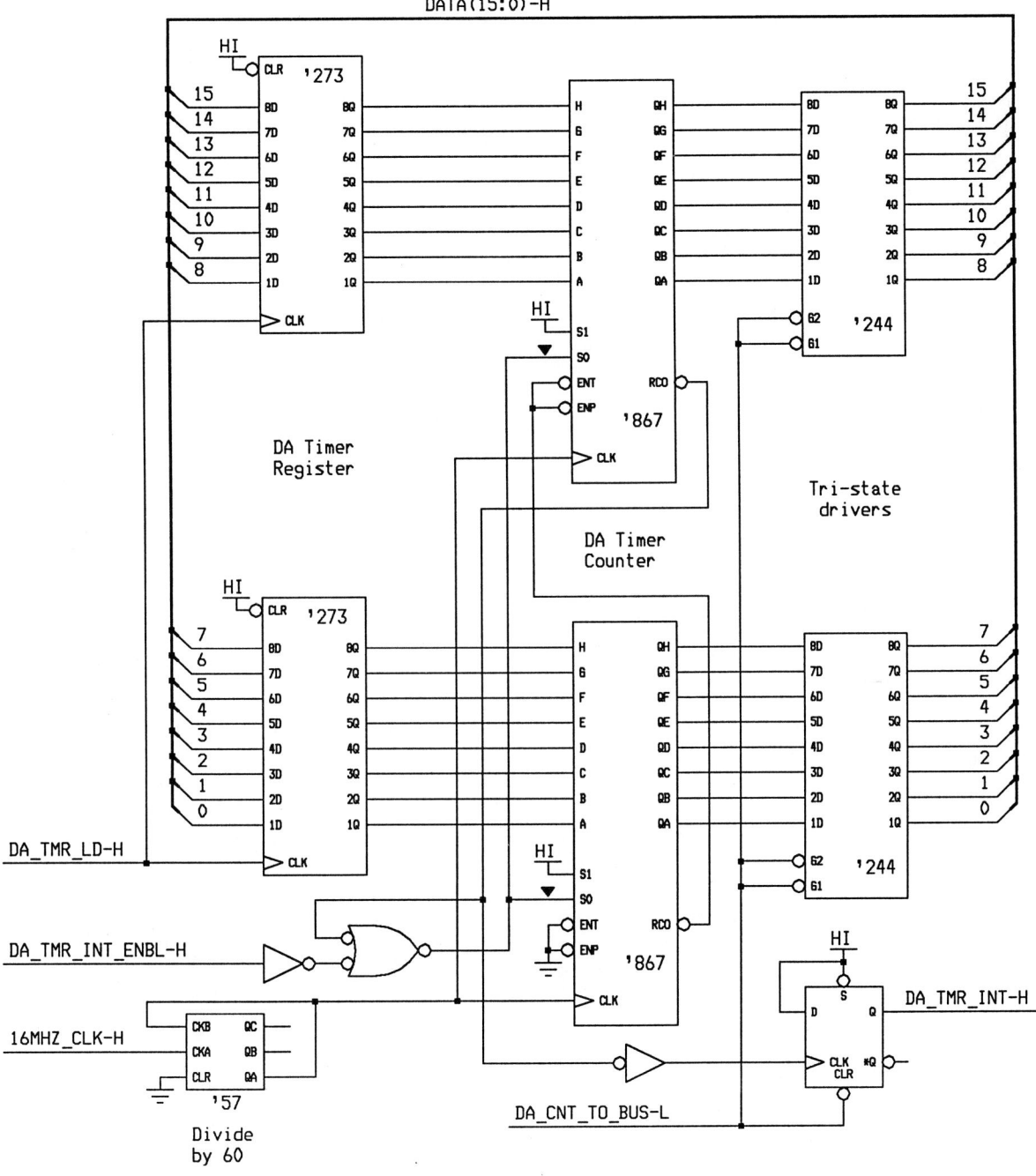

Figure 5.5(b). Additions to Logic of Simple Interface System:
D/A Timer Logic.

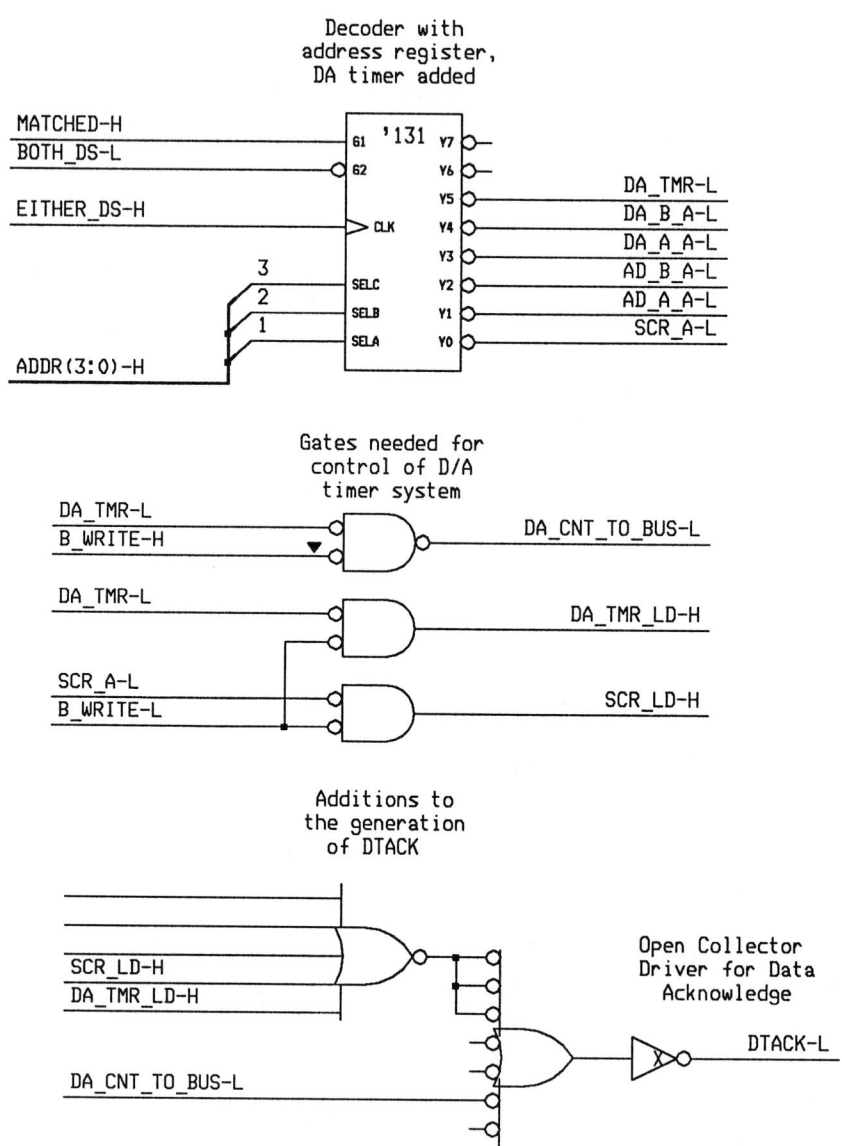

Figure 5.5(c). Additions to Logic of Simple Interface System: Control Logic Changes.

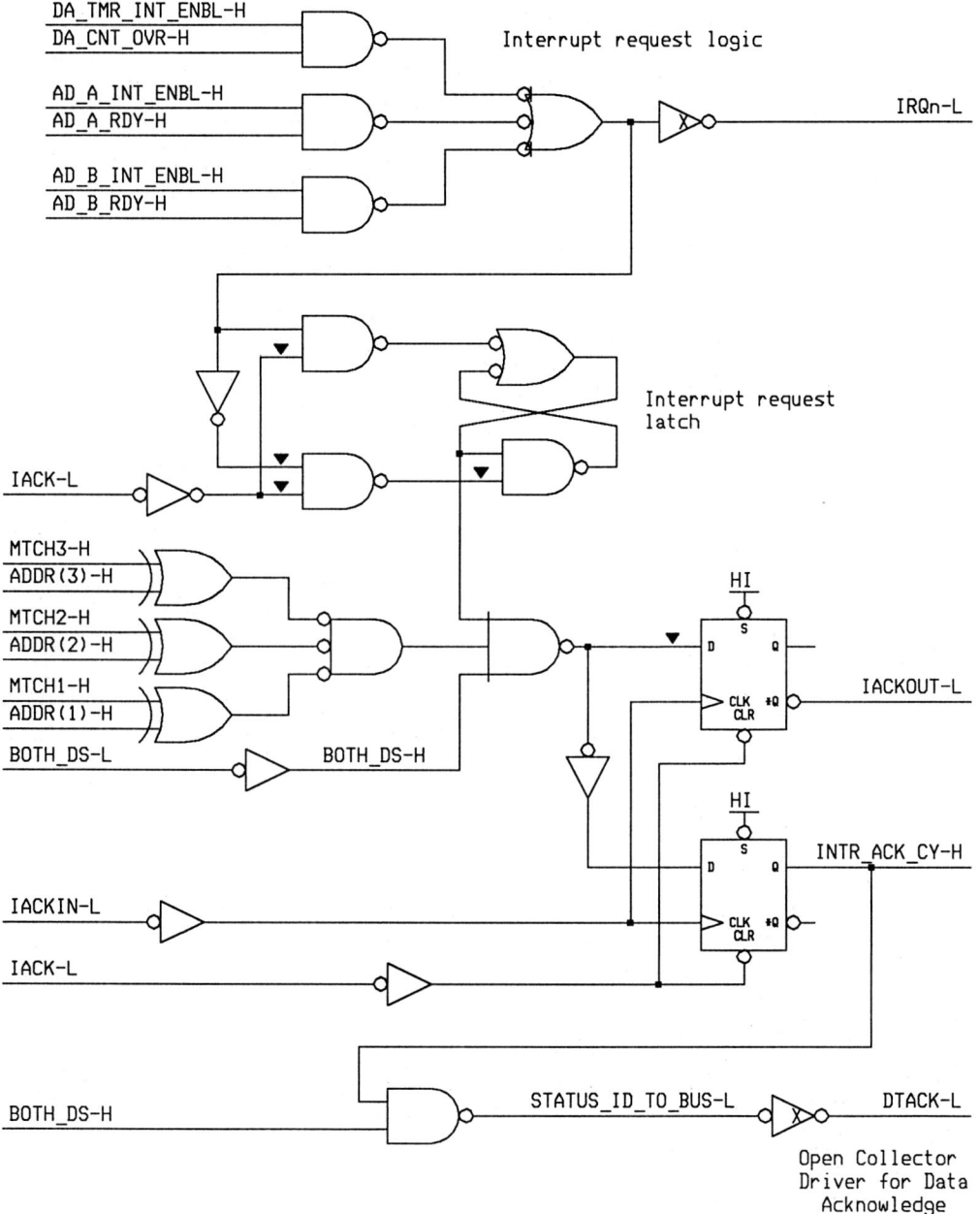

Figure 5.5(d). Additions to Logic of Simple Interface System: Control Logic for Interrupt Acknowledge Action.

The D/A timer logic is shown in Figure 5.5(b). The D/A Timer Register accepts a value from the bus when activated by the load signal (DA_TMR_LD-H). This is the value to which the counter will be initialized when it reaches its maximum value, and when the counter interrupt function is not enabled. When the counter interrupt function is enabled, the counter will increment to maximum value, then reload with the count stored in the register. Each time this process occurs, the DA_TMR_INT flag will be set. This will cause an interrupt to occur, and the action of reading the current value of the counters will reset the flag. Thus, if an interrupt service routine doesn't want the current counter value, it disregards this information, but the act of reading the value has reset the flag. The clock for the counter is derived from a 16 MHz clock signal which is present on the VME bus. This clock has a 62.5 nsec period; counting this clock with a 16-bit counter would give a maximum time between flags of only 4 msec. Therefore, a '57 device is included in the clock line, which will divide the clock signal by 60. This results in a D/A Timer Counter resolution of 3.75 μsec, and a maximum value of almost a quarter of a second. This range appears to be more useful for most applications, but the user can change the clocking arrangements to fit the needs of the system.

The changes to the control logic which are needed to service the new registers are shown in Figure 5.5(c). The decoder functions as before, except that an additional address is used. This address is xxxxxA, and when a write occurs to this address, the D/A Timer Register is loaded; when a read is requested from this address, the current value of the counter is asserted onto the bus. Also, if the D/A timer interrupt flag is set, the read action will reset the flag. The gates which cause this action, as well as the additional gate needed to load the control register, are also included in Figure 5.5(c). Finally, these signals are added to the ORing function which is needed to correctly assert DTACK-L.

The logic which controls the action of the interface during an interrupt acknowledge cycle is shown in Figure 5.5(d). First, the interrupt request signal (IRQn) is asserted when any of three conditions exist. These conditions are that a bit is set in the control register, and the corresponding flag is set. Thus, if the interface is set up to interrupt the system when the A A/D converter has a new value available, the AD_A_INT_ENBL signal will be asserted. When this condition exists and a new value is available (which will be signaled by assertion of AD_A_RDY), then the interrupt line will be asserted. In addition, this assertion level will be present at the input of the interrupt request latch.

The purpose of the interrupt request latch is to present a stable signal during an interrupt acknowledge cycle. There is no basic synchronization between the signals which can cause interrupts (the A/D flags, for example) and the signals on the bus. Therefore, at the beginning of an interrupt acknowledge cycle the current value of the interrupt request line is latched. This is accomplished by the assertion of IACK-L, which is a signal asserted by the interrupt handler at the beginning of an interrupt acknowledge cycle.

The heart of the action shown in Figure 5.5(d) centers around the two flip-flops. Note that when IACK is not asserted, both of these flip-flops will be cleared. When an interrupt acknowledge cycle is initiated, these flip-flops will be able to be set. However, only one of them will be set, since the signal feeding the first is inverted to feed the second. This signal will be asserted if and only if the proper conditions for responding to the interrupt acknowledge cycle exist. These three conditions are the presence of the interrupt request in the interrupt request latch, the matching of the request level with the address lines, and the appropriate width of a request. The width of the transfer will be acceptable when both the data strobes have been asserted, which will be the case when BOTH_DS is asserted. The timing of these signals (guaranteed by the latching of the interrupt request) is such that the data inputs to the flip-flops will be stable prior to the arrival of the IACKIN signal. At that time, one of the two flip-flops will toggle. If the conditions for responding to the cycle have not been met, then the IACKOUT-L signal will be asserted, sending the request to the next device in the daisy chain. If the interface is to respond, then INTR_ACK_CY will be set. The immediate result of the assertion of the flag will be the assertion of the Status/ID bits onto the data bus, as well as the assertion of DTACK. When the cycle is over, the data strobes will be deasserted by the interrupt handler, which will result in the release of the data bus and DTACK. The interrupt handler will also release IACK to restore the bus to its normal functions.

This activity will result in an interrupt cycle for the VME bus, but other buses have similar requirements for interrupt systems. Each bus will define a set of events which must occur during the execution of an interrupt sequence, and any interface which requires an interrupt capability must respond according to the defined sequence. The techniques described here can be used in many of the other interface systems as well.

5.4. Interface with Direct Memory Access

The next step with the interface we have been developing is to add the capability to do Direct Memory Access, or DMA. That is, instead of requiring the processor to control all the individual transfers with the interface, we wish to set up a block of transfers with a single request. In this case, the responsibilities of the processor are quite different from the strictly programmed I/O case, or the interrupt driven case. With a DMA transfer, the processor sets up the transfer by placing in the interface itself the address of the buffer for the data and the number of data values to transfer. It then initiates the transfer, and the interface performs the needed work, usually interrupting the processor when the work is finished.

A block diagram of the modified interface is shown in Figure 5.6. Only the parts of the interface which interest us here are included in the figure. That is, the

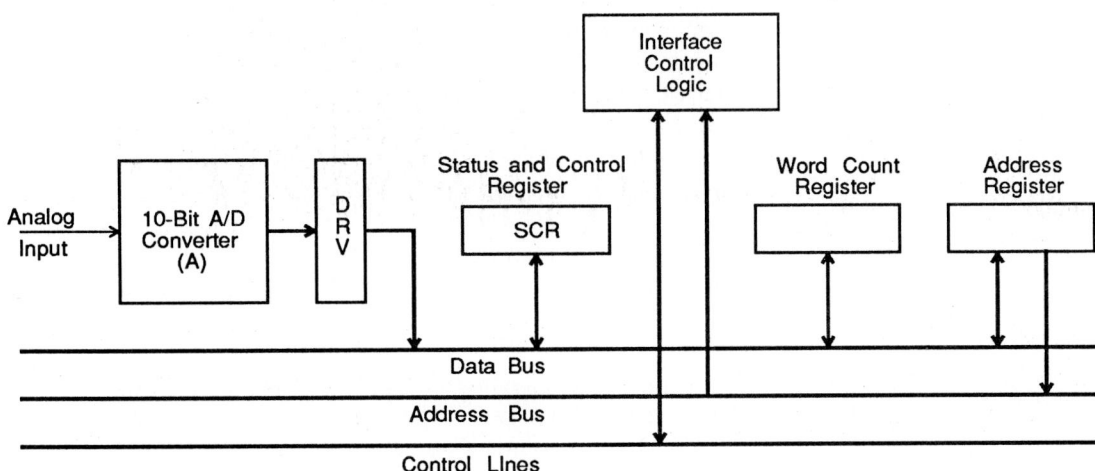

Figure 5.6. Block Diagram of A/D DMA Interface.

D/As and the second A/D are not shown. What is shown demonstrates the methods and mechanisms utilized in this interface system. The Address Register is filled from the data bus, as indicated in the figure. Also, a return path to the data bus is shown. This is not used in normal operation, but rather it provides a mechanism for the processor to interrogate the interface for the next address to be used for data storage. The path used when the interface is performing a DMA transfer is the connection to the address bus. This will be used by the interface to provide an address to the system which identifies the destination of the data. In this interface we are assuming that all transfers are two-byte transfers which take place on the 16 LSBs of the bus, and the address must be set up to reflect that assumption. We are also assuming that the initial address of the buffer area is the lowest address of the buffer, and that the address must be incremented for each successive transfer.

The Word Count Register is set up to identify the number of words (2-byte entities) which are to be transferred. The assumption here is that the processor will place a positive number in this register, and that it will decrement until zero is reached. Another widely used technique is to create a word count function with a count-up counter, and set up the initial value as a negative number. This type of operation calls for the incrementing of the word count until zero is reached. The diagram shown in Figure 5.6 also indicates that the Word Count Register can be read as well as filled. This connection allows diagnostic interrogation of the register; it also allows a processor to determine how many values are left in the current transfer of information into the system. The tri-state drivers used to provide both the Address Register value and the Word Count Register value to the data bus are not, strictly speaking, necessary in the interface. That is, these values are not absolutely necessary for the operation of the system. However, they do provide capabilities which can be very useful in the system.

The Status and Control Register is used to control the action of the system. The only changes in this register from the previous sections is the meaning of the interrupt bit for A/D A and the DMA bit. In this interface, the interrupt will occur when the transfer is complete, rather than on each individual data transfer. Transfers begin when a $0 \rightarrow 1$ transition occurs on the DMA bit; the system assumes that before this occurs the processor has properly initialized the Address Register and the Word Count Register for a transfer of data.

The source of the data for the system in Figure 5.6 is the A A/D converter; the other A/D converter and the D/A converters are not shown, although we are assuming that they are still part of the interface. Information from the A/D converter is directed onto the data bus through a tri-state driver. Thus, one of the required functions of the interface is to enable the data onto the bus at the proper time.

The action of the system is controlled by the block labeled ''Interface Control Logic.'' Under the direction of the control lines, this logic will load the Word Count, Address, and SCR registers, as well as read those registers. Also, this logic is responsible for controlling the interaction of the system with the bus when a DMA transfer has been initiated. When the data path block diagram has been prepared as shown in Figure 5.6, the action of the system can be described with a state diagram.

Before we examine the action of the system with a state diagram, let us identify the changes which have come about because of the expansion of the functions of the system. These can be identified by looking at the expanded Status and Control Register, the addresses required to access the values of the system. The addresses for this system are a modified version of those presented in Table 5.1, and the new set is shown in Table 5.2. Note that initiating action and reading

Chap. 5: Design of I/O Interface Systems

Table 5.2. Addresses for the VME Interface System with DMA.

Address	Read Action	Write Action
xxxxxx00	Read SCR Register	Write SCR Register
xxxxxx02		
xxxxxx04	Read value from A/D B	Initiate sample process for A/D B
xxxxxx06	No action	Write value to D/A A
xxxxxx08	No action	Write value to D/A B
xxxxxx0A	Read value from D/A timer	Write value to D/A timer register
xxxxxx10	Read LSBs of Word Count	Write value to LSBs of Word Count
xxxxxx12	Read MSBs of Word Count	Write value to MSBs of Word Count
xxxxxx14	Read LSBs of Address Reg	Write value to LSBs of Address Reg
xxxxxx16	Read MSBs of Address Reg	Write value to MSBs of Address Reg

values from A/D A have been removed, and that reading and writing to the Address Register and the Word Count Register have been added. Also, this requires more than eight addresses to accomplish, so one additional bit of address decode is necessary, which reduces the number of address bits involved in the generation of the match signal by one.

In addition to the changes in the addresses which are needed, the role of the SCR has been expanded. This is indicated by looking at the bits required, as shown in Table 5.3. Earlier versions of the interface system used the least significant bit of the SCR to indicate the availability of data from the A/D A converter. Since this system will use DMA techniques to move the A/D A data, bit 0 is not used in this version of the interface. Bit 1 is used to indicate that the A/D B converter has data available.

Table 5.3. Function of Bits Used in the Status and Control Register.

Bit Number	Control Register	Status Register
0	No action	No action
1	No action	A/D B has data
2	No action	D/A counter overflow
3	No action	DMA action is proceeding
4	No action	No action
5	No action	No action
6	No action	No action
7	No action	No action
8	Enable A/D A interrupt	A/D A interrupt is enabled
9	Enable A/D B interrupt	A/D A interrupt is enabled
10	Enable D/A B timer interrupt	D/A timer interrupt is enabled
11	Enable DMA action	DMA action is enabled
12	No action	DMA action requested, not started
13	No action	DMA action completed

The status of the D/A counter overflow is available as bit 2 in the SCR; this bit identifies the condition that the timer/counter associated with the D/A system has overflowed. Bit 3 is used to identify the status of the DMA system. When this bit is a 0, the DMA logic is idle; when the DMA logic is active this bit will be a 1.

Bits 8, 9, and 10 enable the interrupts associated with the system. The interrupt mechanism for A/D A is controlled by bit 8. When this bit is cleared, no interrupts will occur because of action associated with A/D A. When this bit is set, an interrupt will occur when the DMA action associated with A/D A has been completed. When bit 9 is set, an interrupt will occur when a new value is available from A/D B; if bit 9 is cleared, then no interrupt will be associated with A/D B. If bit 10 is set, then an interrupt will occur when the D/A timer/counter overflows. If this bit is not set, then no interrupt will be associated with the timer/counter.

The remaining bits in the SCR have to do with the DMA action of the system. DMA action is initiated by causing a $0 \rightarrow 1$ transition on bit 11. Bit 11 will also identify the current status of the DMA system, if it has been started or not. Bit 12 will be a 1 for that brief period of time between request of DMA action and the state machine responding to do the work. In normal operation, bit 12 will almost never be read as a 1; however, if a malfunction occurs in the DMA control logic, this bit can be a useful indication of error. Finally, bit 13 will be read as a 1 when the DMA action requested by a 0 to 1 transition on bit 11 has been completed.

The bits of the SCR, plus the register interaction associated with the addresses accessed by the system, allow programmed I/O instructions to control the action of the interface. For transfers to and from the registers under the control of the processor, the control logic operates as required by the bus protocol and moves the data appropriately. When a DMA action has been requested, then the control logic needed for the bus interaction leaves the idle state and performs the required work.

The state diagram of the DMA action of the control logic is shown in Figure 5.7. Each of the states is labeled with a letter near it, and we will refer to these letters in the process of describing the action of the system. When the interface is not in the process of performing a series of DMA transfers as directed by the processor, the state machine will remain in the idle state, State A. During this time the processor can make whatever preparations which are needed, such as setting up the initial address in the Address Register, placing the number of transfers to make in the Word Count Register, and so forth. When all necessary conditions have been established, the processor signals to the interface that the transfers should commence. This is seen in Figure 5.7 as the ''get data'' condition, which causes the action to leave State A for State B.

State B is used by the interface to await the arrival of data. We are assuming here that the A/D converter has been configured in such a way that an external timing mechanism is being used to let the converter know when to initiate the conversion of data. When that happens, and the conversion process is complete, then new A/D data is available and the system moves to State C.

One of the main differences between this interface and the interfaces presented earlier is that this interface must request ownership of the bus, and when that is granted the interface itself will assert the control lines to move the data. Hence, in State C the interface requests ownership of the bus. The interface will remain in this state until control of the bus is granted to the interface, at which time it will move on. Note that the condition listed on the state diagram is that the bus is granted and that AS is high. The signal AS is the address strobe of the system, and it is asserted low. A high level on this signal indicates that the previous bus master has released the control lines, and hence a new bus master can control the bus lines. When this condition exists (both the bus granted signal and the release of the AS line), the system moves to State D.

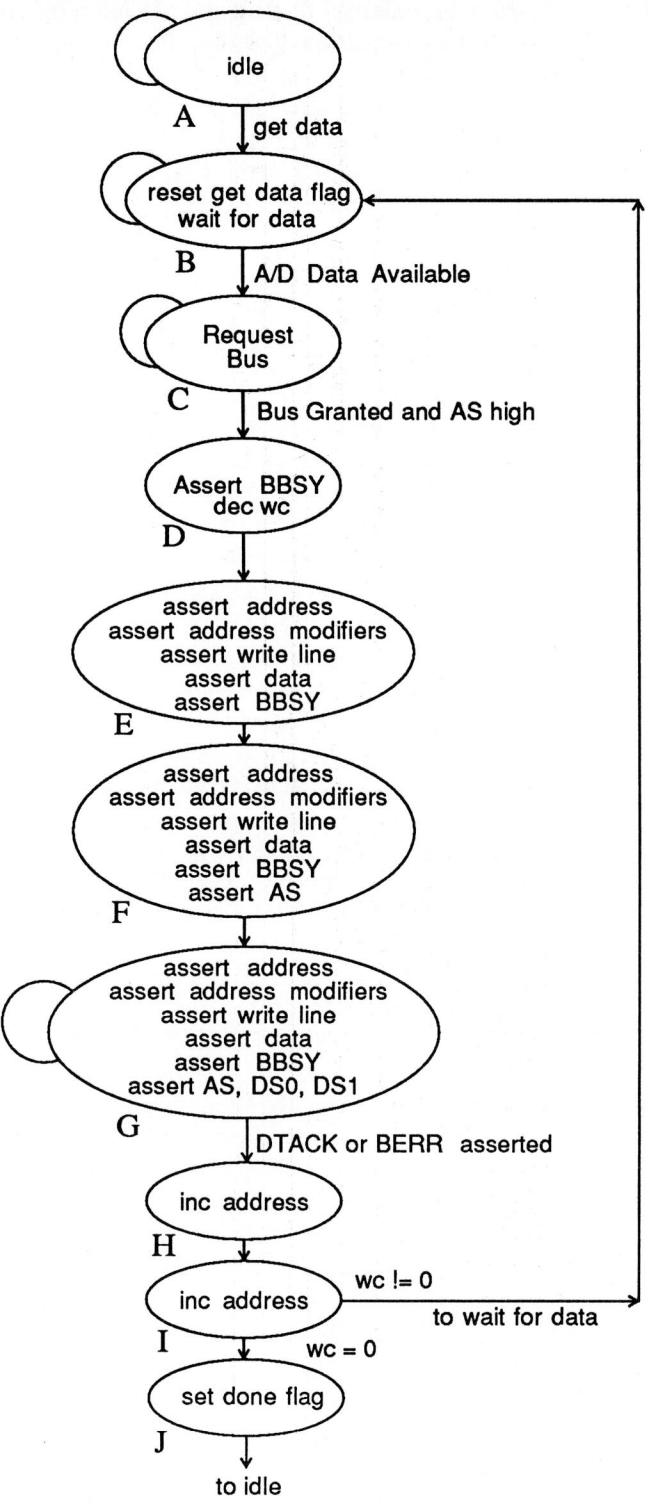

Figure 5.7. Preliminary State Diagram of A/D DMA Interface.

The action of the system in State D is no longer to assert the bus request, but to assert the bus busy line (BBSY). This indicates to the arbitration unit that a successful arbitration was accomplished, and that it (the arbitration unit) can release the bus grant line. The bus busy line will be asserted until the interface is ready to release the bus. This state also is used to decrement the word count. The action of the system then moves to State E.

When the control system of the interface reaches State E the control lines can be asserted. In this state the address lines, the address modifier lines, the write line, and the data are all asserted. Since this interface is used only for writes, the write line and the data are asserted at the same time as the address and address modifiers. In a system designed for both reads and writes, at this time the direction of the transfer would be indicated with the write line, and the data would be asserted for a write, but not asserted for a read. The action then moves to State F.

The signals asserted in State F include all the signals asserted in State E, but adds the address strobe (AS) as well. Asserting the address itself in the previous state and the address strobe in this state guarantees that the address lines will be stable before the address strobe arrives at slave modules. The system then moves to the next state, State G.

State G calls for the assertion of all signals of State F, plus the assertion of the two data strobes, DS1 and DS0. This indicates that a two-byte transfer is taking place since the signal line for a four-byte transfer (LONGWORD) is not asserted. The data strobes act as a request line, asking the addressed slave to respond. In this instance, we assume that the addressed slave is a memory module, and that the response will come as quickly as possible. The desired response is the data acknowledge (DTACK) signal, which indicates that the slave has accepted the data and that the interface can release the bus. Another possible response is the occurrence of the bus error signal (BERR), which indicates that something went wrong. The state diagram of Figure 5.7 shows that upon receipt of either signal the transfer is considered over, and the system moves to State H. Until one of the signals is received the system waits in State G.

All signals relating to the bus are released by leaving State G, and State H calls only for incrementing the address. The next state is always State I. The Address Register is incremented again in State I. The two increments are needed in this configuration since the address is a byte address, and the transfer completed is a two-byte transfer. State I is also used to test the condition of the Word Count Register. If the value in the Word Count Register is zero, then the allotted number of transfers has been completed and the system moves to State J. However, if the value in the Word Count Register is not zero, then the action of the system returns to State B to await the arrival of the next data value.

When the action of the system moves to State J, the done flag is set, which is a bit in the status register accessible to the processor. Then the system returns to the idle state.

This system allows the transfer of information from an A/D converter to a memory with DMA techniques. There are improvements which could be made in the system; we will point out some of them at the end of this section and invite the reader to modify the interface as described here to take care of some of the conditions which have not been enumerated.

The next step in the design process is to create a detailed data path block diagram, on which all the devices which will be used are specified, and all the control signals are identified with their appropriate assertion levels. The reader is invited to perform this process, given the information in Figures 5.6 and 5.7. The counters used to implement this system were '191s, which are up/down counters.

Thus the same type of device (with control lines configured differently) can be used for both the Address Register and the Word Count Register. This device also simplifies the loading of the registers, since the load is asynchronous; many counters require a synchronous load. The other type of device used extensively is a tri-state driver. In this system we use drivers requiring a low-true enable.

With the detailed data path block diagram and the preliminary state diagram, a detailed state diagram can be prepared. This state diagram is shown in Figure 5.8. The lettered states of Figure 5.7 have been identified with the corresponding state numbers. The idle state is State 0, this state assignment chosen so that the clearing of the Present State Register will reset the system and force it to the idle state. To determine when to leave the idle state, the system examines the signal AD_GO, which will be set by the $0 \rightarrow 1$ transition of bit 11 of the SCR. Leaving the idle state, the system moves to State 1. Note that an attempt has been made in the system to number the states so that next states are logically adjacent to one another. This is not absolutely necessary, but will generally result in easier, more straightforward implementations.

The signal CLR_GO is asserted in State 1, which is used to clear the AD_GO flag. Then the system waits for indication of the arrival of new data. The signal which will be asserted when new data arrives is AD_A_RDY; when this line is asserted the system moves to State 3.

In State 3 the bus is requested by asserting the bus request line, BRx-L. The "x" is used because there are four levels of bus request priority, and the user must define which of the levels is required; this applies to the corresponding bus grant signal as well. External to the state machine, the logic of the system ANDs the bus granted condition with the AS high condition, and this signal is called BG_AND_AS. When it is asserted, the system will leave State 3 and go to State 7; until it is asserted the system will remain in State 3.

State 7 of Figure 5.7 corresponds to State D of Figure 5.6. The signal BBSY-L is asserted, as is the signal WCCLK-H. The word count is decremented by WCCLK-H. The bus busy signal will remain asserted through State 12, indicating to other units on the bus that this module has control of the transaction.

State 6 is used to assert ADEN, which is used to enable the address, the address modifiers, and the data onto the bus. Also, the write line (WRITE-L) is asserted at this time.

State 14 adds to the number of asserted signals the assertion of AS. The remaining signals (ADEN, BBSY, WRITE) continue to be asserted.

State 12 continues to assert the above signals, and also causes the assertion of the two data strobes, DS1 and DS0. Logic external to the state machine tests the DTACK and BERR signals, and when one of these signals is asserted, the system moves to state 13.

The move to State 13 releases the lines controlling the bus, and asserts the signal to increment the Address Register (IACLK). State 5 also increments the Address Register, and decides, depending on the status of WCZERO, whether to go to State 1 to await more data, or to go to State 2. WCZERO will be asserted when the Word Count Register reaches zero.

State 2 is used to set the done flag, and then the system returns to the idle state.

The logic of the major portions of the interface is shown in Figure 5.9. Figures 5.9(a) and (b) contain the chips required for the address register. For the full 32 bits, this requires eight '191s; by restricting the requirement to 24 bits or 16 bits, and using a different set of address modifier bits, the number of devices required would be correspondingly reduced. Note that there are two sets of tri-state

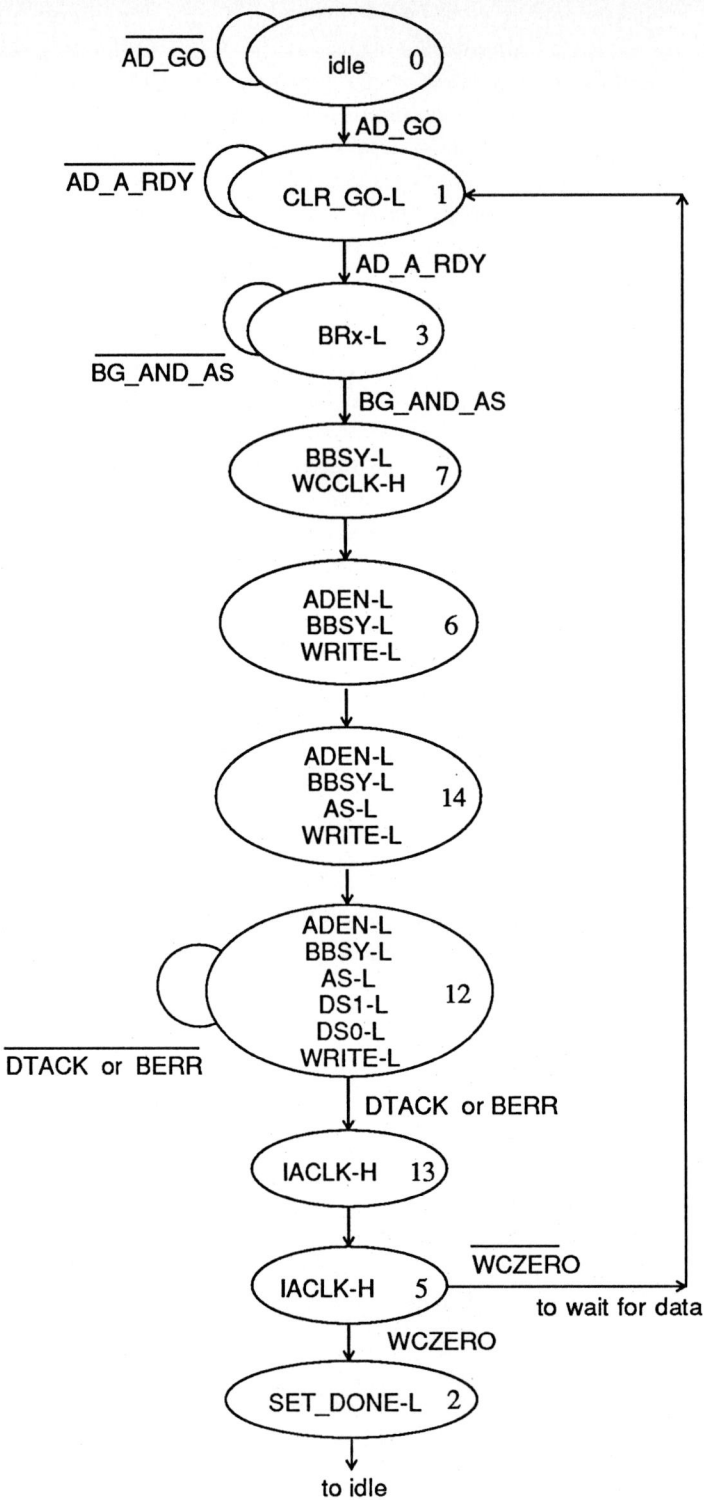

Figure 5.8. Detailed State Diagram of A/D DMA Interface.

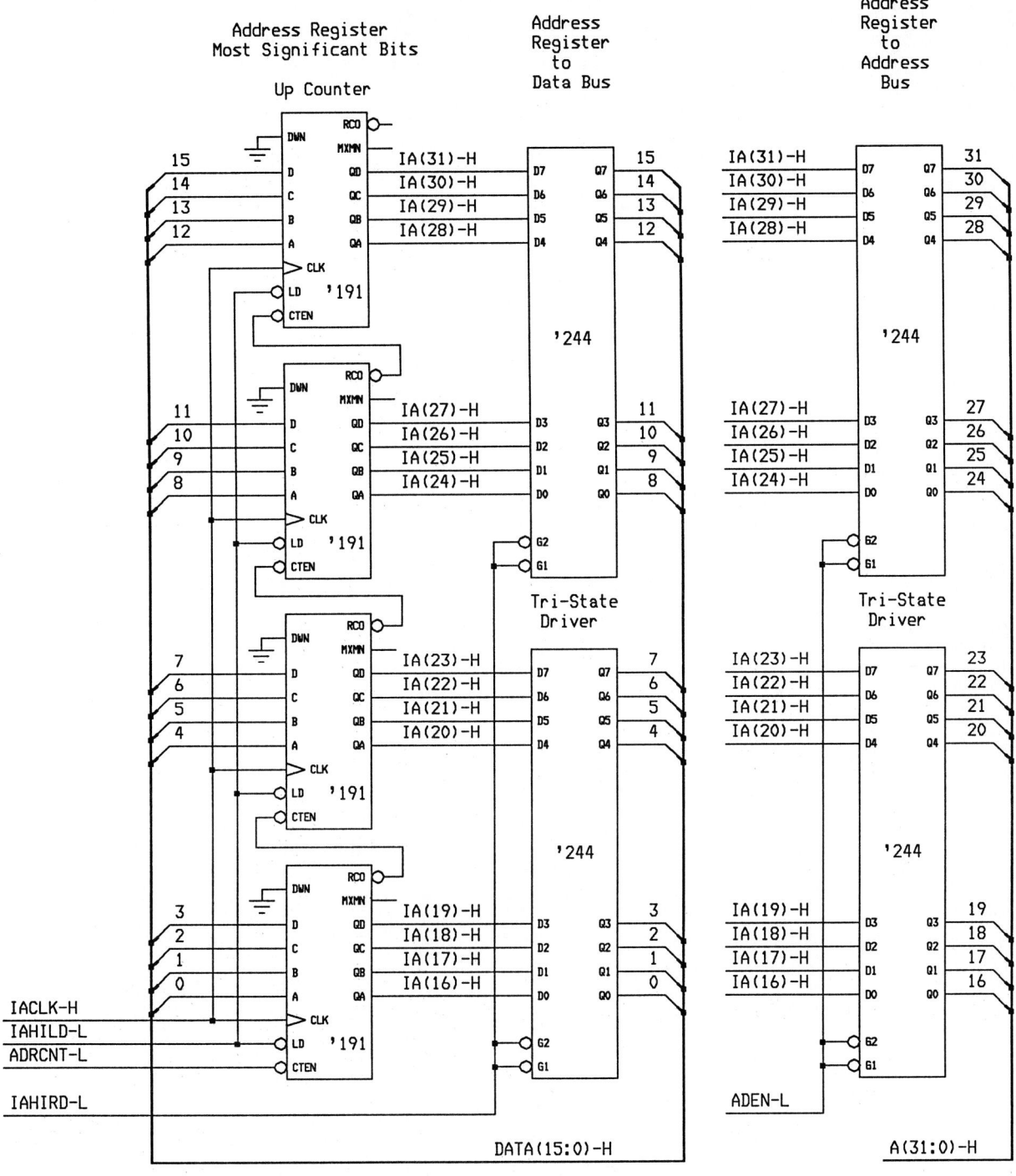

Figure 5.9(a). Logic Diagram of A/D DMA Interface: Address Register Most Significant Bytes.

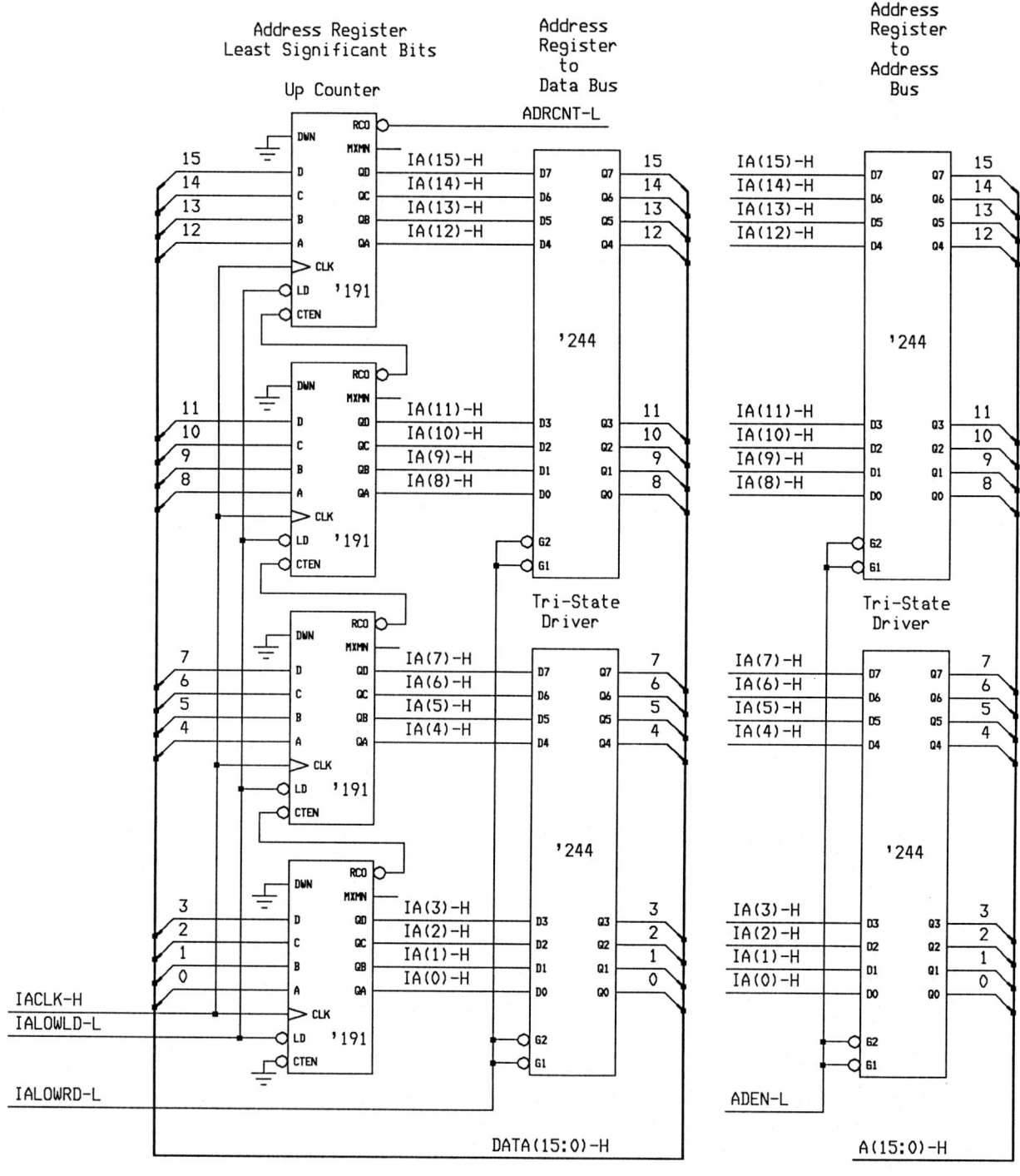

Figure 5.9(b). Logic Diagram of A/D DMA Interface: Address Register Least Significant Bytes.

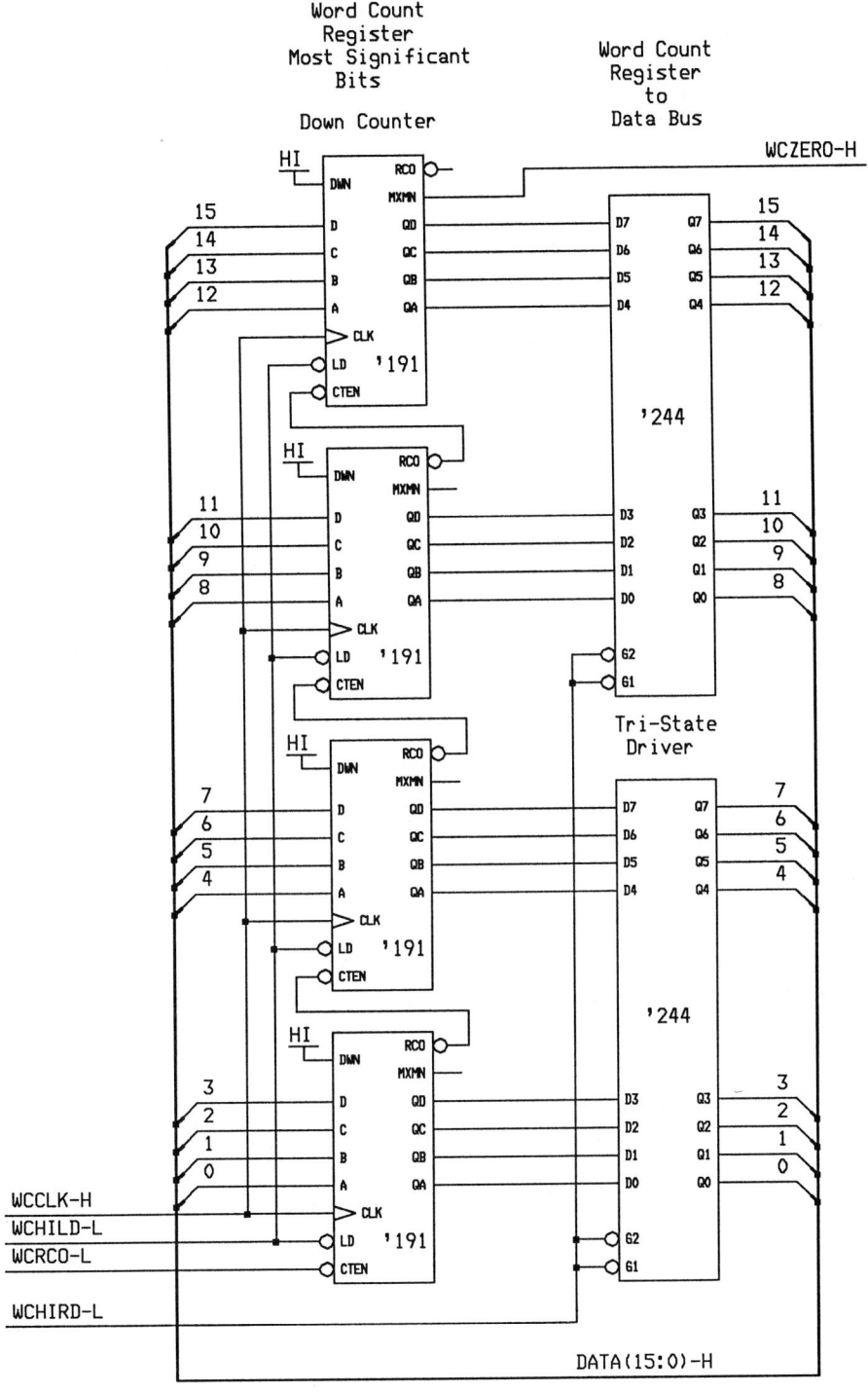

Figure 5.9(c). Logic Diagram of A/D DMA Interface:
Word Count Register Most Significant Bytes.

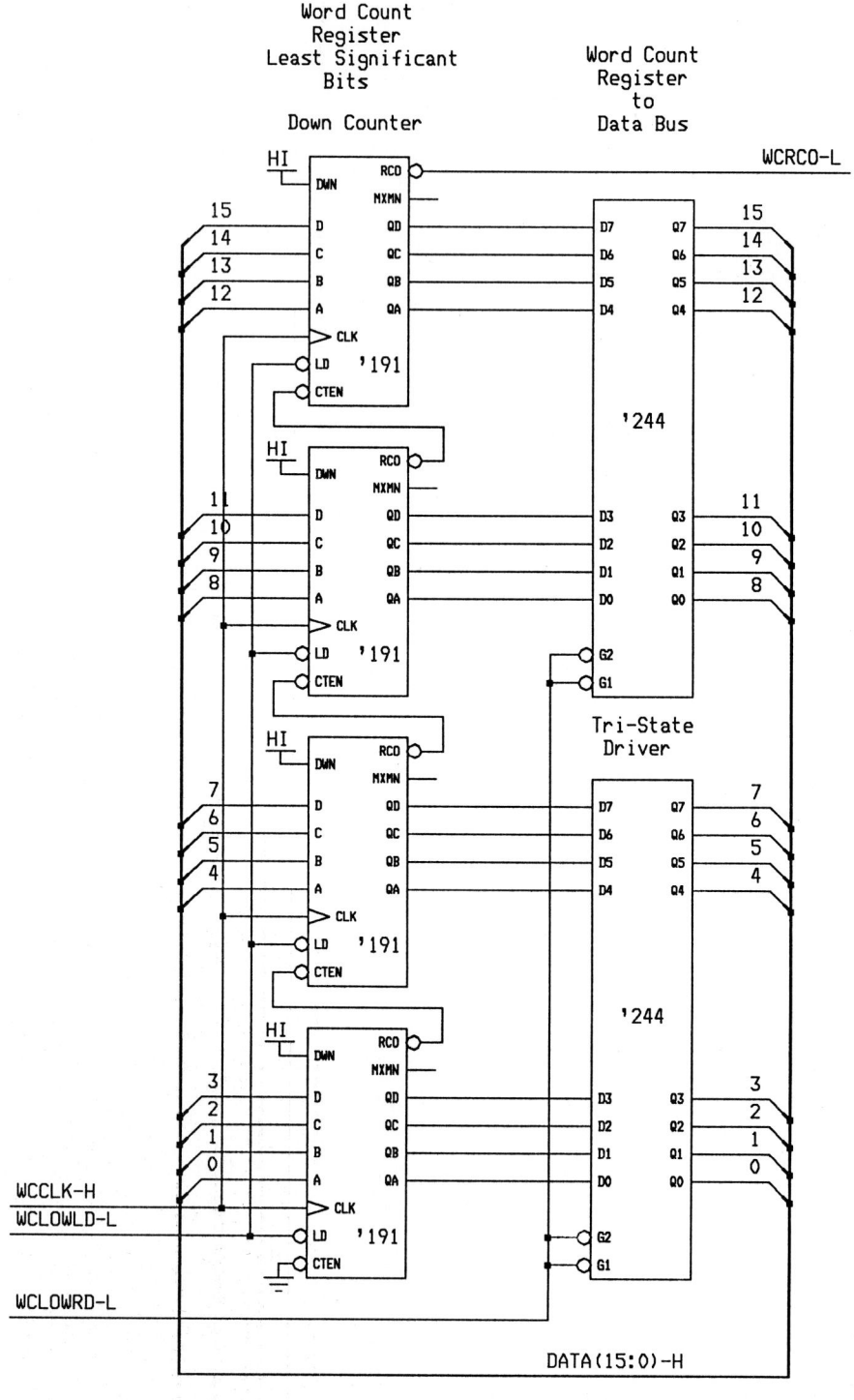

Figure 5.9(d). Logic Diagram of A/D DMA Interface: Word Count Register Least Significant Bytes.

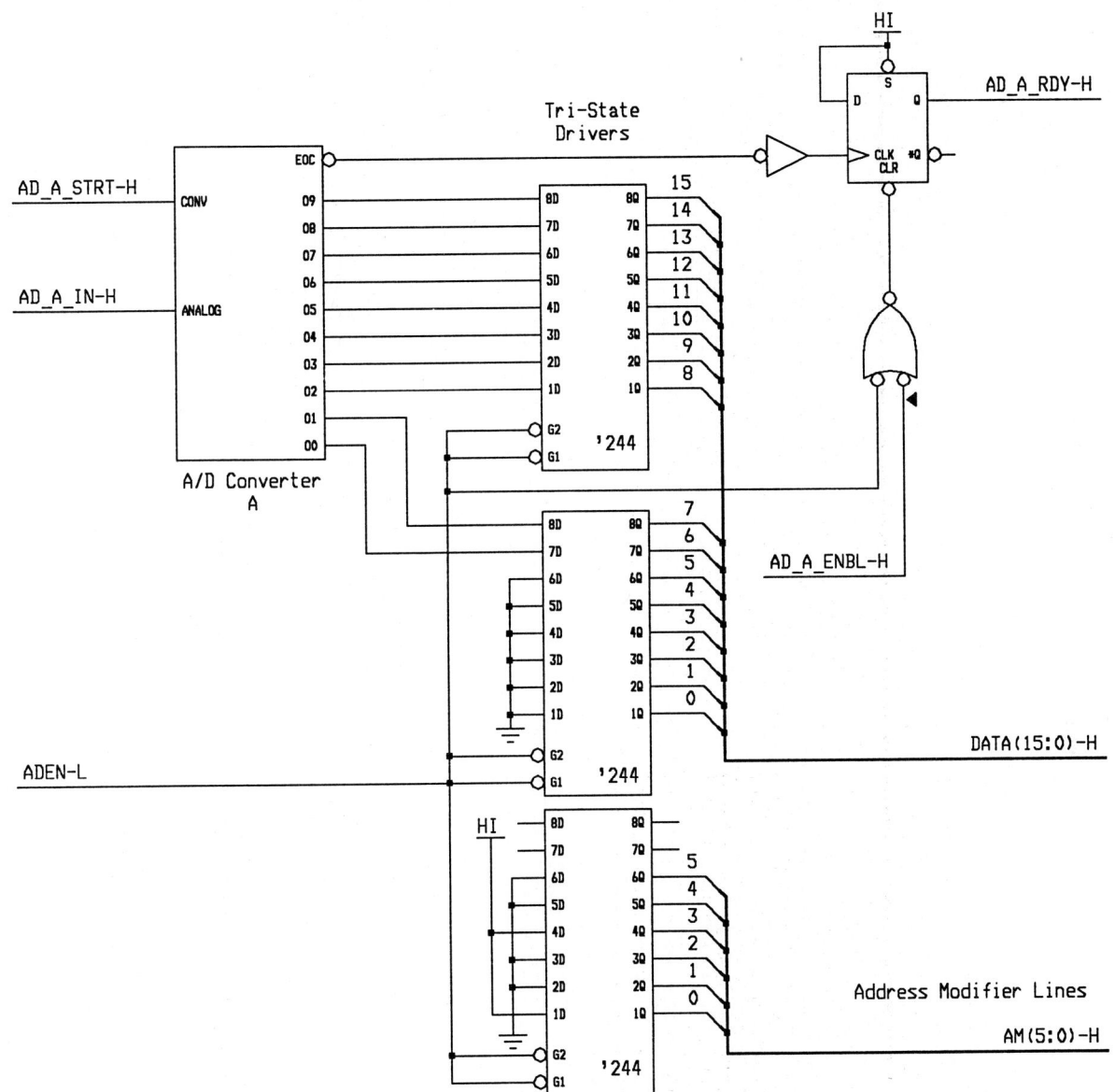

Figure 5.9(e). Logic Diagram of A/D DMA Interface:
A/D Converter and Bus Drivers; Address Modifier Bus Drivers.

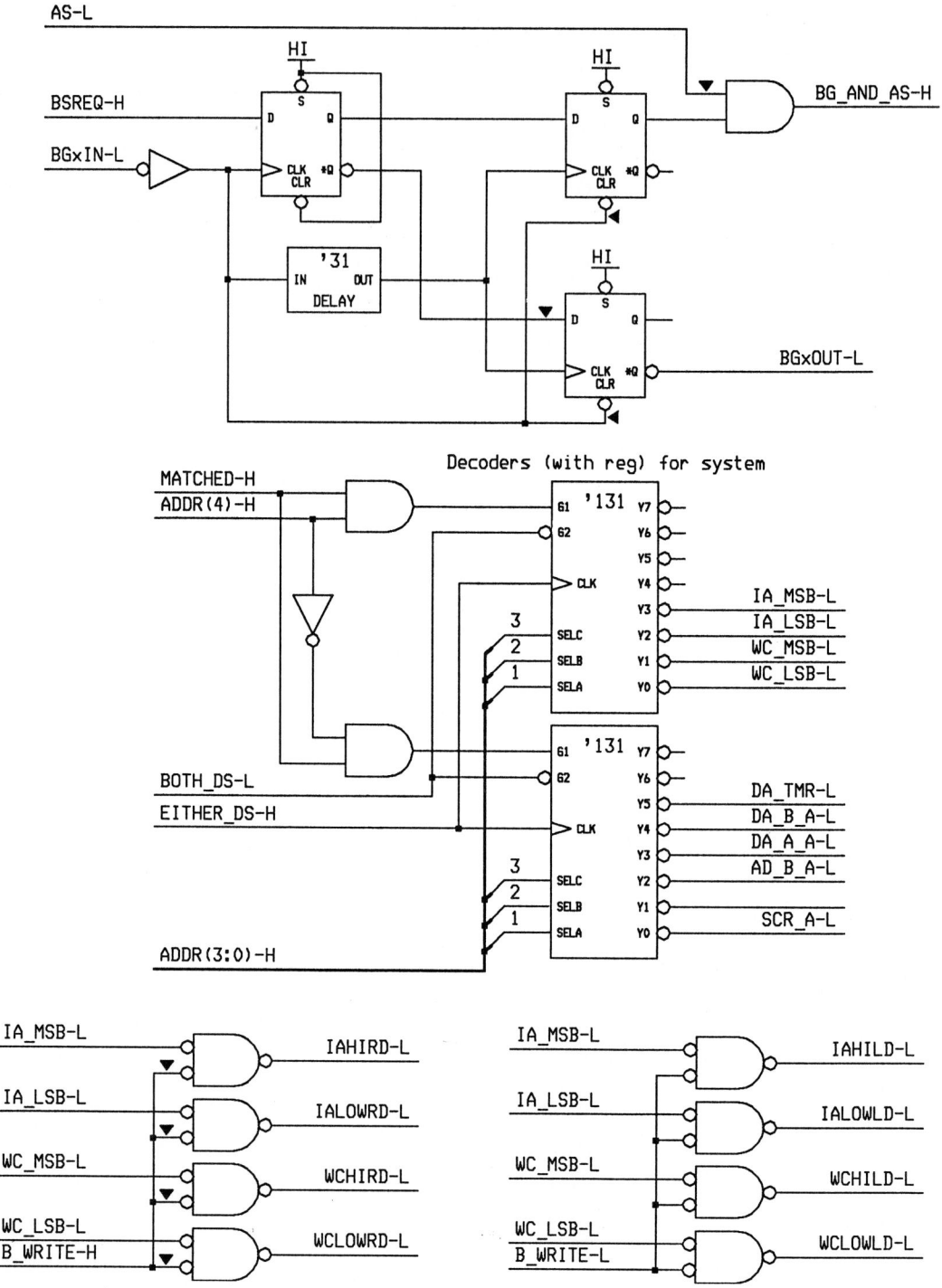

Figure 5.9(f). Logic Diagram of A/D DMA Interface:
Programmed I/O Interface Control Logic.

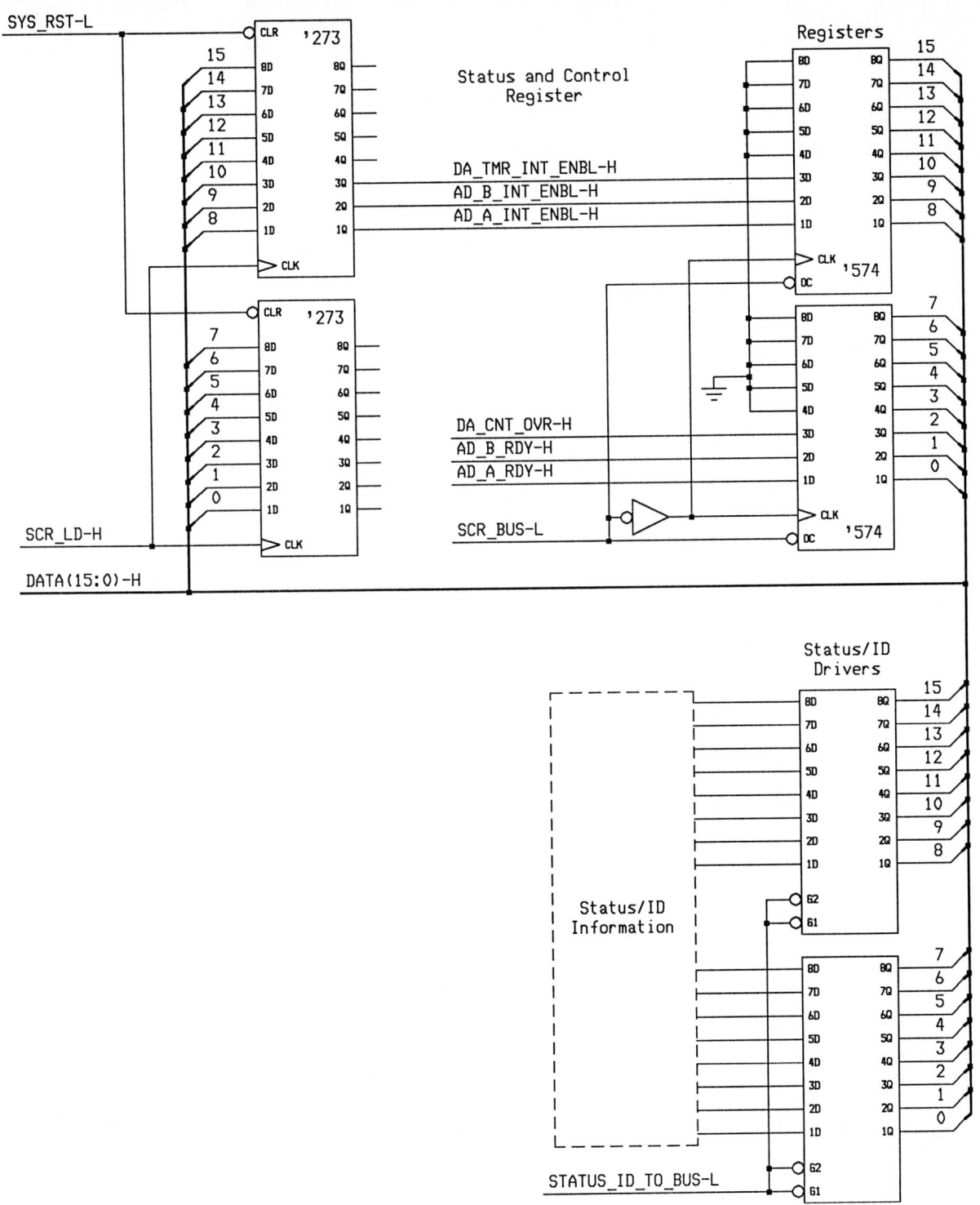

Figure 5.9(g). Logic Diagram of A/D DMA Interface: Status and Control Register.

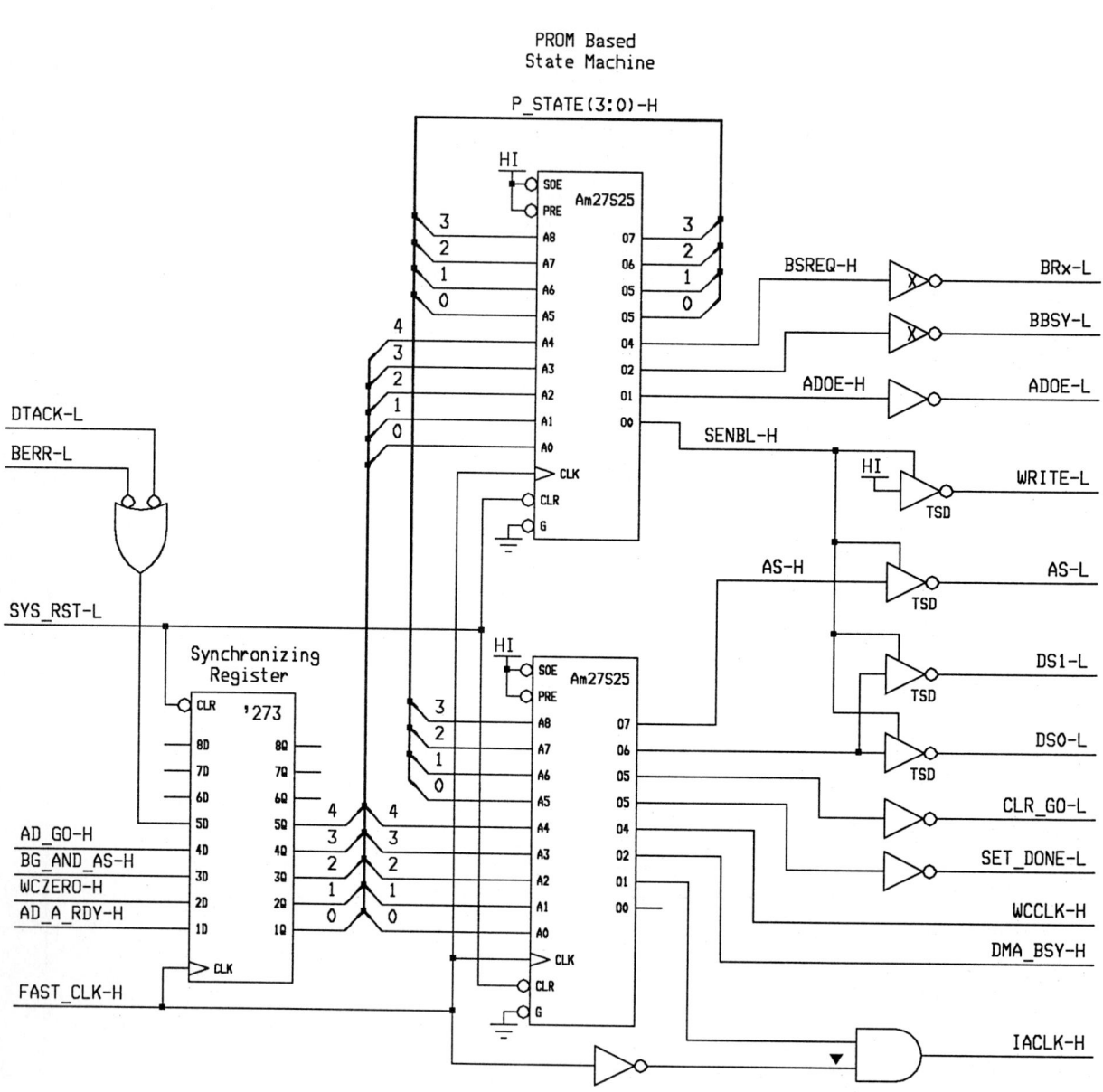

Figure 5.9(h). Logic Diagram of A/D DMA Interface: State Machine and DMA Control Logic.

drivers provided for the interface, one which supplies the address to the data bus, and one which supplies the address to the address bus. As mentioned above, one reason for the choice of the '191s as counters for this system is that they are loaded asynchronously. With this type of a device, the only clock signal required comes from the state machine (IACLK-H).

The devices utilized for the Word Count Register are shown in Figures 5.9(c) and (d). These counters are identical to the Address Register, except that they have been configured to count down instead of counting up. Also, only one set of tri-state drivers is provided, since this information is directed only to the data bus. The zero detection is easily accomplished by logic internal to the '191s, and when the value of the counter is zero, WCZERO is asserted. This counter is actually much bigger than what would be required in almost all systems. That is, the logic shown in Figure 5.9 indicates that 32 bits of count could be used. This would allow a single transfer of 4 GWords, which would be too large for practical applications. Reducing the number of bits in the Word Count Register to 16 would still allow a single transfer of up to 64 KWords, which is still far larger than most systems would use. Making this change would reduce the chip count of the Word Count Register, as well as reduce the size of the address required and some of the gating associated with the programmed I/O logic.

The actual data path for this portion of the interface is included in Figure 5.9(e). Remember that not shown in the various illustrations of Figure 5.9 are the other A/D converter and the two D/A converters. As before, the data from the A/D converter is sent to the data bus through tri-state drivers. Also as before, an edge triggered flip-flop is used to create a flag which will be set when the converter has completed its work. An AND gate used in an ORing function has been added to the clear line of the A/D ready flag. When the A/D action has not been enabled, this flag is forced to a cleared condition. The net effect is that at whatever time the A/D action begins, a new value of the input will be needed. This removes the possibility of an overrun problem between the first and second samples. Also included in Figure 5.9(e) is a tri-state driver configured to deliver the proper pattern to the Address Modifier lines. The pattern shown is for 001001, which is an Extended Non-Privileged Data Access. Other patterns could be established for other types of transfers.

Some of the logic used by the programmed I/O portion of the interface is shown in Figure 5.9(f). The reader is encouraged to identify the additional gating required to provide a complete interface. For example, the generation of the DTACK signal is not shown, nor the changes needed for the modification of the address matching circuitry (the generation of MATCHED). The generation of the signals for filling and reading the Word Count Register and the Address Register are shown. Also included with Figure 5.9(f) is the bus grant circuitry. When the appropriate bus grant in line is asserted (BGxIN-L), an inverter places a rising edge on the clock line of a "D" flip-flop labeled FFA. This is to interrogate the status of the bus request line for this interface. If this module has requested the bus (the interface is in State 3), then the BSREQ signal will be asserted, and this will be accepted by FFA. If the bus request was not generated by this module, then it should be passed to the next module in the daisy chain. This is accomplished by the flip-flop labeled FFC. After the assertion of BGxIN, some time is required for FFA to reach a stable condition. Thus, the clock line used by FFB and FFC is delayed by an amount which allows FFA to become stable. Using ALS technology, this will need to be greater than 18 nsec, and a small delay such as that provided by a 74LS31 could be used for this function. Note also that when the BGxIN signal is not asserted, both FFB and FFC are held in a cleared state. When BGxIN is asserted, one of them will be set, depending on assertion of FFA. If FFA is not set, then FFC will be set, and the BGxOUT signal will be asserted, sending the grant signal to the next module. If FFA is asserted, then BGxOUT will not be asserted, and this interface can obtain control of the bus. When this condition exists, AND the AS signal has been released by the previous bus master, then BG_AND_AS will be asserted, signaling to the state machine that the action can proceed.

The logic associated with the Status and Control Register is shown in Figure 5.9(g). As in the previous section, the register part is created with '273s, and the reading of the status bits is done with '574s. The definition of all bits is given in Table 5.3. Note that the $0 \rightarrow 1$ transition of bit 11 of the register sets a flip-flop (FFD). Before this occurs, the flip-flop for the done flag (FFE) is held in a cleared condition by bit 11 of the register. When the state machine moves out of the idle state it will clear the AD_GO flag with the CLR_GO signal. Finally, when the state machine completes the last transfer and moves back to the idle condition, it will set FFE which asserts AD_A_DONE. All these bits, together with the status of the state machine (DMA_BSY), are available to be read by the system through the '574s.

The actual logic of the state machine is shown in Figure 5.9(h). This is a memory based state machine controller, using Am27S25s. These devices contain a 512×8 PROM and an edge triggered register. Thus, the present state register for the system is contained inside the memory devices. The configuration shown allows for four feedback variables (P_STATE(3:0)) and five input variables. The input variables shown are those called for in the state diagram of Figure 5.8. The clock for the system is simply called FAST_CLK. Associated with the VMEbus is a 16 MHz clock; if this is used as the clock of the system then individual state times are 62.5 nsec. The Am27S25 can operate to about 40 nsec clock times, which is a clock frequency of 25 MHz. Other types of state machine devices, such as registered PALs, can operate even faster.

The signals generated in Figure 5.9(h) are used on the other pages of the Figure 5.9, or are bus signals. The signals BBSY and BRx are open collector signals, and hence the drivers for them are appropriately marked in the figure. The bus control signals (WRITE, AS, DS1, and DS0) are tri-stated, and hence must be driven with tri-state drivers. This will occur when SENBL is asserted, which is the same time that WRITE is called for in the state diagram. One interesting technique is demonstrated by this implementation. Note that the assertion level of all signals coming out of the state machine is high. If low asserted levels are needed, they are created with inverters. There is no reason why this inversion could not be accomplished by inverting all bits (for that signal line) in memory. This reduces the chip count slightly, and may be a very beneficial choice. However, by utilizing high assertion levels at this point, then in the debug phase of the project there is never any question if a signal is asserted or not at the state machine outputs: if it is high it is asserted; if it is low it is not asserted. This can be very helpful in the checkout of the device.

One final note concerns the generation of IACLK. This signal is ANDed with the inverse of the clock to create a signal asserted only in the last half of a cycle. The reason for this is that if this technique were not used, there would only be one active edge, but the signal would be asserted for two states. Hence, the address register would increment by only one, instead of two.

Now that we have described this simple interface, we will make some observations about what we have done, and let the reader explore methods of addressing some of the things which have been oversimplified in this design. First of all, there are some very strict requirements in the definition of the VMEbus for the allowable electrical loads and required driving characteristics of the signals which are provided for the bus. For detailed information, see [Moto85]. The signals involved in this interface need to conform to the definition, and hence the required drivers must be used. In general, this means that care must be taken to choose a logic family which will properly match the requirements.

One of the ways to improve the loading characteristics is to buffer the data bus with a transceiver. The major benefit is that the electrical load is reduced. However, a problem which is introduced is that an additional delay is introduced in the data lines for both read and write operations. What are some of the methods which can be used to deal with this delay?

The state machine shown in Figures 5.7 and 5.8 indicates that the action will continue until the word count reaches zero. However, what changes would be needed to the state diagram (and the hardware) to check the word count before the transfer is initiated? This incorporates an additional feature which will not permit a transaction to begin if the word count is not set to a non-zero value.

Another state machine change involves the bus error condition. In the state diagrams, action moves on with the receipt of a DTACK signal (normal condition) or a BERR condition (abnormal condition). If the bus error condition is to be dealt with in a different fashion than the normal operation, what action makes sense? How can the state diagram (and the hardware) be changed to set an error flag if this occurs?

Still another change in the state machine is required if the desired action of the interface is to assert BBSY and await the release of AS, rather than the current method to await the release of AS before asserting BBSY. This requires an additional state and some logic changes in the state machine. How can this be accomplished?

One of the things not included in Figure 5.9 is the logic required to cause an interrupt when AD_DONE is asserted. What logic is required, and how can this be incorporated into the logic shown in the previous section?

As indicated by these questions, an interface must be designed to account for all of the possible conditions in the bus itself as well as all possible conditions of the device being controlled. The designer must be sure that all of the appropriate conditions are handled by the interface.

5.5. Summary

The design of interface systems for computer systems is a complex task, yet straightforward application of the principles of digital system design will lead to interfaces which perform the necessary tasks. However, prior to embarking on an interface project, it is imperative that the designer understand the requirements of the system, the characteristics of the communication mechanism, and the behavior of the device being controlled.

Most interface systems utilize an asynchronous bus to transfer data and provide the necessary control. An asynchronous protocol calls for a bus master to intiate action by asserting address and control lines (after it has acquired control of the bus from the arbitration system). The actual transfer begins when the bus master asserts the request line; the selected slave responds by asserting the acknowledge line; the master releases the request line; and, finally, the slave releases the acknowledge line. Data is transferred between the master and the slave at appropriate times during the sequence, depending on the type of transaction (read or write). The timing requirements of the transaction and the exact sequence of events will be defined in a bus communication protocol.

A processor can control the action of a device, and the transfer of data, by utilizing the facilities of the bus. Commands can be issued by activating specific addresses, or by sending different patterns to a specific address. Similarly, data can be moved to and from an I/O device by writing to or reading from a specifc address which is assigned to that device. This technique is called programmed I/O, since all control and data transfers are directly controlled by a program.

An interrupt facility can be used in conjuction with programmed I/O. This facility allows a processor to ignore an interface until specific action is needed. At that time, the interface requests an interrupt, and the processor responds in an appropriate fashion by using programmed I/O methods. Interrupt driven I/O leads to somewhat slower data rates, when compared to strictly programmed I/O techniques, but the processor itself can be involved in non-I/O activities.

Data rates can be substantially enhanced by the use of Direct Memory Access techniques. A DMA interface is controlled by the processor by using programmed I/O techniques. The processor uses programmed I/O to fill word count, starting address, and command registers, and to read status information. The interface can notify the processor upon completion of a task by the use of the interrupt facility. Transfers occur when the interface takes command of the bus and moves information directly to or from system memory.

In all of the cases mentioned, the transfer is a sequential process controlled by the interaction of the interface and the bus. For strictly programmed I/O transfers, the sequential nature of the transaction is defined by the sequential behavior of the bus protocol. An interface which is restricted to this type of transfer can perform the necessary action by using gating systems involving the control signals on the bus. Systems which must control the action of a device can make beneficial use of a sequential controller such as those demonstrated in Chapter 4 or Section 5.4.

5.6. Exercises

5.1   Design a programmed I/O interface to send data to an 8 bit D/A converter.

5.2   Using reasonable times for register transfers and memory transfers, determine the maximum data rate for the interface designed in Exercise 5.1.

5.3   Design a byte swap register for a 16-bit asynchronous data bus. This register is filled by writing to address 777650_8, and it is read by accessing the same address. However, the information which returns is in reverse byte order from that which is provided.

5.4   Design a bit rotate register for a 32-bit data bus. This system responds to three addresses: 776540_8, 776544_8, and 776550_8. The first address is used for the data; the second address is used rotated data; and the third address is used for the rotate amount. The rotate amount has a maximum value of 31, so it need be only 5 bits. Writing to 776540_8 fills a register (32 bits wide) with data; reading the address retrieves the same information with no rotate applied. Writing to the second address results in no action; reading the address retrieves the information in the data register rotated by the amount specified in the rotate register. Finally, writing and reading the third address fills and obtains the rotate amount.

5.5   Design an interrupt timer. That is, create a system which will issue an interrupt to the processor every n ×sec, where n is a number written or read under program control.

5.6   ** Obtain the specifications of an ST506 disk drive and create a DMA interface to control it.

5.7. Additional References

[AMD85] Advanced Micro Devices, *Bipolar Microprocessor Logic and Interface Data Book.* Sunnyvale, CA: Advanced Micro Devices, 1985.

[Bart85] Bartee, T. C., *Digital Computer Fundamentals* (6th ed.). New York: McGraw Hill Book Company, 1985.

[Baer80] Baer, J. L., *Computer Systems Architecture.* Rockville, MD: Computer Science Press, 1980.

[Bell70] Bell, C. G., et.al., "A New Architecture for Mini-Computers - the DEC PDP-11," *Proceedings Spring Joint Computer Conference,* 1970, pp. 657-675.

[Chen74] Chen, R. C. H., "Bus Communications Systems," Ph.D. Dissertation. Pittsburg, PA: Department of Computer Science, Carnegie-Mellon University, 1974.

[Clul82] Cluley, J. C., *Minicomputer and Microprocessor Interfacing.* New York: Crane, Russak, 1982.

[Dext86] Dexter, A. L., *Microcomputer Bus Structures and Bus Interface Design.* New York: M. Dekker, 1986.

[Egge83] Eggebrecht, L. C., *Interfacing to the IBM Personal Computer.* Indianapolis, IN: H. W. Sams, 1983.

[Flet80] Fletcher, W. I., *An Engineering Approach to Digital Design,* Englewood Cliffs, NJ: Prentice Hall, 1980.

[IEEE75] Institute of Electrical and Electronics Engineers, "IEEE Standard Digital Interface for Programmable Instrumentation," IEEE Std. 488-1975. The Institute of Electrical and Electronics Engineers, Inc., October 1975.

Intel, *Microsystem Components Handbook.* Intel Corporation, 1984.

[Lang82] Langdon, G. G., Jr., *Computer Design.* San Jose, CA: Computeach Press Inc, 1982.

[Lipo88] Lipovski, G. J., *Single- and Multiple-Chip Microcomputer Interfacing.* Englewood Cliffs, NJ: Prentice Hall, 1988.

[Moto85] Motorola, Inc., *The VMEbus Specification,* October 1985.

[Shiv85] Shiva, S. G., *Computer Design and Architecture.* Boston, MA: Little, Brown, 1985.

[TsSi82] Tseng, C. J., and D. P. Siewiorek, "The Modeling and Synthesis of Bus Systems," Technical Report DRC-18-42-82, Design Research Center. Pittsburg, PA: Carnegie-Mellon University, 1982.

[ThMa79] Thurber, K. J., and G. M. Masson, "Bus Structures," in Distributed Processor Communication Architecture, Lexington, MA: Lexington Books, 1979, pp. 131-174.

[TiLa82] Titus, C. A., J. A. Titus, and D. G. Larson, *STD Bus Interfacing.* Indianapolis, IN: H. W. Sams, 1982.

6

Design of Memory Systems

Memory systems are required in almost all digital computer systems, both for storing the instructions to be executed and for storing the data to be manipulated. There are a number of different types of special memories, such as the memories for a video display, or content addressable memory. This chapter investigates some of the design issues which are involved in fairly standard memory systems. We will begin this investigation by examining one- and two-dimensional techniques for identifying memory to be accessed; then we will demonstrate application of the techniques in both static and dynamic memory systems. Finally, we will design a simple cache memory which can be used in a variety of computer systems. These simple memory systems will demonstrate techniques which can be utilized in many different kinds of memory systems.

6.1. One- and Two-Dimensional Memory Systems

There are many different techniques used in memory systems to specify the memory location (or device, or subsystem, or . . .) which is to be utilized in a transaction. Two of the most easily recognized and most often used techniques are one-dimensional (1-D) and two-dimensional (2-D) decoding schemes. The names refer to the mechanism used to determine the active element. Figure 6.1 gives a block diagram of the 1-D technique for identifying the active element. The basic idea is that one portion of the address is used to select the specific device to be enabled, and that portion of the address is decoded to assert a single enable signal. The decoder, then, accepts N address lines and asserts one of 2^N lines. The assertion can be further modified by additional control lines. That is, each of the 2^N patterns on the address lines will identify a single output line, but the assertion of that line can be further modified by enables or timing signals as called for by the system.

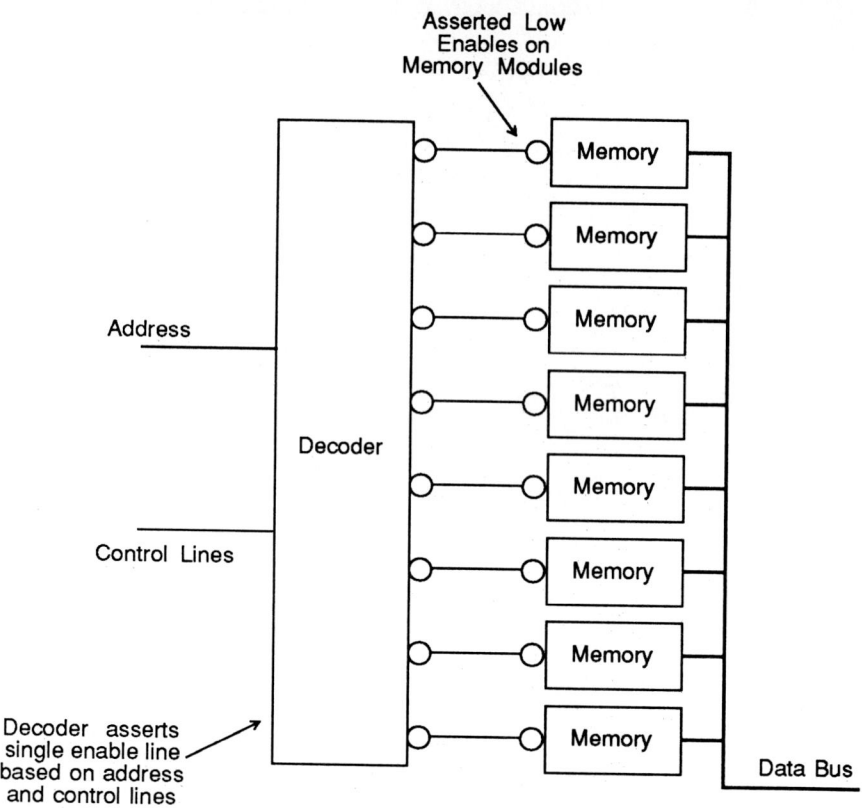

Figure 6.1. One-Dimensional Decoding Technique for Memory Cells.

The 2-D scheme is shown in Figure 6.2. This scheme also accepts N address lines to identify one of 2^N separate locations. However, the N lines are divided into two groups, which are used as inputs to separate decoders. If M lines are used by one of the decoders, then $M-K$ lines will be used by the other decoder. As before, the action of the decoder can be further modified by enable and timing signals. This arrangement requires that the memory elements being activated have two enable lines, where the 1-D scheme required a single enable line per cell. Thus, the 1-D scheme has a more complex decoder with simple cells, while the 2-D scheme has simpler decoders and more complex cells. The schemes can be utilized on silicon substrates, where each individual cell is a single bit, or they can be used at the device level, where each individual cell is a memory device or collection of memory devices.

We will now examine two separate memories to demonstrate the two techniques. In both cases the memory device utilized is a Programmable Read-Only Memory (PROM), which can be used to store programs and data in systems which will not change. The particular device we will refer to is the HN27512G by Toshiba, although other manufacturers have similar devices. This memory is a 64K $\times$ 8 memory (65,536 locations with 8 bits per location). Internal to the device the locations are accessed with a 2-D technique. In addition to the 16 address lines needed for the device, the 28-pin package also contains power and ground, 8 output lines, a chip enable (asserted low), and an output enable (also asserted low). By

Chap. 6: Design of Memory Systems

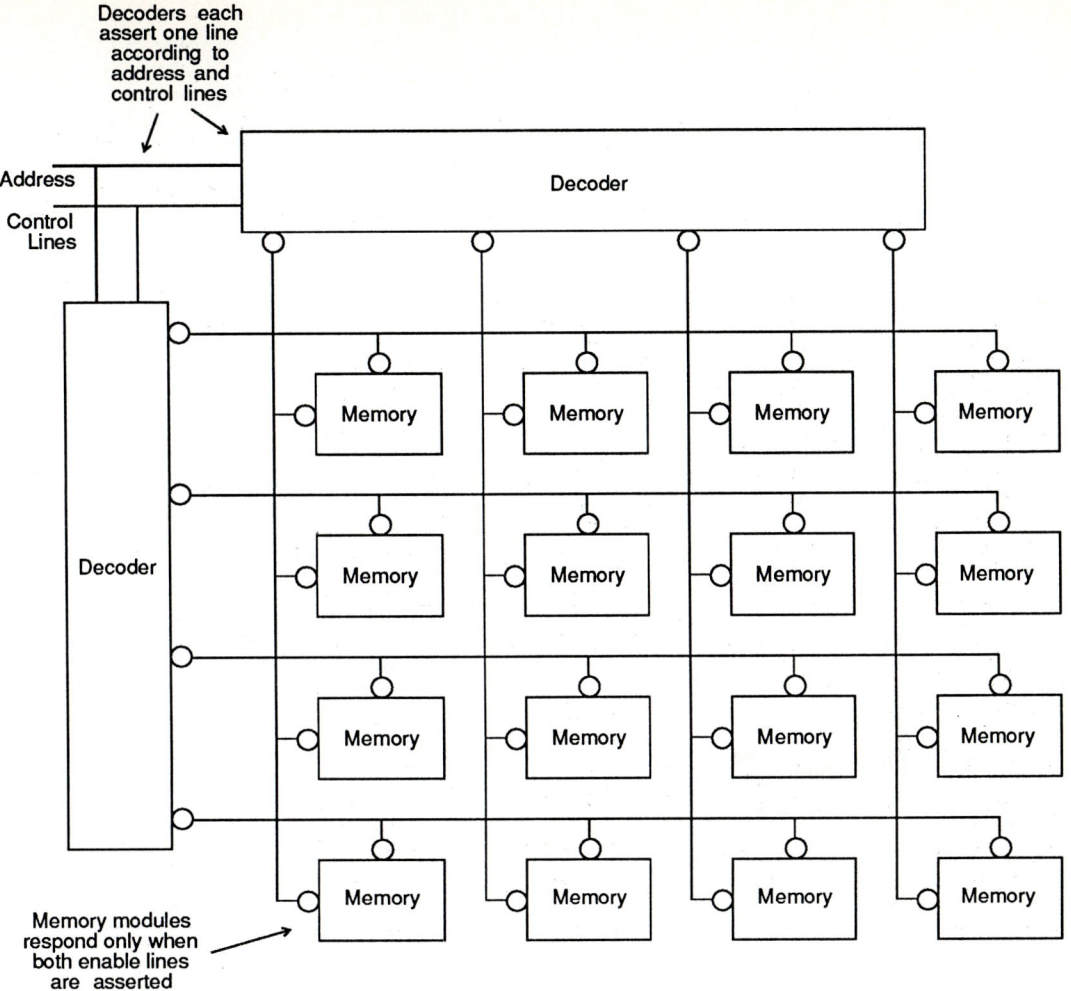

Figure 6.2. Two-Dimensional Decoding Technique for Memory Cells.

using the chip enable and the output enable in reasonable ways both the 1-D and the 2-D techniques can be demonstrated.

The 1-D technique calls for a single decoder to enable the appropriate memory device. This technique is shown in Figure 6.3, where eight 64K × 8 devices are used to create a half-megabyte read-only memory. A single 3-line-to-8-line decoder ('138) is used to select one of the eight memory devices, the output of which will be enabled onto the data bus. The additional control lines indicated in the figure allow disabling the PROMs so some other device can use the data lines.

The characteristics of the memories used in Figure 6.3 allow additional design features for the system. The figure indicates that the decoder outputs are directed to the chip enables of the PROMs. This will permit only one of the PROMs to be active at a time, and that PROM places the information on the data bus. The HN27512-25 has a 250 nsec access time from chip enable, to which must be added the delay time for the decoder, which is about 20 nsec for a 74ALS138. However, if the design is changed to use the output enable lines instead of the chip enable lines, then the response time of the memory can improve. This is because the output-enable-to-data-stable time is 100 nsec, which is better than the 250 nsec chip enable access time. The cost of this additional speed is the difference between the

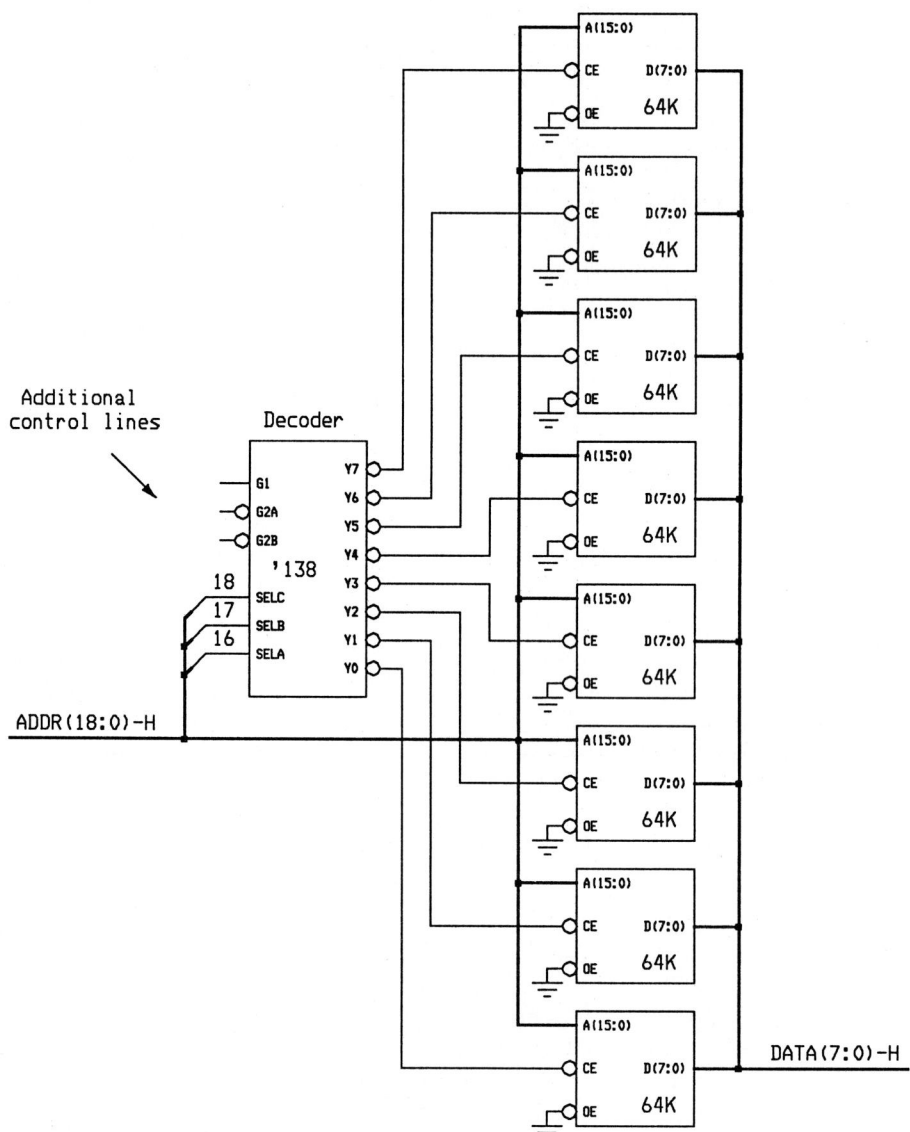

Figure 6.3. PROM Memory System with 1-D Technique.

"active" supply current and the "stand-by" supply current. The active supply current for the device is 45 ma, with the stand-by current much less than that. Thus, the designer can determine which method would best suit the overall system, using the chip enable or the output enable as appropriate. However, using the output enable to select the appropriate memory device instead of the chip enable does not change the fact that the address access time is 250 nsec.

The chip enable and the output enable of the memory used in Figure 6.3 can be used with a 2-D technique as well. Such an arrangement is shown in Figure 6.4, which is the logic diagram for a 1 MByte PROM memory. The decoders used in this system are 2-line-to-4-line decoders ('139s), and each decoder output is directed to four memories. The memory which will provide the information will be the memory which has both output enable and chip enable asserted. Note that the arrangement shown will result in four PROMs which are (chip) enabled at one time, so even though only one of the devices is providing the information, four devices will be drawing active current. Thus, this arrangement may not be advantageous for systems that must conserve supply current.

The 1-D and 2-D techniques can also be applied in other types of memory systems. This applies to selection of an active memory chip, and it also applies to selection to an active device internal to a memory chip.

6.2. Dynamic RAM System

One of the results in the miniaturization of components has been the creation of very large memories. This is especially true in the area of dynamic RAMs, which are currently capable of more than a megabit of information on a chip. This is true because of a combination of the small feature size and the fact that a minimum number of transistors are needed per cell in the dynamic RAM method. The information is maintained as a charge on a capacitor which is created with semiconductor technology. Since the charge will dissipate over time, it needs to be refreshed to maintain the information, and for this reason the memories are called "dynamic."

In this section we will create a design for a 16-MByte dynamic RAM system. The basic memory component of the system is a single dynamic RAM chip which will hold a megabit of information. It is assumed here that we will be creating a memory which will interact with an asynchronous bus system, such as the VME bus of the previous chapter, so the basic timing is the same as indicated there. Some of the details (such as including the address modifier lines) have been omitted from this memory system, and the reader is encouraged to modify the design presented here to conform to all the necessary details required of an actual bus system.

Using a single-bit memory in a system with a 32-bit bus requires multiple copies of the memory chip. The 32-bit width is 4 bytes wide, so the basic memory module to be used here is a bank with 4 megabytes of memory. Four of these banks will be needed for a system with 16 MBytes of memory. These are arranged in a system as shown in Figure 6.5. The 32 data lines are common to all memory banks, and the remainder of the circuitry is used to control the activity and timing associated with the memories. If we assume that there is a 24-bit address, then all address lines are required to identify the specific byte in question. The two most significant lines are used to select which of the four banks are involved in the transaction. The two least significant address lines identify the location of the byte within the 32-bit word. The remaining 20 address lines are used to select the appropriate word, 1 bit in each of 32 data memories.

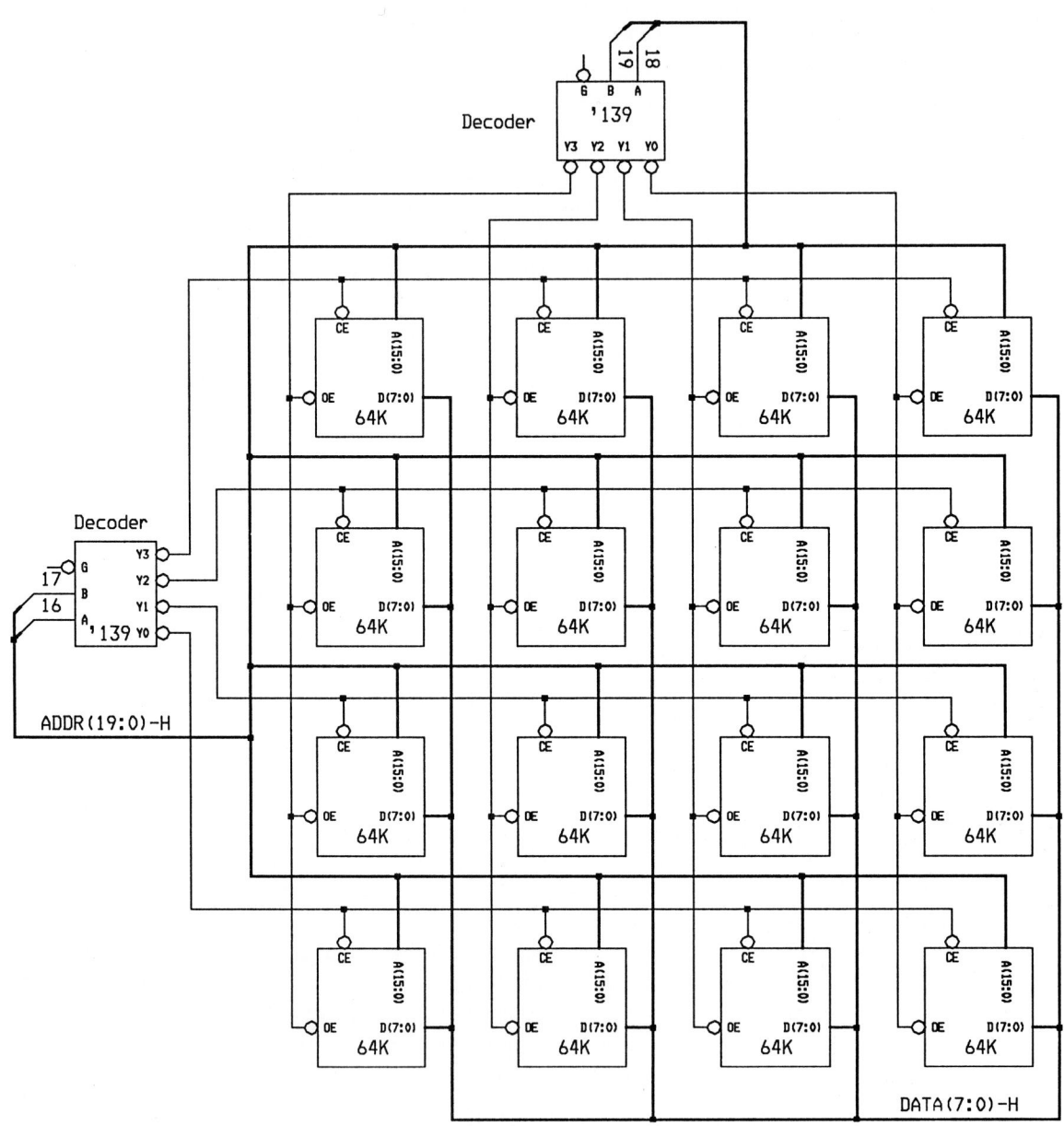

Figure 6.4. PROM Memory System with 2-D Technique.

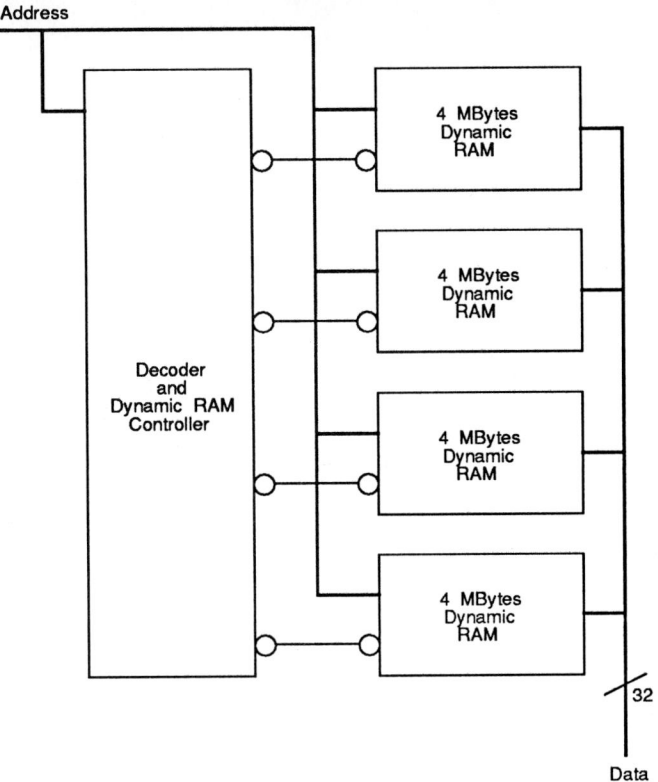

Figure 6.5. Basic Block Diagram of 16 MByte Memory.

Each memory chip, then, requires 20 bits of address. Since 20 address lines are an inordinate number of pins for a chip, dynamic RAMs use a 2-D technique and time multiplex address lines. This method is demonstrated in Figure 6.6. Not all dynamic RAMs will conform to these waveforms, but the mechanism shown in Figure 6.6 is typical of the requirements for DRAMs. The 20 address lines required for a megabit address are time multiplexed on 10 address lines. The designer selects 10 of the lines to form the address of the row in question, and the remainder of the lines form the address of the column in which the bit will be found. Figure 6.6 indicates that these addresses are presented at different times, and each is associated with its own strobe. The row address has associated with it a row address strobe, or RAS, and the column address has a column address strobe, or CAS, associated with it. Both the RAS and the CAS timing will have a setup time and a hold time associated with them, and the control section of the RAM will be responsible for meeting these timing constraints.

The sequence of events for a read operation is shown in Figure 6.6(a). Note that the write line, W-L, is not asserted during this transaction. The transaction begins with the presentation of a valid row address on the address lines. After the address lines have been stable for at least the required setup time, RAS is asserted. This is shown as t_R in the figure. The row address must remain stable for some small period of time after RAS is asserted, and this is the hold time. The 10 address lines can then be changed to the column address, and this address must be stable for some setup time before the assertion of CAS. This time is indicated as t_C in the figure. Again, the lines must be stable for some hold time after the assertion of

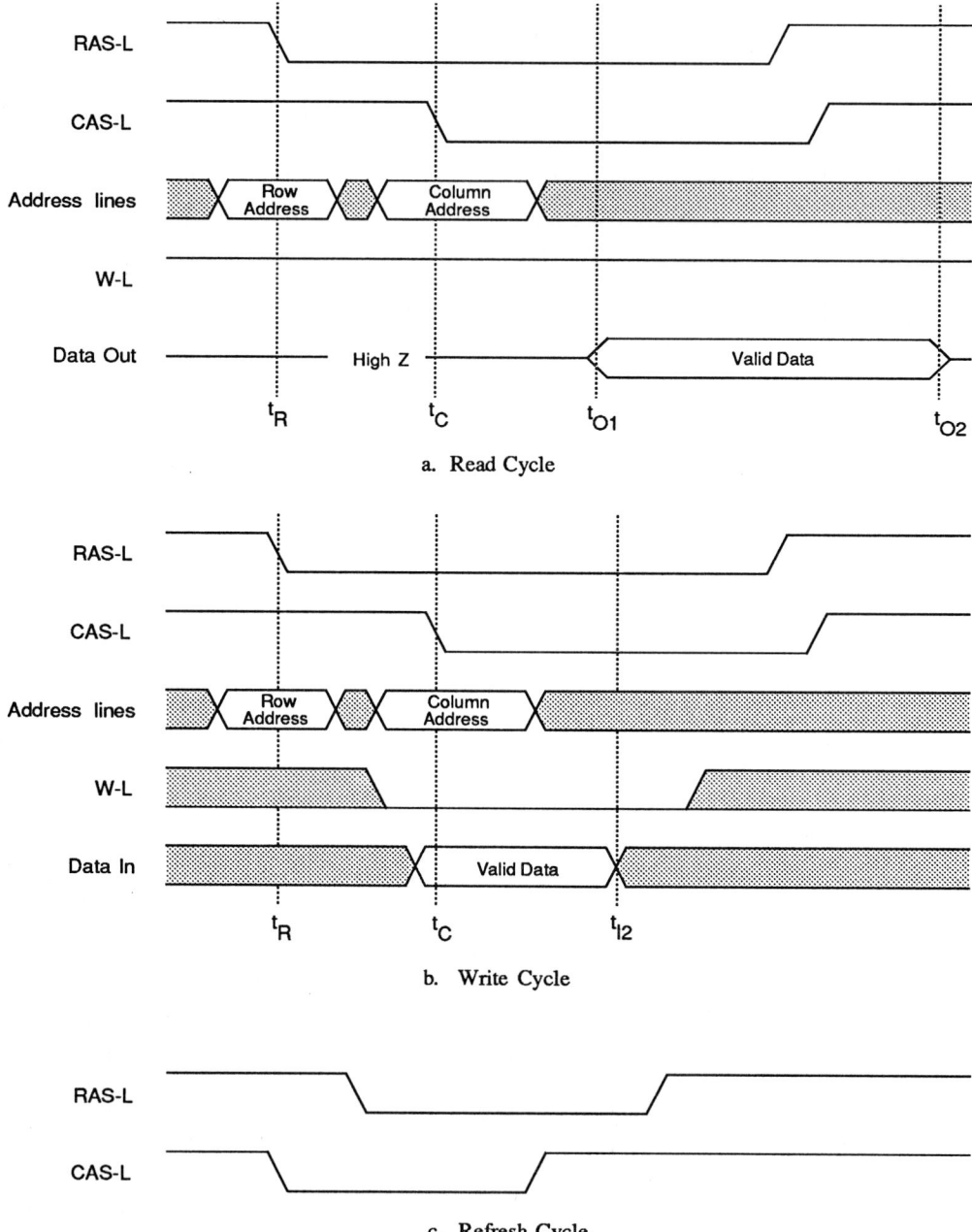

Figure 6.6. Timing Waveforms for Dynamic RAM.

CAS. At some later time, which is a characteristic of the memory itself, the data lines become stable. This time is shown as t_{O1} in Figure 6.6. The data will remain stable until the RAS and CAS lines are deasserted, at which time the output lines return to the tri-state mode (time t_{O2} in the figure).

The write cycle is shown in Figure 6.6(b). The basic addressing mechanism is the same as the read timing: present the row address, assert RAS, change the address lines to the column address, and assert CAS. This time, however, some time before the assertion of CAS the write line must be asserted, and it must remain asserted for a specific amount of time after the assertion of CAS. In our design, however, we will assert the write line for most of the cycle, so this constraint will easily be met. Some time prior to the assertion of CAS the data lines must be driven to a stable state by the circuitry providing the information to be written to the memory. This data must be held on the data lines until the hold time specification of the memory has been met, which is shown as time t_{I2} in the figure. The cycle can now complete, needing only to be sure to meet the minimum assertion time requirements of the system.

The third cycle shown in Figure 6.6 is a refresh action which can be taken. The memory is refreshed by reading a location, since that action recharges the capacitors involved with the information storage. Thus, the action that is required is to periodically read each location. One way this can be done is for the user to provide the necessary addresses; another way is to use address registers which are maintained internal to the memory itself. It is this internal action which is utilized in Figure 6.6(c). This is called CAS before RAS refresh, and the action of asserting CAS first has the effect of incrementing the refresh address as well as using the refresh address internal to the chip for the cycle. Like the other modes, this memory mode places specific timing requirements on the signals. The CAS signal must be asserted a specific time before the RAS signal, and the RAS signal has a minimum assertion time which is a characteristic of the specific chip in question. When these requirements are met, then the memory will be refreshed in a regular way.

The design of this system proceeds just as the systems which have been demonstrated in previous chapters. First, we need a data path block diagram, and the very simple version shown in Figure 6.5 is sufficient for our needs. We will add a few features in the final product, but the basic design is shown in that figure. The information in the basic data path block diagram, combined with the timing information shown in Figure 6.6, can be used to identify the action which must be provided by the control lines. In a previous chapter, we used a state diagram to identify this action. In this case, we identify the work with the flow diagram of Figure 6.7. When the system is ready for a new cycle, one of two courses of action is followed. If a refresh cycle is needed, then the refresh action is taken. If a refresh cycle is not needed, then when a read or write cycle is needed, the read/write action is taken. The only difference, then, between a read and a write cycle, as far as the control system is concerned, is the assertion of the write enable lines on the memory. The timing is identical (in this implementation) for both cycles.

The refresh action begins by asserting CAS. Note that no external address is provided, since the CAS before RAS refresh method uses an internal address for the refresh action. After the assertion of CAS, the controller waits for 15 nsec before asserting RAS. These two lines (RAS and CAS) are the only lines needed in the refresh action. For the specific memories used in this example, a minimum RAS time of 85 nsec is required; hence, the action indicated in Figure 6.7 calls for a wait of 90 nsec. When the time requirement has been met, then the RAS and CAS signals

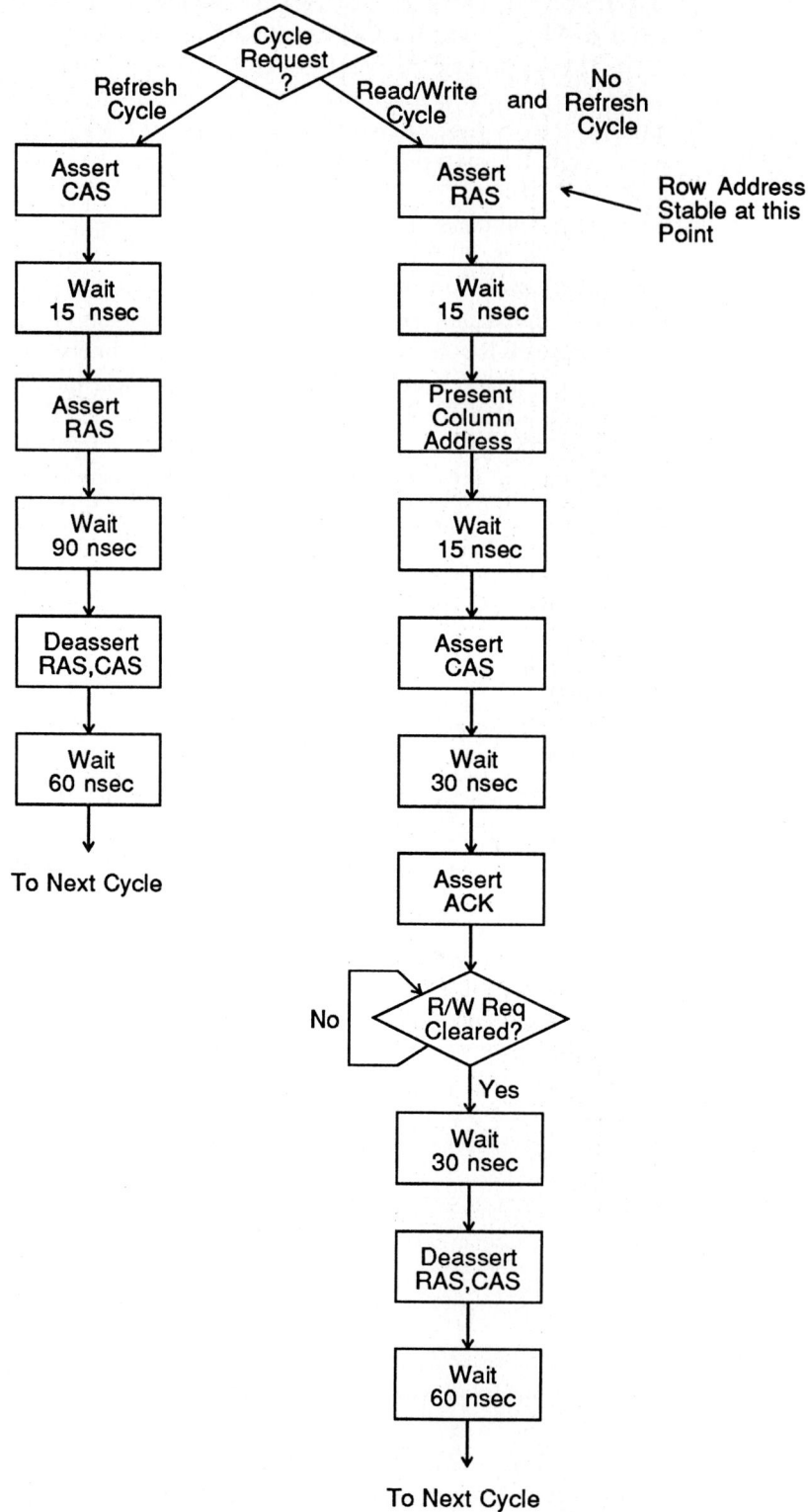

Figure 6.7. Control Flow Diagram for 16 MByte Memory.

are deasserted, followed by an additional wait of 60 nsec. This last wait is to allow the necessary precharge of devices internal to the DRAM, and the timing requirement is identified in the specifications for the memory device.

A read/write action is initiated by assertion of RAS. Due to the nature of the cycle and the address paths as established, the row address will have been stable the required period prior to this time. After the assertion of RAS, the system waits 15 nsec and changes the address for the column address. This is to meet the hold time requirements for RAS. Many DRAMs have a specification which indicates that the column address can become stable at the same time that the CAS signal is asserted, but since different path lengths are used for the two signals, this can be difficult to accomplish. Hence, in Figure 6.7 the address is changed to the column address, and a delay time is inserted to allow the address to stabilize. Then CAS is asserted, and an additional 30 nsec delay is specified before the acknowledge line (ACK) is asserted. This is to allow time for the minimum access time from the assertion of CAS and RAS. Once ACK is asserted, the system waits for the request to go away, which is indicated by the release of READ_WRITE. When the request has been removed, then a 30 nsec delay is followed by the deassertion of RAS and CAS, and finally a 60 nsec delay before allowing the next cycle.

Each different DRAM will have its own specification of parameters for timing. In the timings shown in Figure 6.7 and described above, all timings are multiples of 15 nsec. This was done purposely to permit ease of control design. More complicated designs can match timings more exactly, with a corresponding increase in the complexity of the control system.

A set of logic diagrams for the 16 MByte memory is shown in Figure 6.8. The memory elements themselves and some of the needed buffers are shown in Figure 6.8(a). The principal element in the figure is a symbol representing a megabyte of memory:

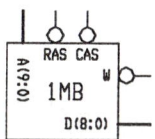

This symbol indicates that a memory module requires 10 address lines as input, as well as RAS and CAS and the write enable. Each memory module also has 9 lines for data [D(8:0)]. The 9 lines are for 8 bits of data and a single parity bit. With this arrangement the system shown in Figure 6.8(a) requires 144 separate memory chips. One way of reducing the area required by a memory of this nature is to package the memories in single-in-line packages (SIPs), with 9 memories to a SIP for a data byte plus parity. Thus, the memories shown in the figure can be packaged in 16 SIPs.

There are a number of buffer elements also shown in the figure. Each group of 4 MBytes of memory has a buffer in the address lines. This buffer needs to be capable of driving the address lines of 36 separate memory devices. Although the current drive required for this is not excessive, the combination of input capacitance and trace capacitance can cause severe problems with the address and control lines, and hence care must be taken to provide a sufficient drive capability to handle the situation. Various manufacturers provide drivers with high current capability, as well as trapezoidal output drive characteristics to reduce the under- and overshoot problems associated with high capacitance loads.

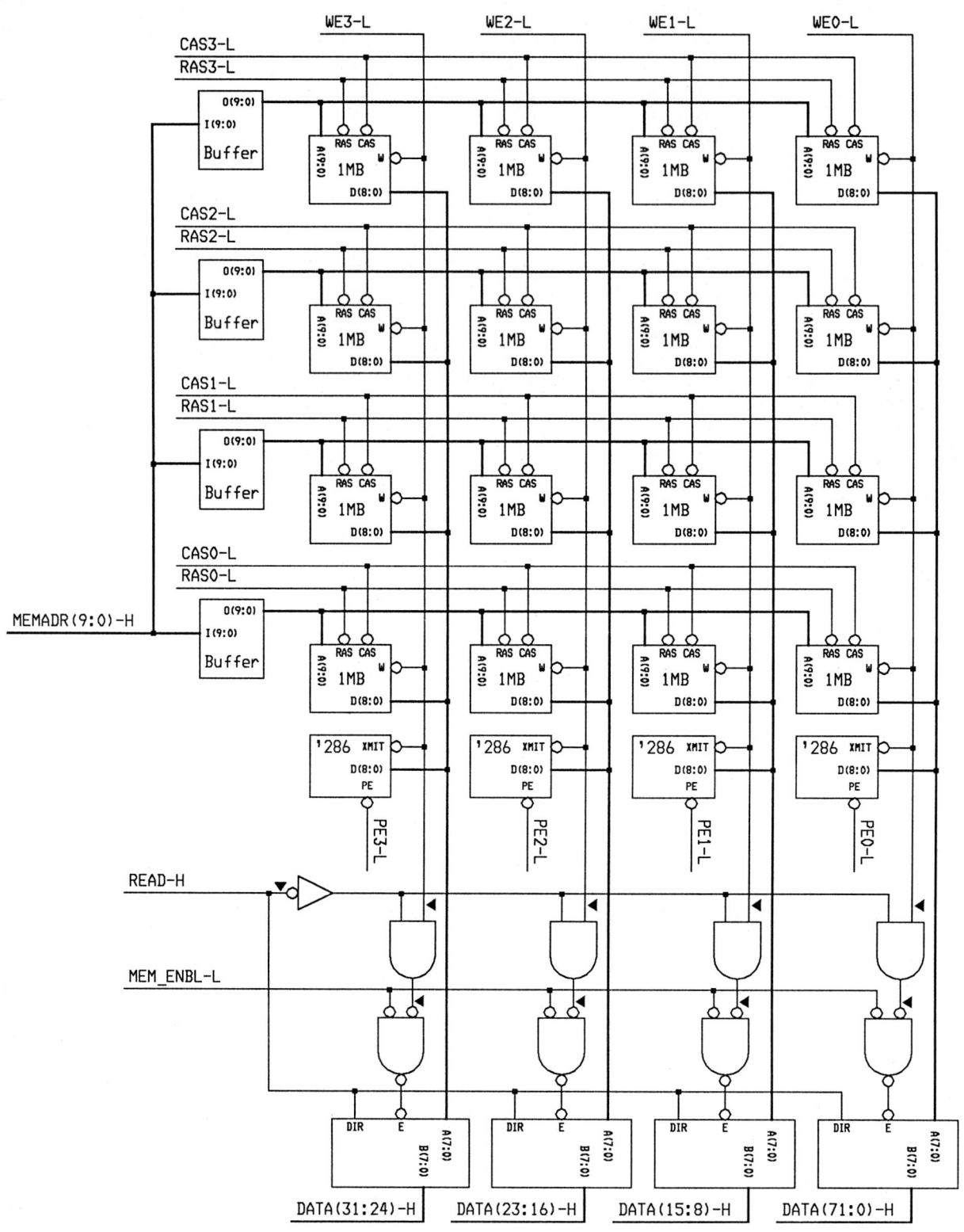

Figure 6.8(a). 16 MByte Memory: Memory Array and Buffers.

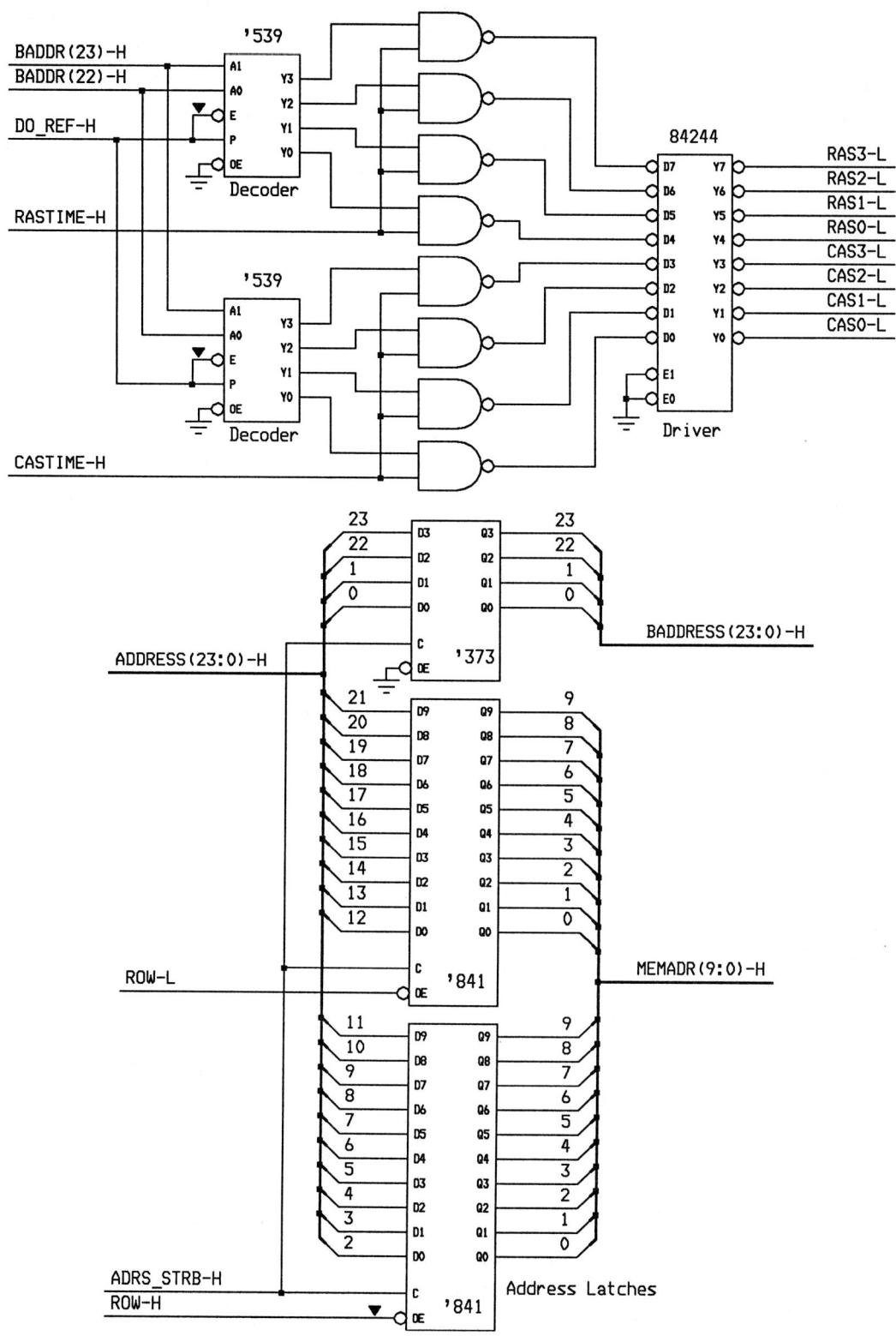

Figure 6.8(b). 16 MByte Memory: Address Buffers and RAS and CAS Logic.

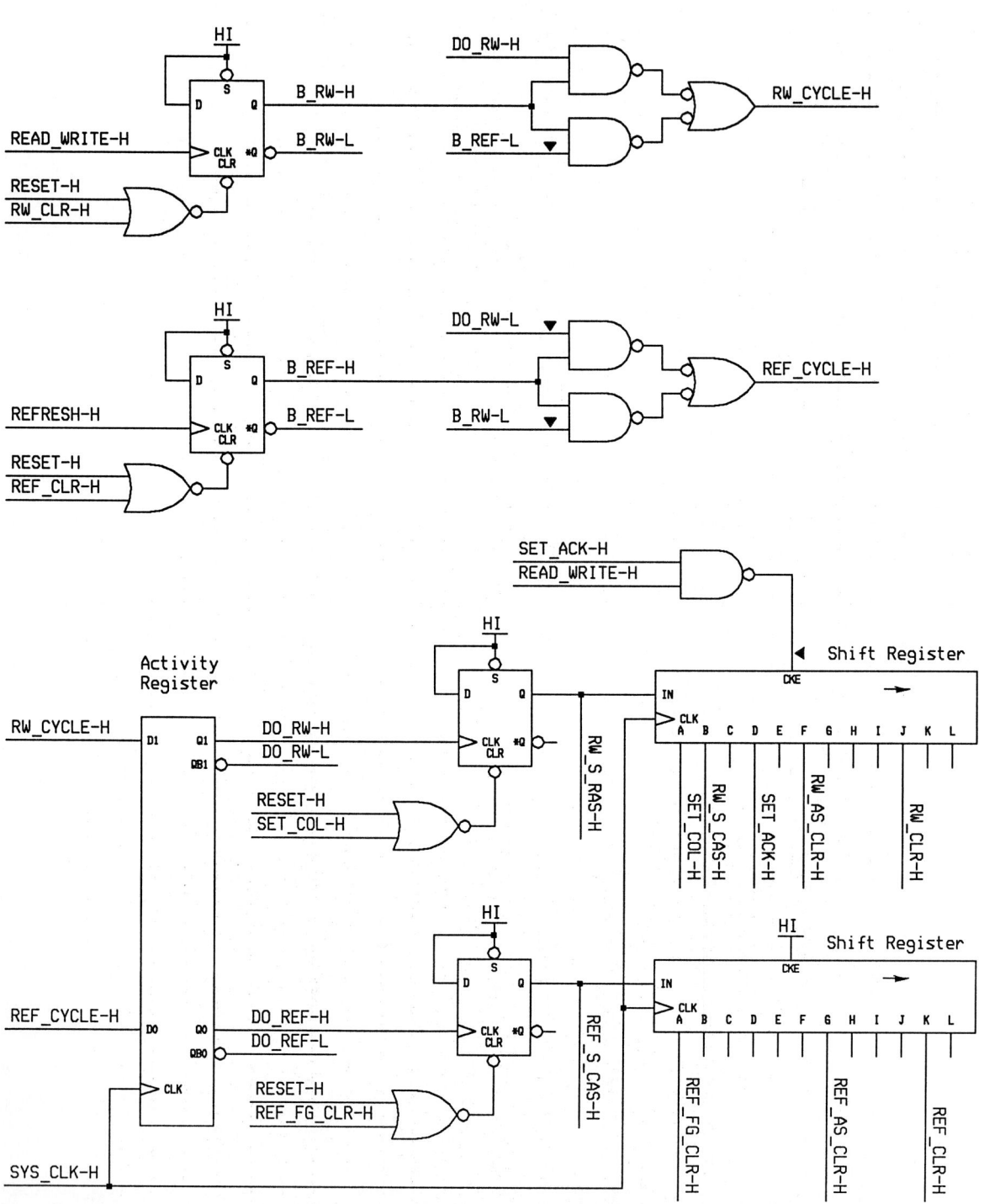

Figure 6.8(c). 16 MByte Memory: Control Logic.

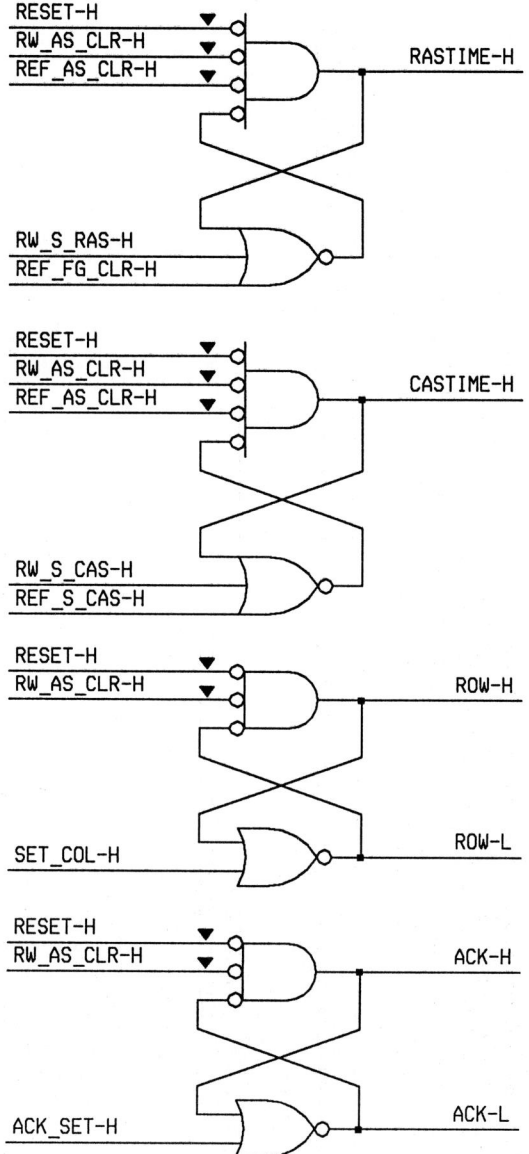

Figure 6.8(d). 16 MByte Memory: Generation of Timing Signals.

Also shown in the figure are transceivers on the data lines, one transceiver for each byte. The transceivers are individually enabled, although the direction line is common to all four devices. Individual write enable signals are provided, one for each byte position. It is assumed that these signals are provided in a timely fashion to meet the write enable setup and hold time requirements for the memories. The reader is invited to create a feasible gating system to generate the four write enables from the address lines and data strobes associated with an asynchronous bus. When the transaction is a read from memory, all the transceivers are enabled, and the direction of data flow is from the memory chips to the bus. For writes to memory, only the transceivers associated with the bytes involved in the transaction are enabled, and the direction of data flow is from the bus to the memory devices. The reason is that, when RAS and CAS are asserted in the manner that they will be for both read and write, the activity of the I/O lines of the memory chips will be determined by the level on the write enable line. If it is not asserted, then information stored at the location identified by the address presented to the device will be asserted on the I/O lines. Hence, the transceiver of a device not involved in the write transaction must be disabled, so that the memory and the transceiver will not both attempt to assert the same wire at the same time.

The other facility included in Figure 6.8(a) is the use of parity to check for errors in the memory. The capability shown is merely a parity bit for each byte, generated on write and checked on read. If the transaction is a read and a parity error is detected, then the appropriate parity error line will be asserted (PE3-L, PE2-L, PE1-L, PE3-L). If error correction is desired as well as detection, then additional error bits and appropriate circuitry must be added. Double error detection and single error correction can be provided across 32 bits by using seven error bits. However, when a single byte is to be written in the 32-bit word, using a code for 32 bits, then the other 3 bytes of the word must be read from memory, a new set of 7 error bits generated, and the word with the new byte and a new set of error bits can be placed in the memory.

The logic shown in Figure 6.8(b) includes the buffers needed to hold the address [ADDRESS(23:0)-H] for the transaction. These latches accept the information on the address lines when the address strobe is asserted (ADRS_STRB-H). The address has been grouped into three parts, corresponding to three different latching devices. The first device holds 4 bits, and the outputs are called BADDRESS(23:0)-H, or buffered address lines. These 4 bits include the two most significant and the two least significant bits. The two least significant are not involved in the selection of the word in memory, but can be used to identify a specific byte within a word, and hence are needed in some operations. The two most significant bits are used to select the appropriate 4 MByte block. The remaining 20 bits are used to select one of 2^{20} words. These bits are divided into a 10-bit row address and a 10-bit column address, and are alternately enabled onto the 10 address lines common to all DRAMs [MEMADDR(9:0)-H]. When the ROW-L signal is asserted, the row address is provide (this is the normal condition), and when it is not asserted, the column address is provided.

Also included with Figure 6.8(b) is the circuitry for the generation of the row address strobes (RAS3-L, RAS2-L, RAS1-L, RAS0-L) and the column address strobes (CAS3-L, CAS2-L, CAS1-L, CAS0-L). These signals are provided by a high current driver, an 84244. This drive capability is needed to drive the control lines of the memory in a timely fashion. The appropriate RAS signal is asserted whenever the signal RASTIME-H is asserted. Likewise, the appropriate CAS signal is asserted when the signal CASTIME-H is asserted. The other signal shown in the figure is the DO_REF signal, which is asserted during a refresh cycle. This has the effect of forcing all outputs of the decoder ('539) high, so that when the RAS and CAS generation signals are asserted all RAS lines and all CAS lines are asserted, instead of just one.

The generation of the refresh, RAS, and CAS signals is shown in Figures 6.8(c) and (d). The very sequential nature of the events needed in the DRAM cycle suggests a timing chain method to the controller design, and this assumption was used in the creation of the timing specification of Figure 6.7; the timings shown in these figures occur at multiples of 15 nsec.

One of the input signals for the controller is REFRESH-H, which is derived from an external timer circuit. The memory specification indicates that we must refresh each location every 8 msec, which can be accomplished by doing a refresh cycle every 15.625 μsec. This assumes that two rows are refreshed simultaneously, a fairly common feature for DRAMs. When REFRESH is asserted, a flag is set, which is used by the controller. Whenever a refresh flag is available, the next transaction which will be performed is a refresh cycle, which will occur either immediately or when the current read/write cycle has been completed.

The second input signal is READ_WRITE, which is derived from the control lines of the asynchronous bus. When the request line is asserted, a transaction is ready to commence. Thus, this signal is provided as input to the clock of a D-type flip-flop set up as a flag. When the flag is set (READ_WRITE), the bus is requesting either a read or a write transaction with the memory.

The gating which follows the flags is used to allow assertion of only one of the RW_CYCLE and REF_CYCLE signals. These signals are then directed to the Activity Register. One assumption which has been made here is that READ_WRITE and REFRESH occur synchronously with SYS_CLK, and in a timely fashion so that there are no race conditions between the generation of RW_CYCLE and REF_CYCLE and the system clock. If this is not a valid assumption, then some kind of synchronization register is required before the signals can be utilized.

The activity register identifies the start of a read/write cycle with the assertion of DO_RW, or the start of a refresh cycle with the assertion of DO_REF. The assertion of DO_REF sets a flag which is used, in turn, to begin the assertion of CAS (with the REF_S_CAS-H signal). This signal is presented to a shift register, and at the next occurrence of the system clock, a "1" is clocked into the first position of the shift register. This action asserts REF_FG_CLR, which clears the REF_S_CAS signal, and presents a "0" to the serial input of the shift register for the next occurrence of the system clock. The REF_FG_CLR signal is also used to begin the assertion of RAS. Six clock cycles later, or 90 nsec later with a 15 nsec clock time on the system clock, the signal REF_AS_CLR-H is asserted, which causes the deassertion of both RAS and CAS. Then, four clock cycles or 60 nsec later, REF_CLR-H is asserted to clear the refresh flag (B_REF-H), which action allows a read/write cycle to proceed when necessary. Note that this timing follows exactly the flow chart shown in Figure 6.7.

When a read/write cycle is started, the DO_RW signal in the Activity Register will be started. This sets a flag flip-flop asserting the signal RW_S_RAS-H. This causes the assertion of the RAS signal, and also sets up a condition such that a "1" will be presented to the serial input of the shift register for the next system clock. When the next clock cycle does occur, the SET_COL-H signal will be asserted. This clears the RW_S_RAS flag and also disables the row address and enables the column address. This is followed 15 nsec later by the assertion of RW_S_CAS-H, which asserts CAS. Two clock cycles later, or 30 nsec, the acknowledge line is set (SET_ACK). The assertion of this signal also disables the shift action of the register, since both READ_WRITE and RW_S_CAS will be asserted. Thus, the register will not shift any further until the master removes the read/write request (deassertion of READ_WRITE-H), at which point the enable line of the shift register will again be high, and the shift action will return. When this happens, RW_AS_CLR will follow by 30 nsec, and 60 nsec after that RW_CLR-H will be asserted to clear the read/write request cycle.

Figure 6.8(d) contains four set-reset flip-flops created from NOR gates. These flip-flops are used to create the actual signals used in the logic of Figures 6.8(a) and (b), as well as the acknowledge signal. The logic shown in Figure 6.8 is not complete for a bus system, and the reader is encouraged to add the necessary gating and data paths to match a specific bus system.

The memory presented in this section follows the method described in previous chapters for creation of digital systems. The basic problem is defined, and the requirements of the system are ascertained. The characteristics of the component parts are also identified, and a design strategy selected. The data path block diagram is developed, and the timing required of the system is specified. The timing specification in this case was indicated in a flow diagram, rather than a state diagram. Then the logic design proceeds, mapping the requirements of the data path and the timing specification into actual hardware.

6.3. Dynamic Memory System with Cache

For our last design in this chapter we will modify the design of the previous section to include a cache system. This will be a direct mapped, write-through, 128 KByte cache, which minimizes some of the design problems. (For a more thorough treatment of caches in general, see [Poll90].) The basic idea is that the information will reside in the cache memory most of the time, and this memory will be constructed to be as fast as possible, hence allowing the processor to execute at speed as much as possible.

The data path block diagram for a memory system using this simple cache is shown in Figure 6.9. The processor accesses memory in the normal manner, and the information can either come from the cache memory or the dynamic RAM. If the information is available in the cache, then is is presented quickly, and the processor continues. If the information is not in the cache, then it must be obtained from the dynamic RAM, which takes longer.

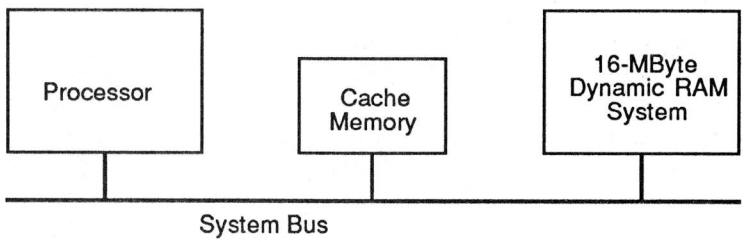

Figure 6.9. Memory System with Cache.

The direct mapping characteristic means that if the information is in cache, there is only one location in which that information can be found. The cache is organized in small quantities called lines, and in this design a line is 16 bytes. Other line sizes are possible, but this system will make use of four word transfers to fill a line. This can be a useful mechanism, as we shall see. Thus, we will take the 24-bit address and break it up as follows: 4 LSBs identify location within a line, next 13 bits identify line (there are 8192 lines), 7 MSB form tag, which identifies which of the 128 possible lines is actually in the cache. This is for a 24-bit address; to expand the system to a 32-bit address would require more tag bits.

The write through characteristic of the cache indicates that all writes will write directly to main store. If the address also happens to be in the cache, then the cache will be updated as well.

The block diagram of Figure 6.9 identifies how the system will function in a general sense, but there is not enough information to identify the signals which need to be asserted. A data path block diagram of the cache only is shown in Figure 6.10. In addition to the block diagram of Figure 6.10, the signals associated with the dynamic RAM of the previous section must also be considered. The system in this design consists of a data section which is 128K × 32 bits, and a tag system which is 8K × 8 bits. Since there are four 16-bit quantities per line, one tag is matched with four locations in the cache RAM.

The data lines of the cache RAM are connected to the data bus of the system; this path is used to fill the cache as well as to remove data from the cache. The address to the cache RAM is obtained from ADR(16:2), which is 15 bits, while above we stated that 13 bits would be used to identify a location in the cache. Indeed, 13 bits identify the line within the cache, and the next bits are used to identify in which 4-byte location within the line the desired byte address can be found. Thus, the two LSBs supply a special function, and one of the blocks in the diagram of Figure 6.10 is labeled "LSB Control." This block contains a counter and a multiplexer. During normal operation, the two LSBs are provided through the multiplexer from the address. However, when a line of the cache is being loaded from the DRAM, the two LSBs are provided from a counter which increments the address through four locations. The address available to the cache on the 2 LSBs is the address supplied by the system unless the signal CNTR-H is asserted, at which time the address is supplied by the counter portion of the unit. The counter is loaded to agree with the address provided by the CTR_LD signal.

The remaining control lines of the cache data area are those which are expected in a RAM system. Data is written into the cache when CACHE_WE-L is asserted, the actual RAM chip is enabled by CACHE_CE-L, and data is output from the RAM chip when CACHE_OE-L is asserted. Similar signals are used in the tag portion of the cache system. The tag RAM is a special device containing both a memory and a comparison system. The information stored at the address identified on the address lines is compared to the information present on the data lines, and if they match, MATCH-H is asserted. The lines in Figure 6.10 which are connected to the data lines of the tag system are the seven MSB of the address (assuming a 24-bit address). What should also be shown, but which is assumed in the figure since it is a constant condition, is one line which is tied high. The purpose for this is to provide a mechanism for initialization. The assertion of RESET-H will clear the contents of the tag RAM, so regardless of the address provided on the input, a match will not occur until a value has been placed in the tag, since "0" (the forced-clear value) will not match the "1" which is present on the data input. This essentially provides an identification of empty locations in the cache. The address lines used in the tag RAM are the 13 address lines which identify the line in question. Finally,

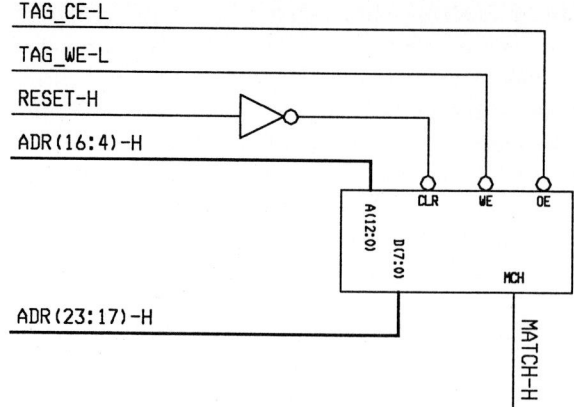

TAG_CE-L

TAG_WE-L

RESET-H

ADR(16:4)-H

ADR(23:17)-H

Tag RAM contains current line
identification. One tag for each
line in cache; tag RAM is 8K x 8

MATCH-H

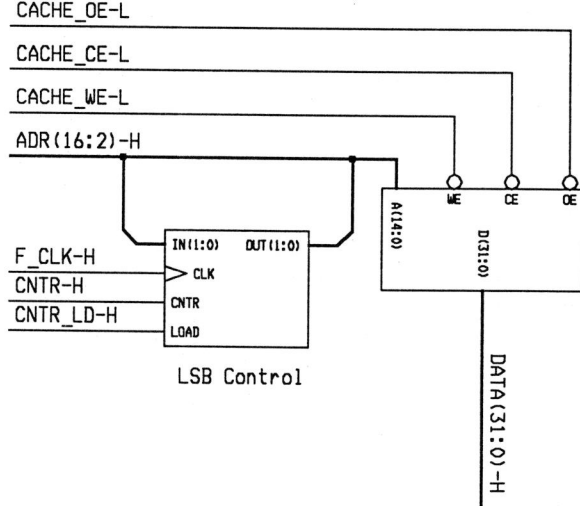

CACHE_OE-L

CACHE_CE-L

CACHE_WE-L

ADR(16:2)-H

F_CLK-H

CNTR-H

CNTR_LD-H

LSB Control

Cache memory holds recent data.
Cache is 32 bits wide by 32Kwords.
Each line contains 4 words (16 bytes).

DATA(31:0)-H

Figure 6.10. Block Diagram of Cache System.

data is written (the address available on the data lines of the tag RAM) to the tag
memory when TAG_WE-L is asserted, and the function of the unit is enabled when
TAG_CE-L is asserted. What is now required is a mechanism for asserting these
signals, as well as the dynamic RAM signals, in a reasonable and regular fashion.

In the struggle to understand the function of the system, a designer can utilize
many tools. We have demonstrated the use of state diagrams to describe the desired
behavior of a sequential device. Another mechanism utilized often is a timing
diagram, or a set of timing diagrams. In the creation of these diagrams, a designer
must incorporate various pieces of information, characteristics of the devices being
used in the system. We will present four timing diagrams in this section, one for
each of the four different cache conditions: read-hit, read-miss, write-hit, and
write-miss. As with the dynamic system of the previous section, the activity will be
reported in increments of 15 nsec. Also, even though not all the signals are used in

each instance, all the signals are included in each timing diagram, so that no confusion should be present concerning signals which are not activated in a cycle.

The four timing diagrams are contained in Figure 6.11. The first signal included (Address) represents the set of address lines directed to both the dynamic RAM and the cache system. The second signal (Data) represents the data lines, which in this case are obtained from the cache system. The third signal (Read-H) represents the read line, which identifies whether the transaction is a read or a write transaction. The fourth signal (Request-H) is shown asserted high, and represents a request line associated with an asynchronous bus. This signal may be derived from the appropriate control lines of a bus system. The next signal (B_RW-H) is the buffered read/write signal of the dynamic RAM system of the previous section, and the signal below that (DO_RW-H) represents the initiation of a read/write cycle, also from the previous section. Next follow signals for the RAS time and CAS time, associated with the dynamic RAM system, as well as the Row-H signal identifying when the row address is presented to the dynamic RAM modules. The next three signals are the write enable, chip enable, and output enable of the cache system. This is followed by the match signal from the tag section of the cache, and the write enable (Tag_WE-L) of the tag section. Finally, the acknowledge line (ACK-H) identifies a handshake protocol activity to coordinate the cycle with a bus master.

The simplest cycle is the read-hit cycle, shown in Figure 6.11(a). In this cycle the address and read lines are asserted, followed by the assertion of Request-H. When this happens, the cache system is signaled to the fact that a read operation has been requested, so it asserts the necessary signals to determine if the information is in the cache. The chip enable and the output enable are asserted to place the data on the bus if it is there. The tag memory is checked, and MATCH-H indicates, by 22 nsec into the transaction, (this is a figure that comes from the specification of the cache tag RAM, and will vary from device to device) that indeed the line stored at the location identified by ADR(16:4) matches the current address. We are assuming that the actual RAM devices used here provide the data within 40 nsec, so the acknowledge is asserted at the time of the third system clock, or 45 nsec into the transaction.

The assertion of the request line causes the buffered read/write signal (B_RW-H) to be asserted, but the DO_RW-H signal will be asserted only if the system is not doing a refresh cycle. When the DO_RW signal is asserted, then the acknowledge signal can follow. Once the acknowledge signal has been asserted, the system will remain in that configuration until the request line is released. This starts the termination of the read transaction, releasing the data lines and the acknowledge signal.

The most interesting activity occurs when there is a read-miss, as shown in Figure 6.11(b). At this time, the processor (or other device on the bus) requests information, and the information is not in the cache. The transaction begins as before: the address becomes stable, the read line indicates that this is a read, and the request line is asserted. However, the tag of the line in the cache does not match the tag area of the address provided, as shown by the MATCH-H line at the time of the third clock cycle. Since that is the case, the cache output enable is deasserted, and a dynamic RAM cycle is initiated. The initiation is indicated by the assertion of the Rastime-H signal. This is followed by the deassertion of Row-H, and then the assertion of Castime-H. After this activity on RAS and CAS, data will become available on the data lines after the appropriate delay. This data must be present a certain amount of time prior to the active edge (the low-to-high transition) of the cache write enable line. Hence, at the time labeled 165 in Figure 6.11(b) one fourth of the line (one 32-bit word) is retrieved from the dynamic RAM and written into

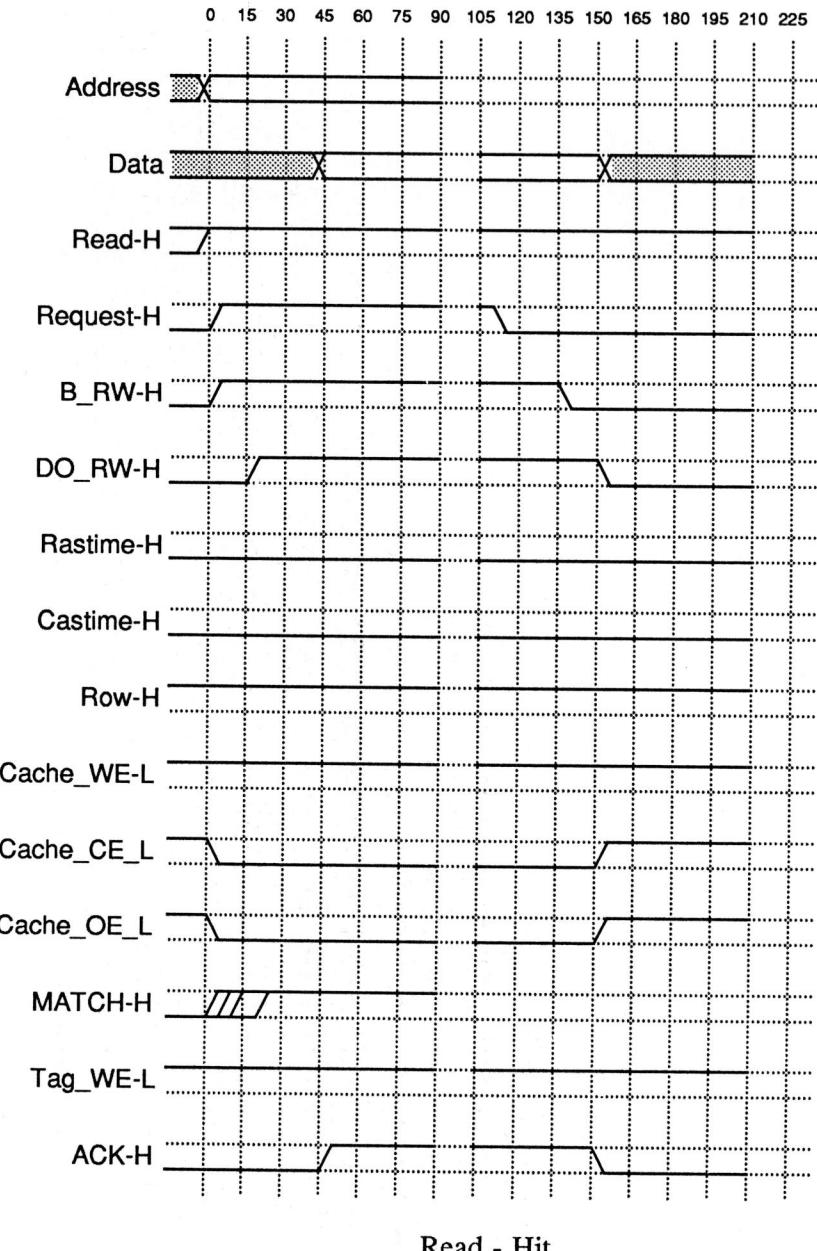

Figure 6.11(a). Timing Diagram for Memory System with Cache: Read-Hit.

Chap. 6: Design of Memory Systems

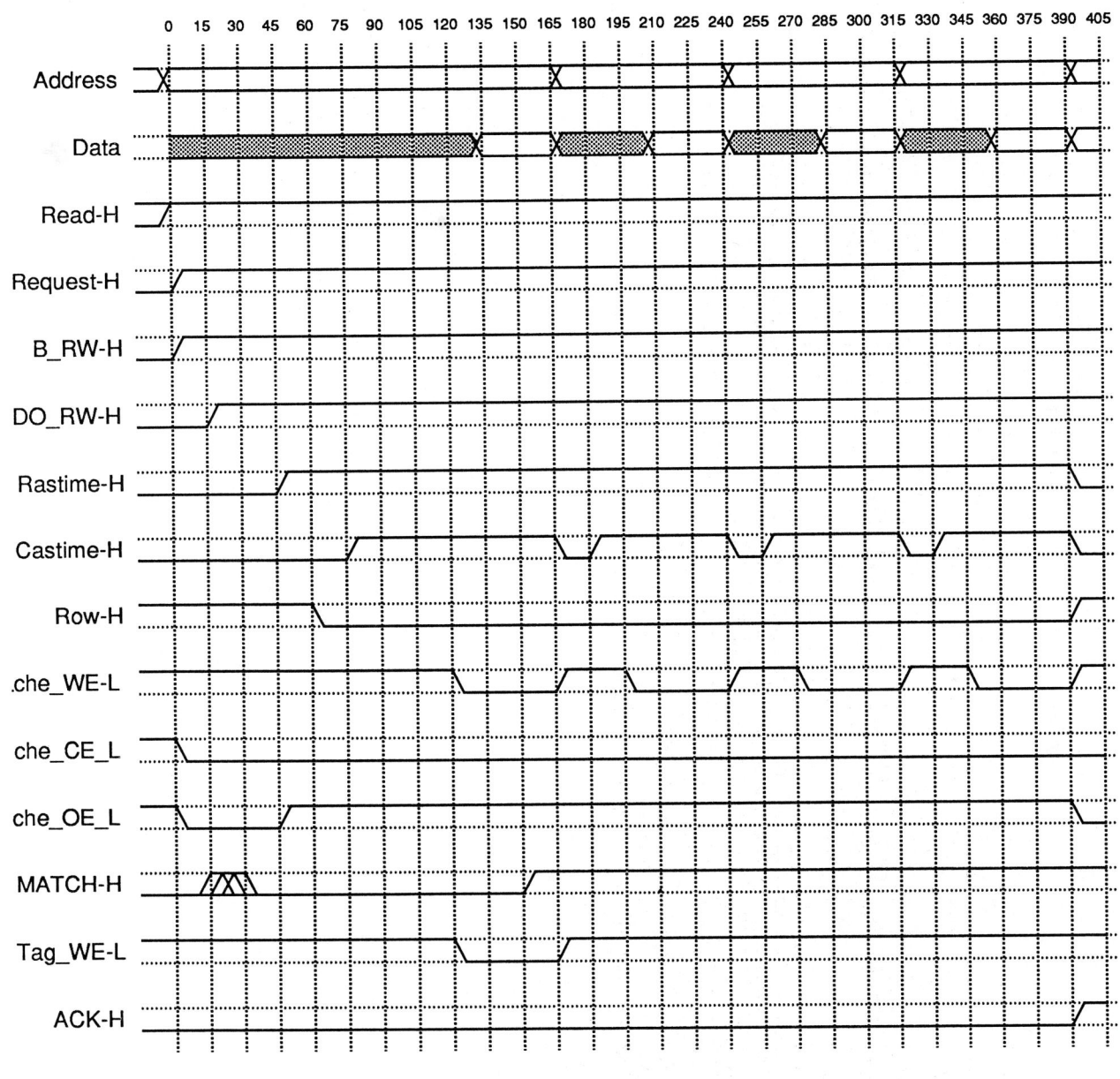

Figure 6.11(b). Timing Diagram for Memory System with Cache: Read-Miss.

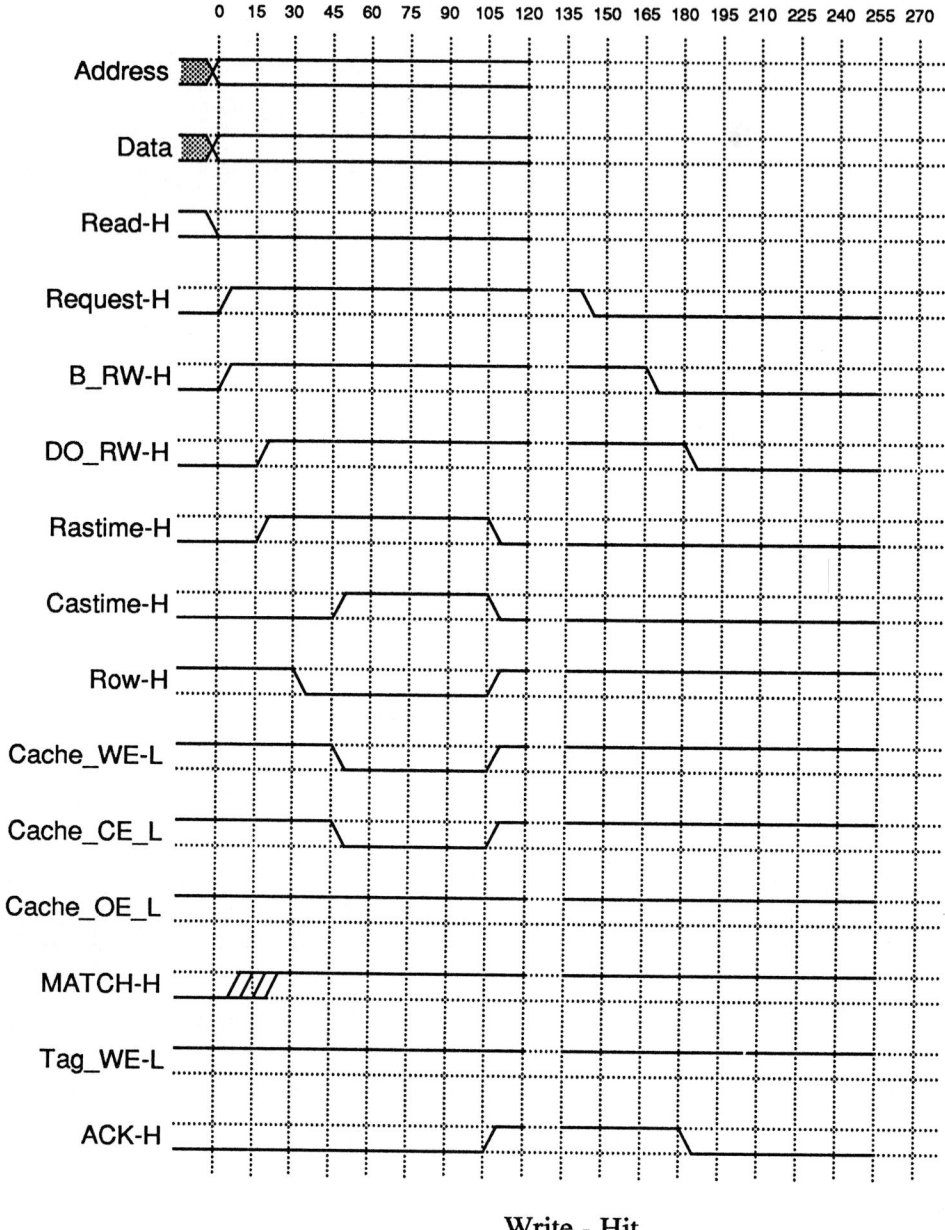

Figure 6.11(c). Timing Diagram for Memory System with Cache: Write-Hit.

Write - Hit

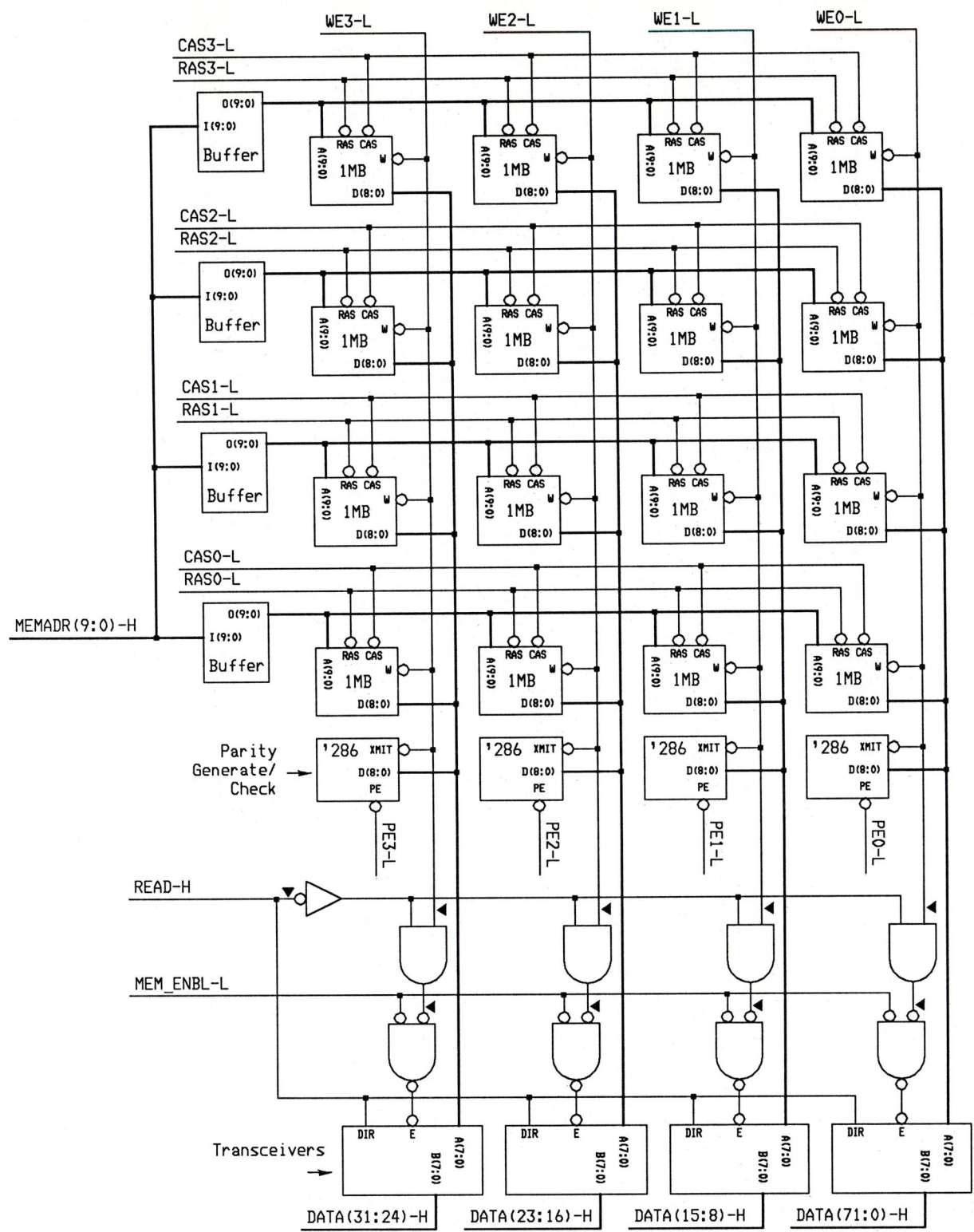

Figure 6.12(a). Memory System with Cache: Memory Array and Buffers.

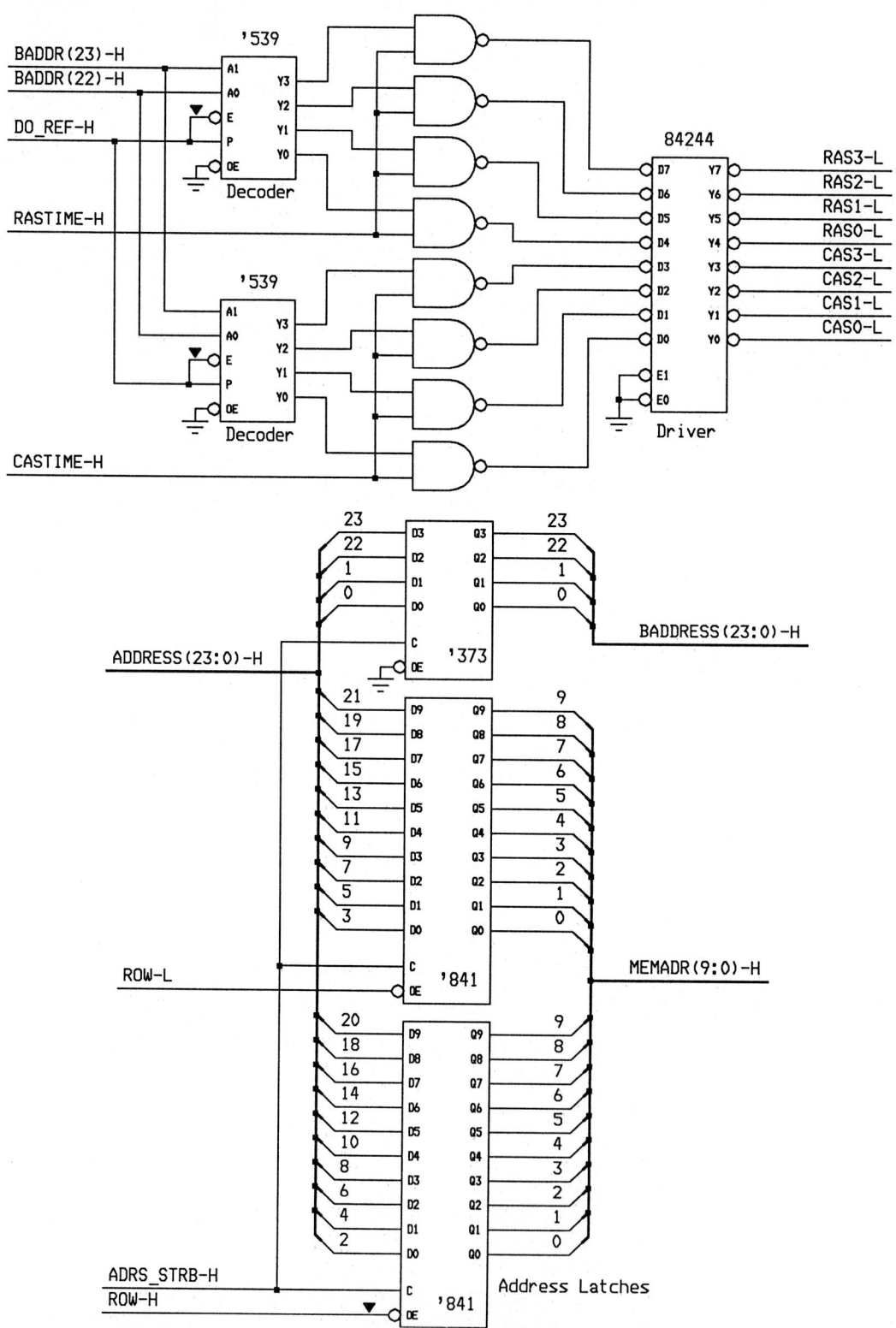

Figure 6.12(b). Memory System with Cache: Address Buffers and RAS and CAS Logic.

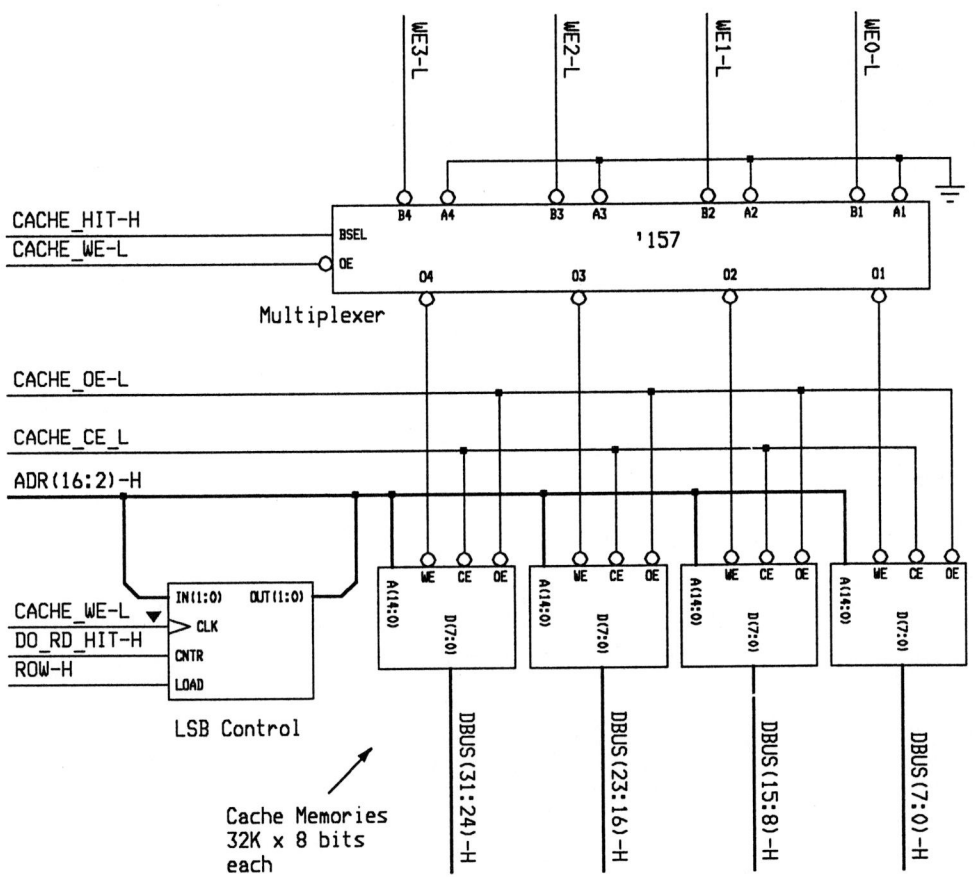

Figure 6.12(c). Memory System with Cache: Cache RAMs.

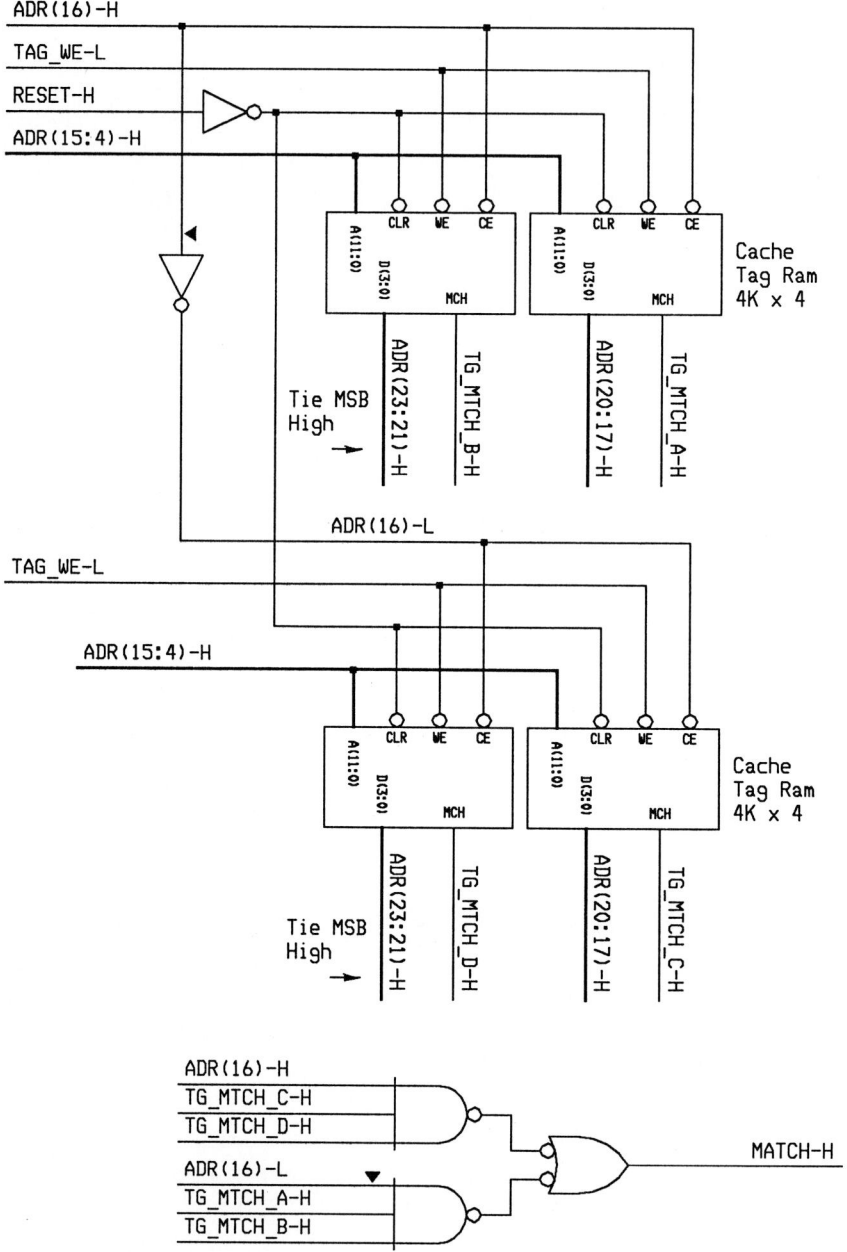

Figure 6.12(d). Memory System with Cache: Cache Tag RAMs.

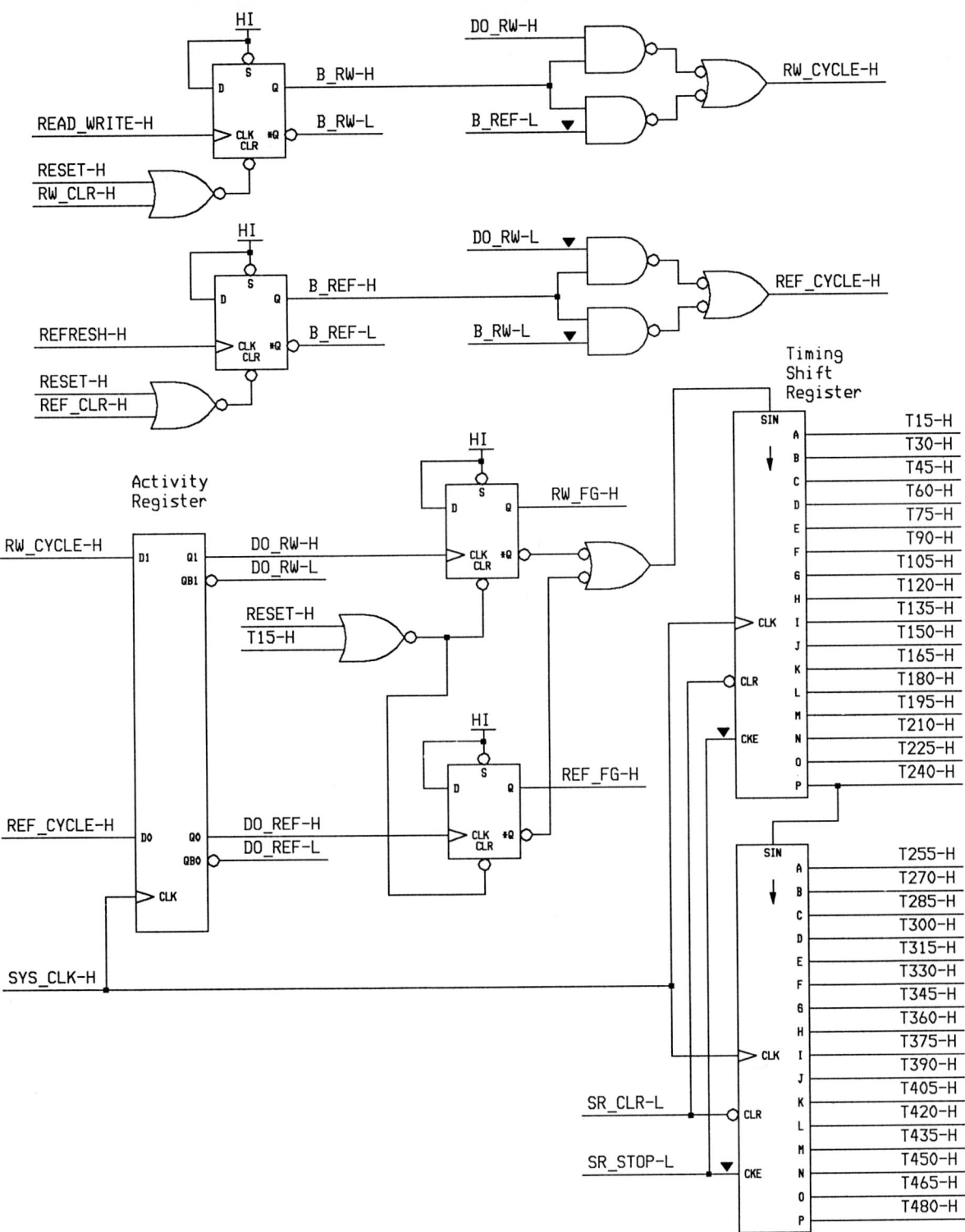

Figure 6.12(e). Memory System with Cache: Timing Chain Generation.

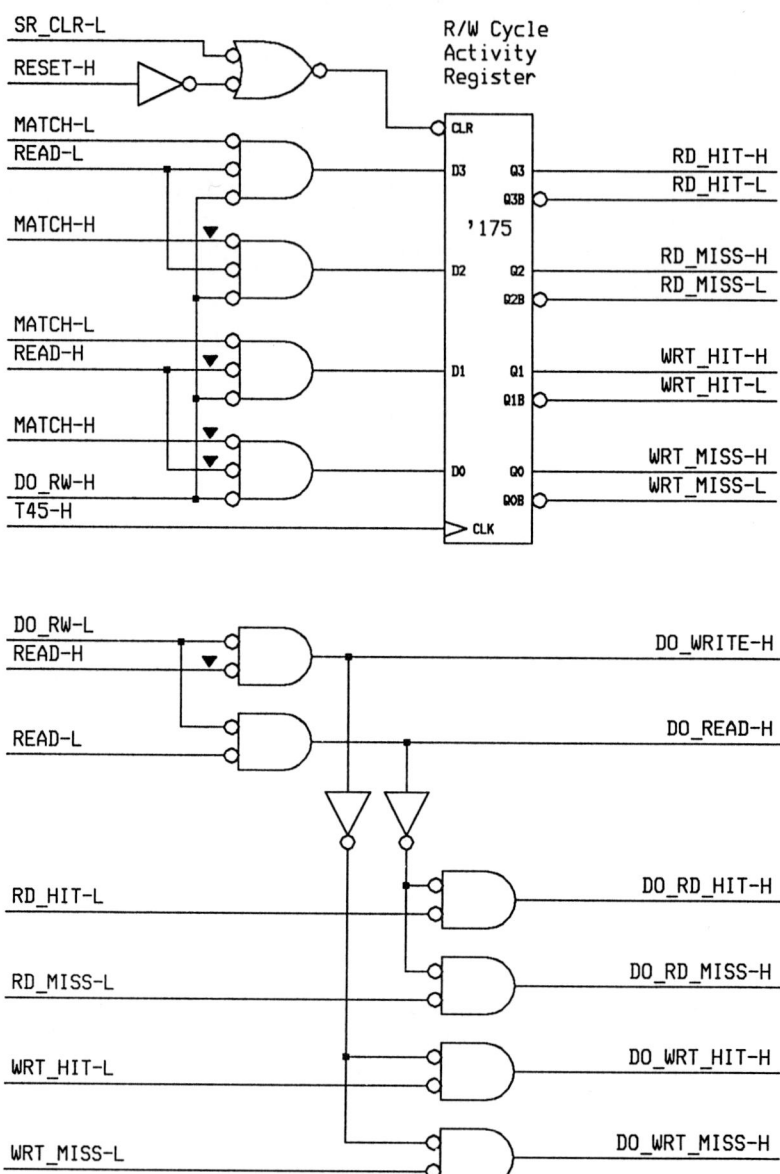

Figure 6.12(f). Memory System with Cache: R/W Cycle Activity Logic.

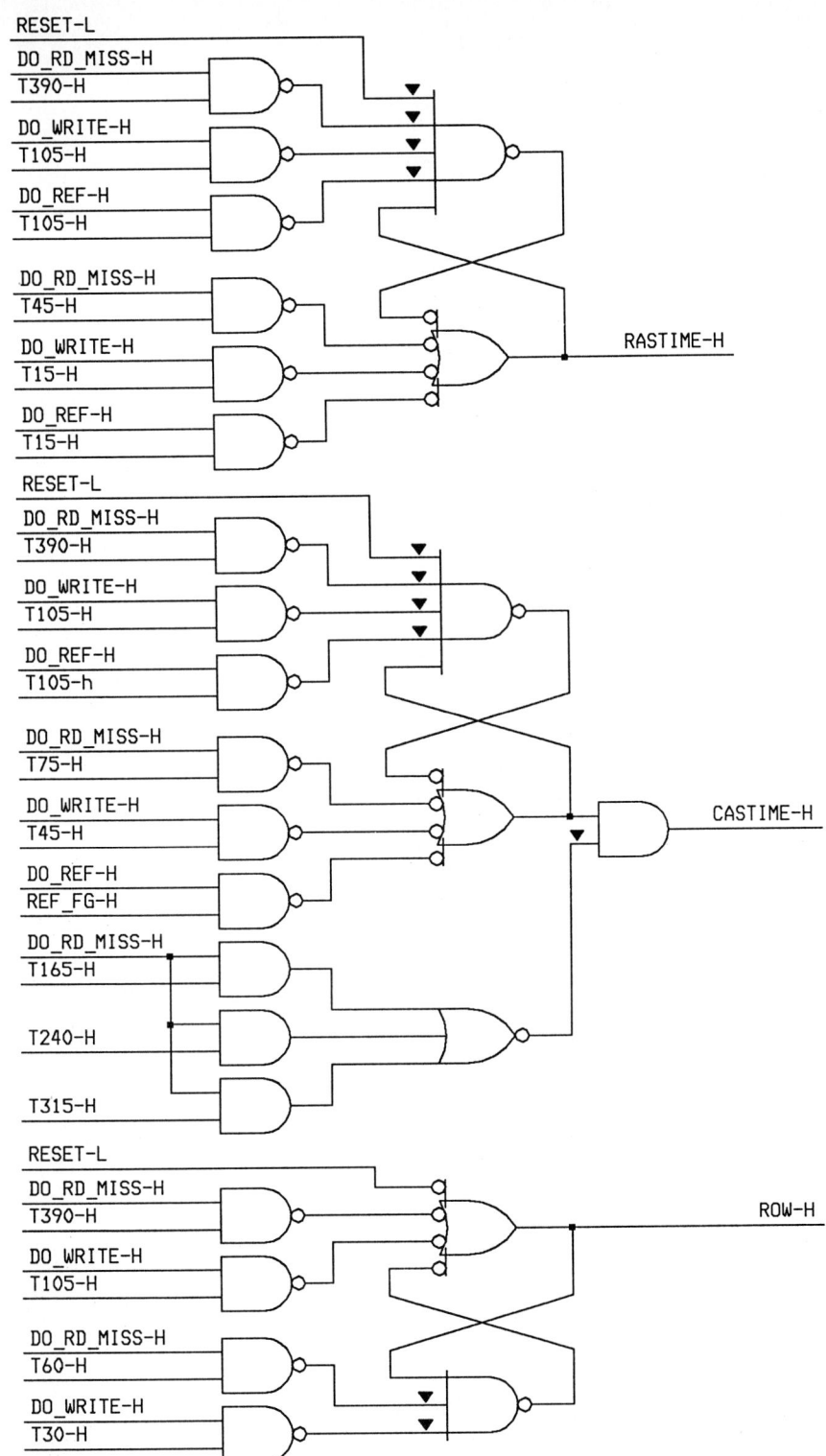

Figure 6.12(g). Memory System with Cache: Logic for DRAM Control.

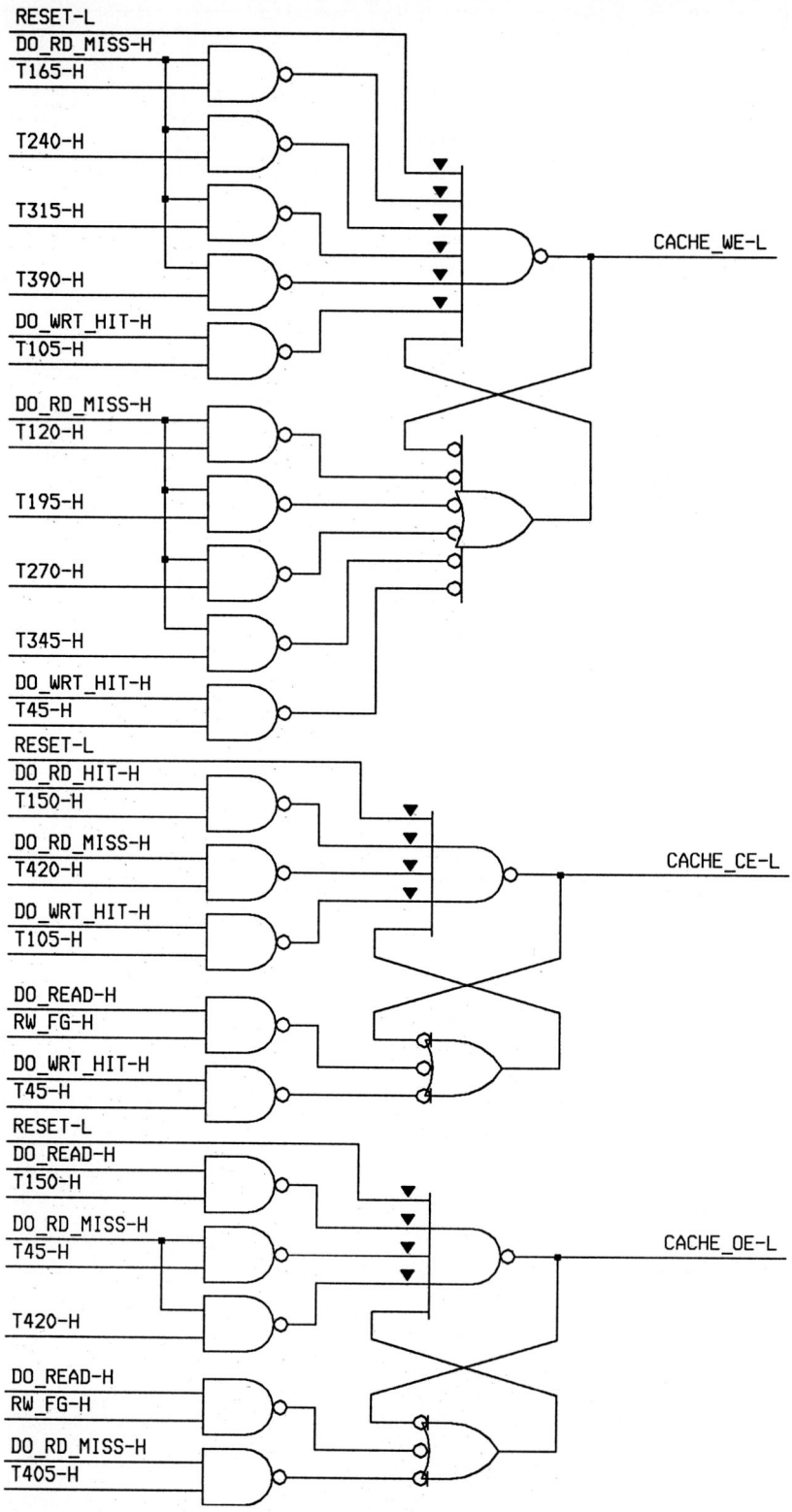

Figure 6.12(h). Memory System with Cache: Cache Control Signals.

Chap. 6: Design of Memory Systems

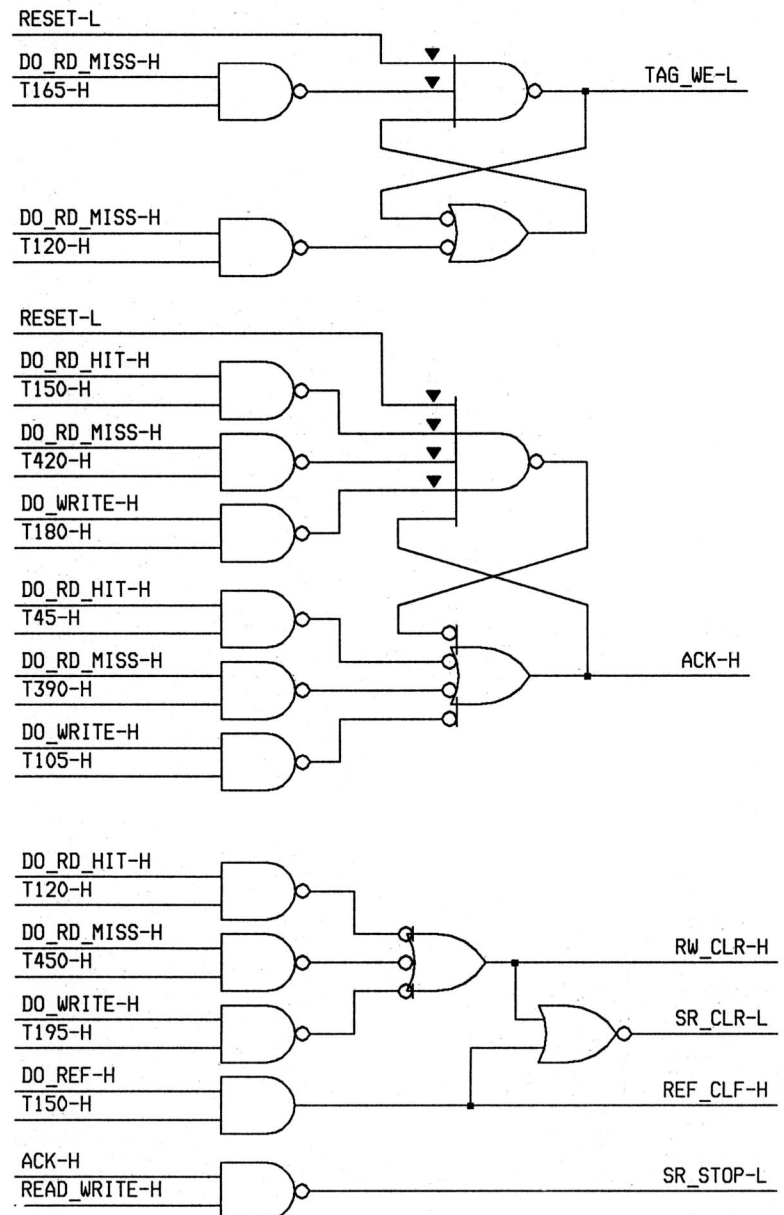

Figure 6.12(i). Memory System with Cache: Miscellaneous Timing Signals.

output enables are tied together, as are the chip enables, since those functions can be shared by all four memory devices. However, the write enables are separate, and are directed to the appropriate cache memory device through a multiplexer. When there is a cache hit, the write enable lines which should be utilized are provided by the system, via the WExx signals; hence only those lines are asserted. Otherwise, all four lines will be enabled at CACHE_WE time. Finally, there is a block labeled LSB Control. This unit is used to provide the two least significant bits of the address. For a cache hit, either read or write, these bits come from the address bus. For a cache miss, the two bits are provided by a counter, cycling through the four combinations to identify the correct four words as the bits are retrieved from the dynamic RAMs in nibble mode.

The tag memories are contained in Figure 6.12(d). The devices used for this design are 4K × 4 memories; hence, for 8K lines two sets are needed, and for 8 bits wide, a pair of devices is needed. Address line 16 is used to select the appropriate bank, as well as to enable the appropriate match gates. Note that if the match output of the tag RAMs can be wire-ORed together, then the gating itself will not be necessary. Information is written into the tag RAM when TAG_WE-L is asserted, as shown in the timing diagram of Figure 6.11(b).

The control circuitry for the system begins in Figure 6.12(e) with the generation of the timing circuitry. Much of the circuitry is exactly the same as for the corresponding function in Figure 6.8(c), since the function is the same. That is, the flags for starting a read/write cycle or a refresh cycle are the same, as is the circuitry to guarantee that only one of the cycles proceeds at a time. Also, the activity register performs the same function. The basic difference is the inclusion of a single, longer shift register which is used to generate the timing signals. When either a read/write cycle or a refresh cycle is initiated, a ''1'' is made available at the input of the serial shift registers, and with each system clock, the pulse moves down the shift register. The names of the signals identify when they occur within the total cycle, pulse T105-H occurring 105 nsec after the cycle started. The shift registers have three control lines. The first is the clock, which is connected to the system clock and activated every 15 nsec. The second is a clear line, which will be activated when the current cycle has completed its function, removing the pulse from further activity in the system. The third control line is an enable, allowing the pulse to stop when needed. Hence, when the acknowledge has been asserted and the system is waiting for the release of the request which initiated the action, the shift action of the system is disabled.

The logic included in Figure 6.12(f) identifies the type of cycle which is underway when a read/write cycle has been requested. If it is a read cycle, then DO_READ will be asserted; if it is a write cycle, then DO_WRITE will be asserted. Note that if the current cycle is a refresh cycle, then none of these signals will be asserted. After 45 nsec, the match condition can be tested; hence the signal T45-H is used to clock the condition into an activity register. Only one of the four outputs will be asserted, and the one which is asserted will determine the activity which will follow. This activity has been defined by the timing diagrams of Figure 6.11, and is implemented in the remaining logic diagrams.

The logic for the generation of RASTIME, CASTIME, and ROW is included in Figure 6.12(g). These signals are created by cross-coupled NAND gates, and upon reset all the signals are set in the unasserted condition. (Note that ROW is asserted, but this is the desired condition for this signal.) After initialization, the signals are set and reset by the appropriate pulses from the timing shift register, enabled by the current condition. Hence, on a read-miss cycle, RASTIME is asserted at time T45 and deasserted at time T390. On a write cycle, whether hit or miss, RASTIME is asserted at time T15 and deasserted at time T105. Finally, on a refresh cycle, RASTIME is asserted at time T15 and deasserted at time T105. The same type of logic applies to the creation of CASTIME. However, there is one gate beyond the flip-flop in this case. This gate is used to deassert CASTIME for the appropriate times on a read-miss condition, as shown in Figure 6.11(b).

The logic for the generation of the cache control signals is contained in the flip-flops of Figure 6.12(h). As before, these flip-flops are set and reset at the appropriate times by the timing signals from the shift register enabled by the appropriate condition cycle.

The only tag control signal which is timing dependent is the write enable, which is one of the control lines contained in Figure 6.12(i). The only time when the tag RAM is updated is during a read-miss condition, when a new line is being brought into the memory. Hence, on a read-miss, this signal is asserted at T120 and deasserted at T165. The acknowledge line is asserted at different times according to the type of cycle in progress, as shown by the timing diagrams of Figure 6.11. The acknowledge action of the timing diagrams is implemented by the flip-flop action of the logic in Figure 6.12(i). Finally, the clear and enable signals for the timing shift register are generated. The clearing of the B_RW and B_REF flip-flops of Figure 6.12(e) is done enough time after the release of RASTIME (when asserted) to guarantee the proper precharge time of the DRAMs. In any case, when one of these flags is cleared, timing bit is cleared out of the shift register in preparation for the next cycle. Otherwise, two timing pulses could be present in the shift register at the same time, and cause undesirable signal behavior. The final gates in the figure make sure that the shift register shifts until an acknowledge has been asserted and the corresponding release of the request has not been detected.

Having presented the memory system shown in Figure 6.12, some comments need to be made concerning the design. First of all, the design is not complete as shown. That is, before it could be made a real design, a specific bus would have to be chosen, and the interface with that bus completed. This would identify which control signals must be tri-state, and when they need to be enabled, and so on. The selection of a particular protocol would identify the correct assertion levels of the signals, as well as any specific control lines which are needed but not included in Figure 6.12.

An implementation of the design of Figure 6.12 must also consider the real effects of real devices in the system. The stray capacitances, as well as the actual input and output capacitances of the devices, must be dealt with in a reasonable manner, and any timing adjustments introduced as needed. Also, the current drive needed by the various components will vary with different memory devices and different technologies of implementation. All these real effects must be considered in the creation of an actual implementation.

Another problem with the design of Figure 6.12 is the timing aspect. That is, we have presented the design as if all the gates are perfect, with minimal delay time and maximum drive capability. Real gates will not have those characteristics, and so appropriate steps must be taken to be sure that the system will function. In a TTL implementation, FAST gates and functional modules should be fast enough to function, as should AS (advanced schottky) devices. The system could also be implemented in other technologies, or in a system internal to a semiconductor chip.

The gating systems presented in Figure 6.12 contain a number of repetitions which, when eliminated, would reduce the gate count. Also, more efficient gating systems can be devised which combine some of the functions in a number of ways. However, the timing chain mechanism which is demonstrated by the design can be a viable method for creating a digital system, and should not be discarded as a design technique.

One final note about the system concerns the implementation mechanism. Certainly, if all the gates shown in the logic diagrams are placed in the system using standard gates, the number of devices would be very large. However, if the system is implemented with a PLA or other programmable logic device, then the number of physical devices required can be kept to a minimum.

The memory design presented in this system demonstrates a method for adding a cache to a memory system. The design also demonstrates the shift register method for generating timing sensitive control signals. The method is not as efficient in some ways as a state machine approach to control design, but it can provide capabilities in control system design which are not available with other techniques.

6.4. Summary

Memory system designs must be created with careful attention to the details of the requirements as well as to the methodology of the overall approach. One- and two-dimensional techniques have been demonstrated in this chapter, and higher dimensions are possible by extending the ideas presented here. The method selected for implementation should be selected only after consideration of the implications of the design techniques. The one- and two-dimensional techniques can apply to memory systems created within a chip or to systems created by combining memory chips together.

In creating a memory system design, all the appropriate engineering criteria must be considered. We have presented some of the techniques which can be utilized to create systems, but designers must consider the use of critical system resources, and create a system which makes use of all system resources in reasonable ways. This is a very vague statement, but each memory system will be applied in a different set of circumstances, and the criteria for evaluation will vary accordingly.

Some memory systems, such as the 1-D and 2-D systems of the first section, can be created without control timing constraints. However, for large, dynamic RAM systems, the implementation of the memory system must include consideration of the size of the memory system as well as the timing of the control signals. This chapter presented control systems implemented with a shift-register approach to timing control, but any appropriate method for generation of the control signals could be utilized.

6.5. Exercises

6.1 Given that you have a 32K × 8 PROM memory with two low-true chip enables, design a 2-D memory system which will provide 2 megabytes of read-only memory to a 16-bit bus.

6.2 Given that you have a 64K × 8 static RAM memory with a low true output enable, chip enable, and write enable, design a 1 megabyte memory to interface to an 8 bit-bus.

6.3 The dynamic RAM memory of Section 6.2 utilized a serial-in, parallel-out shift register with a clock enable. Design such a shift register with 74xx74 ''D'' type flip-flops.

6.4 The dynamic RAM memory of Section 6.2 utilized a serial-in, parallel-out shift register with a clock enable. Design such a shift register by utilizing a 74F674 shift register device.

6.5 The dynamic RAM memory of Section 6.3 utilized a serial-in, parallel-out shift register with a clock enable and a direct clear. Design such a shift register using appropriate devices, either individual flip-flops or complex integrated circuits.

6.6 The dynamic RAM memory of Section 6.3 utilized a block labeled LSB Control in Figure 6.12(c). The three control lines for this system are a clock, a select line, and a load line. So long as the select line is high, the 2 output bits will be copies of the 2 input bits. When the select line is low, the 2 output bits will be provided by a 2-bit counter. The counter is filled whenever the load line is asserted (high), and will increment at the low-to-high edge of the clock. Design such a counter system.

6.7 Create a flow chart similar to Figure 6.7 which represents the timing specifications of Figure 6.11. Also include the refresh cycle timing.

6.8 Implement the control section of the dynamic RAM system of Section 6.2 with a state machine controller.

6.9 Implement the control section of the dynamic RAM system of Section 6.3 with a state machine controller.

6.10 Expand the addressing mechanisms of the RAM system of Section 6.3 to handle a 32-bit address.

6.11 Modify the write-through cache system of Section 6.3 to be a write-back cache system.

6.6. Additional References

[Baer80] Baer, J. L., *Computer Systems Architecture.* Rockville, MD: Computer Science Press, 1980.

[Bart85] Bartee, T. C., *Digital Computer Fundamentals* (6th ed.). New York: McGraw Hill Book Company, 1985.

[HaVr78] Hamacher, V. C., Z. G. Vranesic, and S. G. Zaky, *Computer Organization.* New York: McGraw Hill Book Company, 1984.

[Haye88] Hayes, J. P., *Computer Architecture and Organization,* (2nd ed.). New York:

McGraw Hill Book Company, 1988.

[Lang82] Langdon, G. G., Jr., *Computer Design.* San Jose, CA: Computeach Press Inc., 1982.

[LaFr89] Langholz, G., J. Francioni, and A. Kandel, *Elements of Computer Organization.* Englewood Cliffs, NJ: Prentice Hall, 1989.

[Poll90] Pollard, L. H., *Computer Design and Architecture.* Englewood Cliffs, NJ: Prentice Hall, 1990.

[Shiv85] Shiva, S. G., *Computer Design and Architecture.* Boston, MA: Little, Brown, 1985.

Appendix A

Boolean Minimization Program

Minimization of Boolean systems can be very cumbersome for a large number of variables. However, this process can be automated in many respects, and the program which is listed in this Appendix has been created to generate appropriate equations. The input to the program is a table with two columns. The first column contains a character for each input variable, and the second column is a character representing the desired output for that combination of input variables. The permissable characters are ''0,'' ''1,'' ''x,'' and ''X.'' The ''x'' and ''X'' representations are for the don't care condition.

The program has two arguments, one required and one optional. The first argument is required, and is the name of the file in which the input specification is found, although the standard input file can be specified by using a hyphen on the command line. The second argument is an optional string of characters to use to represent the input variables. Without this argument, the program assigns the letter ''a'' to the first column of the input specification, ''b'' to the second column, and so forth.

The output of the program is a Boolean equation in which the AND terms are implied and the OR terms are specified with a ''+'' sign. Complemented variables in the equation are indicated by a trailing apostrophe. For example, consider the system specified by the following input file:

```
0000xxxxx   0
00010xxxx   1
00011xxxx   0
001xxxxxx   1
011xx0xxx   1
011xx1xxx   0
100xxxx1x   x
100xx0x0x   x
100xx1x0x   x
101xxxxx0   x
101xxxxx1   x
110xxx0xx   x
110xxx1xx   x
111xxxxxx   x
```

The 14 lines form the specification for a Boolean system with nine inputs. As can be seen, don't care conditions can be included in both the input and output specifications. Using this file as an input to the program results in the following output:

```
9 bits provided.

   After selection of prime-implicants:
        b'de' + b'c + cf'
```

As mentioned above, the characters representing the various inputs can be changed if desired. This program was used to generate and check many of the equations used in Chapters 2 and 4, but no guarantee is made concerning its general operation.

```
/*
 * Copyright (C) 1988, 1989 by W. Tait Cyrus (cyrus@pprg.unm.edu)
 *
 * The author does not make any warranty expressed or implied, or assumes any
 * liability or responsibility for the use of this software.
 *
 * Should the program prove defective, you assume the cost of all necessary
 * servicing, repair or correction.
 *
 * In no event, unless required by applicable law, will the author and/or any
 * other party be liable to you for damages, including any lost profits, lost
 * monies, or other special, incidental or consequential damages arising out
 * of the use or inability to use (including but not limited to loss of data
 * or data being rendered inaccurate or losses sustained by third parties or
 * a failure of the program to operate with any other programs) this program,
 * even if you have been advised of the possibility of such damages, or for
 * any claim by any other party.
 *
 * This software or its use shall not be: sold, rented, leased, traded, or
 * otherwise marketed without the expressed written permission of the author.
 *
 * All rights not granted by this notice are reserved.
 */

/*
 * Note: as of September 1989, a copy of this program can be obtained via
 *       anonymous ftp from pprg.unm.edu (/pub/misc/find_boolean_primes.c).
 *       No guarantee, though, on how long this program will continue
 *       to be made available.
 */

/*
 * This program reads in boolean minterms, expressed as binary digits,
 * and determines the prime-implicants.  It then chooses those
 * prime-implicants that give an expression with the least number of
 * literals.  This method is know as the Quine-McQluskey method.
 * More information can be found in:
 *
 * Quine, W. V., "The Problem of Simplifying Truth Functions."
 *      Am. Math. Monthly, Vol. 59, No. 8 (October 1952), 521-31.
 * McCluskey, E. J., Jr., "Minimization of Boolean Functions."
 *      Bell System Tech. J., Vol. 35, No. 6 (November 1956), 1417-44.
 * Mano, M. M., "Digital Logic and Computer Design."
 *      Prentice-Hall, ISBN 0-13-214510-3, pages 102-10.
 *
 * Initially written 11/22/88 by W. Tait Cyrus
 * Last modified 8/29/89 by W. Tait Cyrus
 */

#include <stdio.h>
#include <ctype.h>

struct numbits {
        struct term     *term;
        };

#define numberdashes    output

struct term {
        struct term     *next;
        char            *value;
        char            checked;     /* Checked off of not.  */
        char            output;      /* What the output is.  */
        int             count;       /* General counter.     */
        };

#define DEFAULT_ELEMENTS "abcdefghijklmnopqrstuvwxyzABCDEFGHIJKLMNOPQRSTUVWXYZ"

#define MAX_LEN 1000

char *variables;                    /* Pointer to string of variables names. */
int real_number_bits;               /* # of bits across we are dealing with. */
char *program_name;                 /* Pointer to program name (how called). */
char buffer[MAX_LEN];               /* Temp buffer for the char strings.     */

struct numbits **numbits;
int number_minterms;
struct term *sorted_primes, *sorted_minterms, *sterms, *prev;
```

```
#define usage() {\
    fprintf( stderr, "Usage: %s inputfile [variable_string]\n", program_name);\
    exit(1);\
    }

main(argc,argv)
    int argc;
    char *argv[];
    {
    int i;
    int pass;
    int num_bits;
    struct term *add_expand();
    char output, c, get_output();
    FILE *infile;

    program_name = argv[0];

    if( (argc < 2) || ( argc > 3) ) usage();

    if( argv[1][0] == '-' ) infile = stdin;
    else {
        if( (infile = fopen( argv[1], "r" )) == NULL ) {
            usage();
            }
        }

        /*
         * To figure out how many bits are available, we have to read
         * the first line and count them.
         */
    fgets( buffer, MAX_LEN, infile );
    real_number_bits = find_data_length( buffer );
    printf( "%d bits provided.\n", real_number_bits);
    initialize();

    if( argc == 3 ) {
        if( strlen( argv[2] ) < real_number_bits ) {
            fprintf( stderr, "Error: %d to few variables specified\n",
                    real_number_bits - strlen( argv[2] ) );
            exit(1);
            }
        bcopy( argv[2], variables, real_number_bits );
        }
    else {
        if( real_number_bits > strlen(DEFAULT_ELEMENTS) ) {
            fprintf( stderr, "Error: Internal limit of %d variables reached,\n",
                strlen(DEFAULT_ELEMENTS) );
            fprintf( stderr, "        change DEFAULT_ELEMENTS in program to increase\n" );
            exit(1);
            }
        else bcopy( DEFAULT_ELEMENTS, variables, real_number_bits );
        }

/*
 * PHASE 1 -- Determine of prime-implicants.
 */

        /*
         * Read in the minterm values and add them to the first, zeroth,
         * column.
         */
    output = get_output( buffer );
    if( output != '0' ) add_expand( buffer, output, 0, 0 );
    while( fgets( buffer, MAX_LEN, infile ) != NULL ) {
        output = get_output( buffer );
                /*
                 * Add term to first column of terms.
                 */
        if( output != '0' ) {
            buffer[real_number_bits] = 0;
            add_expand( buffer, output, 0, 0 );
            }
        }
    fclose( infile );
```

```
                /*
                 * Now go through creating the next column by comparing those
                 * minterms that differ by only one bit position.
                 */
        for( pass = 1; pass <= real_number_bits; pass++ ) {
            find_primes( pass );
            }


                /*
                 * Now we need to go through looking for those terms that were
                 * not checked off.  We link them together as we do this so we
                 * are able to remove any duplicate terms again.  It also makes
                 * printing out the equation a LOT easier.
                 */
        find_unchecked();

/*
 * PHASE 2  -- minimize prime-implicants.
 */

                /*
                 * We need to find out the number of initial minterms used.
                 * We can't just count them as they are read in because we
                 * can't assume there will NOT be duplicates.  Therefore, we
                 * walk though the first column of the table counting.
                 * As we are doing this, we link together the linked lists
                 * representing each of the number of bits per minterm.
                 * We don't include any terms where the output is a don't
                 * care.
                 */
        count_minterms();
                /*
                 * If there are no terms, then we are done.
                 */
        if( number_minterms == 0 ) {
            sorted_primes = 0;
            goto done;
            }


                /*
                 * We need to find out how many of the newly found primes will
                 * produce the SAME initial minterm.
                 */
        mark_minterms();


                /*
                 * And now, the main work.  Look at the comment for
                 * 'reduce_primes' to gain a GOOD understanding of what it
                 * does.
                 */
        reduce_primes();
done:
        printf( "\n After selection of prime-implicants:\n\t" );
        print_terms( sorted_primes, 1 );
        printf("\n");
        }

initialize() {
        int i;
        char *bool_alloc();

                /*
                 * Allocate space for the string representing the variables.
                 */
        variables = (char *)bool_alloc( real_number_bits );

                /*
                 * Make the two dimensional array.  Given 'n' bits of input
                 * data, we need a 2D array of (n+1)x(n+1).
                 */
        numbits = (struct numbits **)bool_alloc( sizeof(struct numbits) *
                                        (real_number_bits + 1) );
        for( i = 0; i <= real_number_bits; i++ ) {
            numbits[i] = (struct numbits *)bool_alloc( sizeof(struct numbits) *
                                        (real_number_bits + 1) );
            }
        }
```

```
/*
 * This routine allocates ALL memory.  It checks to see that the memory
 * is available and that once it is, makes sure the memory is
 * initialized to zero (0).
 */
char *
bool_alloc( size )
   int size;
   {
   char *tmp;

   if( (tmp = (char *)malloc( size )) == NULL ) {
      fprintf( stderr, "%s: can't alloc requested %d bytes of memory.  Bye\n",
         size, program_name );
      exit(1);
      }
   else {
      bzero( tmp, size );
      return( tmp );
      }
   }

/*
 * This routine expands don't cares in the input into 0's and 1's
 * and then adds them.
 */
#define is_x(c) ((c == 'x') || (c == 'X'))
#define is_not_x(c) (!is_x(c))

struct term *
add_expand( buffer, output, pass, start )
   char buffer[];
   char output;
   int pass;
   int start;
   {
   int i;
   char *s;

   if( start == real_number_bits ) {
      add( buffer, output, pass );
      return;
      }

   s = buffer;
   for( i = start; (i < real_number_bits) && is_not_x(buffer[i]); i++ );
   if( i < real_number_bits ) {
      buffer[i] = '0';
      add_expand( buffer, output, pass, i + 1 );
      buffer[i] = '1';
      add_expand( buffer, output, pass, i + 1 );
      buffer[i] = 'x';
      }
   else add( buffer, output, pass );
   }

/*
 * This routine adds the passed minterm to the specified column (pass).
 */
add( buffer, output, pass )
   char buffer[];
   char output;
   int pass;
   {
   int num_bits;
   int i;
   struct numbits *bits;
   register struct term *terms;
   register struct term *t;

   num_bits = find_number_bits( buffer );
   bits = &numbits[pass][num_bits];
   t = bits->term;
         /*
          * Before adding new entry, check to see that it does not
          * already exist.  This is a gross way to do things and
          * should really use a hash function.
          */
```

```
        while( t ) {
           if( bcmp( t->value, buffer, real_number_bits ) == 0 ) return;
           t = t->next;
           }
        terms = (struct term *)bool_alloc(sizeof(struct term));
        terms->value = (char *)bool_alloc( real_number_bits );
        terms->next = bits->term;
        bits->term = terms;
        bcopy( buffer, terms->value, real_number_bits );
        terms->output = output;
        }


/*
 * This routine just makes a duplicate of the node passed in
 * and returns the address of the new node.
 */
struct term *
copy_node( n )
   struct term *n;
   {
   struct term *ret;

   ret = (struct term *)bool_alloc(sizeof(struct term));
   bcopy( n, ret, sizeof(struct term) );
   ret->next = NULL;
   return( ret );
   }


/*
 * This routine returns the number of ones (1) or dashes (-) seen in
 * the specified buffer.
 */
find_number_bits( buffer )
   char buffer[];
   {
   int i, num_bits;

   num_bits = 0;
   for( i = 0; i < real_number_bits; i++) {
       if( (buffer[i] == '1') || (buffer[i] == '-') ) num_bits++;
       }
   return( num_bits );
   }



/*
 * This routine looks in the previous column and compares those
 * groups that differ by only one bit.
 */
find_primes( pass )
   int pass;
   {
   register struct term *terms1, *terms2;
   int bits;

   for( bits = 0; bits < real_number_bits ; bits++ ) {
       terms1 = numbits[pass-1][bits].term;
       terms2 = numbits[pass-1][bits+1].term;
       while( terms1 ) {
          compare( terms1, terms2, pass );
          terms1 = terms1->next;
          }
       }
   }

/*
 * This routine compares to values to see if there is only one bit
 * different between them.  If there is, then the bit position that
 * differs is replaced with a dash (-) and this new value is
 * added to the specified column (pass).
 */
compare( term1, term2, pass )
   register struct term *term1, *term2;
   int pass;
   {
   register int index;
   register char *a, *b;
   register int i;
```

```
            while( term2 ) {
                /*
                 * This code looks for the index of the bits that are different
                 * between the two strings.  Iff there is more than one bit different,
                 * then index = -1, else index contains the index of the differing bits
                 */
                index = -1;
                a = term1->value;
                b = term2->value;
                for( i = 0; i < real_number_bits; i++ ) {
                    if( *a++ != *b++ ) {
                        if( index == -1 ) index = i;
                        else {
                            /*
                             * More than 1 bit differs.
                             */
                            index = -1;
                            break;
                        }
                    }
                }
                if( index != -1 ) {
                    bcopy( term1->value, buffer, real_number_bits );
                    buffer[ real_number_bits ] = 0;
                    buffer[ index ] = '-';
                    add( buffer, 1, pass );
                    term1->checked = 1;
                    term2->checked = 1;
                }
                term2 = term2->next;
            }
        }


/*
 * This routine walks through all of the columns looking for unchecked
 * values.  All unchecked primes are linked together to form a single
 * linked list.
 */
find_unchecked()
    {
    register struct term *terms;
    int pass;
    int bits;

    prev = 0;
    for( pass = 0; pass <= real_number_bits; pass++ ) {
        for( bits = 0; bits <= real_number_bits; bits++ ) {
            terms = numbits[pass][bits].term;
            while( terms ) {
                if( terms->checked == 0 ) {
                    /*
                     * We kind of cheat here.  We need 'sorted_primes'
                     * to point to the first primes & this is the
                     * easiest way to do it.
                     */
                    if( sorted_primes == 0 ) sorted_primes = copy_node( terms );
                    /*
                     * Since we are linking everything together, we need
                     * to remember the last link looked at.
                     */
                    if( prev == 0 ) prev = sorted_primes;
                    else {
                        prev->next = copy_node( terms );
                        prev = prev->next;
                    }
                }
                terms = terms->next;
            }
        }
    }
}

print_function( buffer )
    char buffer[];
    {
    int i;
    int number_ones;
```

```
        number_ones = 0;
        for( i = 0; i < real_number_bits; i++) {
            if( buffer[i] == '-' ) number_ones++;
            else {
                putchar( variables[i] );
                if( buffer[i] == '0' ) putchar( '\'' );
                }
            }
        if( number_ones == real_number_bits ) putchar( '1' );
        }

print_terms( t, flag )
    struct term *t;
    int flag;
    {
    register struct term *terms;
    int print_plus;

    if( t == 0 ) {
        putchar( '0' );
        return;
        }
    print_plus = 0;
    terms = t;
    while( terms ) {
        if( terms->checked == flag ) {
            if( print_plus ) {
                printf( " + " );
                print_plus = 0;
                }
            print_function( terms->value );
            print_plus = 1;
            }
        terms = terms->next;
        }
    }


/*
 * This routine just counts the number of minterms found in the
 * first column of the table as well as construct a single
 * linked list of initial minterms.  One thing we do, though, is
 * exclude all minterms whose output is a don't care.
 */
count_minterms() {
    register struct term *terms;
    int bits;

        /*
         * Initialize number of minterms to zero.
         */
    number_minterms = 0;
        /*
         * Since we are going to relink everything in column 0,
         * we need to remember the last link looked at so we can
         * connect everything together.
         */
    prev = 0;
    for( bits = 0; bits <= real_number_bits; bits++ ) {
        terms = numbits[0][bits].term;
        while( terms ) {
                /*
                 * We are ONLY interested in those minterms that
                 * have an output of 1 and not x or X.
                 */
            if( terms->output == '1' ) {
                /*
                 * We kind of cheat here.  We need 'sorted_minterms'
                 * to point to the first minterm & this is the easiest
                 * way to do it.
                 */
            if( sorted_minterms == 0 ) sorted_minterms = copy_node( terms );
                /*
                 * Since we are linking everything together, we need
                 * to remember the last '1' link looked at.
                 */
            if( prev == 0 ) prev = sorted_minterms;
            else {
                prev->next = copy_node( terms );
```

```
                        prev = prev->next;
                        }
                    number_minterms++;
                    }
                terms = terms->next;
                }
            }
        }

    /*
     * We need to find out how many of the newly found primes will
     * produce the SAME initial minterm.
     */
    mark_minterms() {
        register struct term *terms, *sterms;
        int i;
        int match;

        terms = sorted_minterms;
        while( terms ) {
            sterms = sorted_primes;
            while( sterms ) {
                match = 1;
                for( i = 0; i < real_number_bits; i++ ) {
                    if( sterms->value[i] != '-' ) {
                        if( sterms->value[i] != terms->value[i] ) match = 0;
                        }
                    }
                if( match ) {
                    terms->count++;
                    }
                sterms = sterms->next;
                }
            if( terms->count == 0 ) {
                printf( "Hmm, strange, but %s didn't get reduced into anything\n",
                    terms->value);
                printf( "This should NOT happen, if it does, then there is a bug.\n");
                }
            terms->checked = 0;
            terms = terms->next;
            }
        }

    /*
     * This is were all the work of minimizing the number of
     * primes is done.
     */
    reduce_primes() {
        register struct term *terms, *primes;
        int i;
        int match;

            /*
             * The first thing we do is to go through ALL initial minterms
             * checking off those that result from a SINGLE prime.  We
             * also "check" the prime so we won't look at it again.
             */
        terms = sorted_minterms;
        while( terms ) {
                /*
                 * If the currently looked at minterm has NOT been
                 * checked off AND it results from only ONE prime, then
                 * we look to see which prime it results from.
                 */
            if( (terms->checked == 0) &&
                (terms->count == 1) &&
                (terms->output == '1' ) ) {
                primes = sorted_primes;
                while( primes ) {
                    match = 1;
                        /*
                         * If a prime that has NOT already been checked.
                         */
                    if( primes->checked == 0 ) {
                        for( i = 0; i < real_number_bits; i++ ) {
                            if( primes->value[i] != '-' ) {
                                if( primes->value[i] != terms->value[i] ) match = 0;
                                }
```

```
                     }
                             /*
                              * Found a prime that produced the currently
                              * looked at minterm.  We now need to mark ALL
                              * minterms that result from this prime.
                              */
                     if( match ) mark_terms( primes );
                         }
                 primes = primes->next;
                     }
             }
         terms = terms->next;
         }

         /*
          * While there are still unchecked minterms, we use the prime
          * that produces the most unchecked minterms.
          */
     while( number_minterms > 0 ) {
        register struct term *p;

        primes = sorted_primes;
          /*
           * For each of the remaining unchecked primes, we count the
           * number of minterms that are produced by this prime.
           */
        p = 0;
        while( primes ) {
           if( primes->checked == 0 ) {
              number_matches( primes, 0 );
              if( p == 0 ) p = primes;
              else {
                 if( (primes->count > p->count) ||
                     ((primes->count == p->count) &&
                      (primes->numberdashes > p->numberdashes) ) ) {
                    p = primes;
                    }
                 }
              }
           primes = primes->next;
           }
        number_matches( p, 1 );
        }
     }

/*
 * Found a prime that produced the currently
 * looked at minterm.  We now need to mark ALL
 * minterms that result from this prime.
 */
mark_terms( prime )
    register struct term *prime;
    {
    register struct term *terms;
    int match;
    int i;

    terms = sorted_minterms;
    while( terms ) {
       match = 1;
       for( i = 0; i < real_number_bits; i++ ) {
          if( prime->value[i] != '-' ) {
             if( prime->value[i] != terms->value[i] ) match = 0;
             }
          }
       if( match ) {
          if( terms->checked == 0 ) number_minterms--;
          terms->checked = 1;
          prime->checked = 1;
          }
       terms = terms->next;
       }
    }

number_matches( p, flag )
    register struct term *p;
    int flag;
    {
```

```
                    int i;
                    int match, matches;
                    register struct term *t;
                    int number_dashes;

                    matches = 0;
                    t = sorted_minterms;
                    p->numberdashes = 0;
                    number_dashes = 0;
                    while( t ) {
                        if( t->checked == 0 ) {
                            match = 1;
                            number_dashes = 0;
                            for( i = 0; i < real_number_bits; i++ ) {
                                if( p->value[i] != '-' ) {
                                    if( p->value[i] != t->value[i] ) match = 0;
                                    }
                                else number_dashes++;
                                }
                            p->numberdashes = number_dashes;
                            if( match ) {
                                matches++;
                                if( flag == 1 ) {
                                    t->checked = 1;
                                    p->checked = 1;
                                    number_minterms--;
                                    }
                                }
                            }
                        t = t->next;
                        }
                    p->count = matches;
                    }

#define is_valid(c) ( \
        (c != '0') && \
        (c != '1') && \
        (c != 'x') && \
        (c != 'X') ? 0 : 1)

find_data_length( buffer )
    char *buffer;
    {
    char *s;
    int len;

    s = buffer;
    for( len = 0; (len < MAX_LEN) && *s; len++ ) {
        if( (*s == ' ') || (*s == '\t') ) return( len );
        if( !is_valid(*s) ) {
            fprintf( stderr, "%s: Error; found '%c' but expected one of 0, 1, x or X.\n",
                program_name, *s );
            exit(1);
            }
        s++;
        }
    if( !*s ) {
        fprintf( stderr, "%s: Error; program only understands %d bits\n",
            program_name, MAX_LEN);
        exit(1);
        }
    return( -1 );
    }

char
get_output( buffer )
    char buffer[];
    {
    char *s;

    s = &buffer[real_number_bits];
    while( isspace( *s ) ) s++;
    return( *s );
    }
```

Appendix B

Memory Contents for PROM Based Controller for Floating Point Multiplier of Chapter 4

The contents of the memory which are needed for the PROM based controller implementation of Chapter 4 are listed below. The address is made up of bits from the present state register and inputs. The contents of the location specified by the combination of present state and inputs identify the next state and the signals which will be asserted in that state.

| Present State Bits | Inputs | | | | | PROM Addr | Outputs | | | | | | | | |
	DO MULT	FLOW ERR	FNL ERR	PROD MSB	TEN		Next State	RND	PROD SHFT	PROD OUT	NO PN	EXCEP TION	IDLE CK	PROD CK	CLK PL
000	0	0	0	0	0	0	000	0	0	0	0	0	1	0	0
000	0	0	0	0	1	1	000	0	0	0	0	0	1	0	0
000	0	0	0	1	0	2	000	0	0	0	0	0	1	0	0
000	0	0	0	1	1	3	000	0	0	0	0	0	1	0	0
000	0	0	1	0	0	4	000	0	0	0	0	0	1	0	0
000	0	0	1	0	1	5	000	0	0	0	0	0	1	0	0
000	0	0	1	1	0	6	000	0	0	0	0	0	1	0	0
000	0	0	1	1	1	7	000	0	0	0	0	0	1	0	0

Present State Bits	Inputs					PROM Addr	Next State	Outputs							
	DO MULT	FLOW ERR	FNL ERR	PROD MSB	TEN			RND	PROD SHFT	PROD OUT	NO PN	EXCEP TION	IDLE CK	PROD CK	CLK PL
000	0	1	0	0	0	8	000	0	0	0	0	0	1	0	0
000	0	1	0	0	1	9	000	0	0	0	0	0	1	0	0
000	0	1	0	1	0	10	000	0	0	0	0	0	1	0	0
000	0	1	0	1	1	11	000	0	0	0	0	0	1	0	0
000	0	1	1	0	0	12	000	0	0	0	0	0	1	0	0
000	0	1	1	0	1	13	000	0	0	0	0	0	1	0	0
000	0	1	1	1	0	14	000	0	0	0	0	0	1	0	0
000	0	1	1	1	1	15	000	0	0	0	0	0	1	0	0
000	1	0	0	0	0	16	110	0	0	0	0	0	0	1	1
000	1	0	0	0	1	17	110	0	0	0	0	0	0	1	1
000	1	0	0	1	0	18	110	0	0	0	0	0	0	1	1
000	1	0	0	1	1	19	110	0	0	0	0	0	0	1	1
000	1	0	1	0	0	20	110	0	0	0	0	0	0	1	1
000	1	0	1	0	1	21	110	0	0	0	0	0	0	1	1
000	1	0	1	1	0	22	110	0	0	0	0	0	0	1	1
000	1	0	1	1	1	23	110	0	0	0	0	0	0	1	1
000	1	1	0	0	0	24	001	0	0	0	0	1	0	0	0
000	1	1	0	0	1	25	001	0	0	0	0	1	0	0	0
000	1	1	0	1	0	26	001	0	0	0	0	1	0	0	0
000	1	1	0	1	1	27	001	0	0	0	0	1	0	0	0
000	1	1	1	0	0	28	001	0	0	0	0	1	0	0	0
000	1	1	1	0	1	29	001	0	0	0	0	1	0	0	0
000	1	1	1	1	0	30	001	0	0	0	0	1	0	0	0
000	1	1	1	1	1	31	001	0	0	0	0	1	0	0	0
001	0	0	0	0	0	32	101	0	0	1	0	0	0	0	0
001	0	0	0	0	1	33	101	0	0	1	0	0	0	0	0
001	0	0	0	1	0	34	101	0	0	1	0	0	0	0	0
001	0	0	0	1	1	35	101	0	0	1	0	0	0	0	0
001	0	0	1	0	0	36	101	0	0	1	0	0	0	0	0
001	0	0	1	0	1	37	101	0	0	1	0	0	0	0	0
001	0	0	1	1	0	38	101	0	0	1	0	0	0	0	0
001	0	0	1	1	1	39	101	0	0	1	0	0	0	0	0
001	0	1	0	0	0	40	101	0	0	1	0	0	0	0	0
001	0	1	0	0	1	41	101	0	0	1	0	0	0	0	0
001	0	1	0	1	0	42	101	0	0	1	0	0	0	0	0
001	0	1	0	1	1	43	101	0	0	1	0	0	0	0	0
001	0	1	1	0	0	44	101	0	0	1	0	0	0	0	0
001	0	1	1	0	1	45	101	0	0	1	0	0	0	0	0
001	0	1	1	1	0	46	101	0	0	1	0	0	0	0	0
001	0	1	1	1	1	47	101	0	0	1	0	0	0	0	0

Present State Bits	Inputs					PROM Addr	Next State	Outputs							
	DO MULT	FLOW ERR	FNL ERR	PROD MSB	PROD TEN			RND	PROD SHFT	PROD OUT	NO PN	EXCEP TION	IDLE CK	PROD CK	CLK PL
001	1	0	0	0	0	48	101	0	0	1	0	0	0	0	0
001	1	0	0	0	1	49	101	0	0	1	0	0	0	0	0
001	1	0	0	1	0	50	101	0	0	1	0	0	0	0	0
001	1	0	0	1	1	51	101	0	0	1	0	0	0	0	0
001	1	0	1	0	0	52	101	0	0	1	0	0	0	0	0
001	1	0	1	0	1	53	101	0	0	1	0	0	0	0	0
001	1	0	1	1	0	54	101	0	0	1	0	0	0	0	0
001	1	0	1	1	1	55	101	0	0	1	0	0	0	0	0
001	1	1	0	0	0	56	101	0	0	1	0	0	0	0	0
001	1	1	0	0	1	57	101	0	0	1	0	0	0	0	0
001	1	1	0	1	0	58	101	0	0	1	0	0	0	0	0
001	1	1	0	1	1	59	101	0	0	1	0	0	0	0	0
001	1	1	1	0	0	60	101	0	0	1	0	0	0	0	0
001	1	1	1	0	1	61	101	0	0	1	0	0	0	0	0
001	1	1	1	1	0	62	101	0	0	1	0	0	0	0	0
001	1	1	1	1	1	63	101	0	0	1	0	0	0	0	0
010	0	0	0	0	0	64	000	0	0	0	0	0	1	0	0
010	0	0	0	0	1	65	000	0	0	0	0	0	1	0	0
010	0	0	0	1	0	66	000	0	0	0	0	0	1	0	0
010	0	0	0	1	1	67	000	0	0	0	0	0	1	0	0
010	0	0	1	0	0	68	000	0	0	0	0	0	1	0	0
010	0	0	1	0	1	69	000	0	0	0	0	0	1	0	0
010	0	0	1	1	0	70	000	0	0	0	0	0	1	0	0
010	0	0	1	1	1	71	000	0	0	0	0	0	1	0	0
010	0	1	0	0	0	72	000	0	0	0	0	0	1	0	0
010	0	1	0	0	1	73	000	0	0	0	0	0	1	0	0
010	0	1	0	1	0	74	000	0	0	0	0	0	1	0	0
010	0	1	0	1	1	75	000	0	0	0	0	0	1	0	0
010	0	1	1	0	0	76	000	0	0	0	0	0	1	0	0
010	0	1	1	0	1	77	000	0	0	0	0	0	1	0	0
010	0	1	1	1	0	78	000	0	0	0	0	0	1	0	0
010	0	1	1	1	1	79	000	0	0	0	0	0	1	0	0
010	1	0	0	0	0	80	000	0	0	0	0	0	1	0	0
010	1	0	0	0	1	81	000	0	0	0	0	0	1	0	0
010	1	0	0	1	0	82	000	0	0	0	0	0	1	0	0
010	1	0	0	1	1	83	000	0	0	0	0	0	1	0	0
010	1	0	1	0	0	84	000	0	0	0	0	0	1	0	0
010	1	0	1	0	1	85	000	0	0	0	0	0	1	0	0
010	1	0	1	1	0	86	000	0	0	0	0	0	1	0	0
010	1	0	1	1	1	87	000	0	0	0	0	0	1	0	0
010	1	1	0	0	0	88	000	0	0	0	0	0	1	0	0
010	1	1	0	0	1	89	000	0	0	0	0	0	1	0	0
010	1	1	0	1	0	90	000	0	0	0	0	0	1	0	0
010	1	1	0	1	1	91	000	0	0	0	0	0	1	0	0
010	1	1	1	0	0	92	000	0	0	0	0	0	1	0	0
010	1	1	1	0	1	93	000	0	0	0	0	0	1	0	0
010	1	1	1	1	0	94	000	0	0	0	0	0	1	0	0
010	1	1	1	1	1	95	000	0	0	0	0	0	1	0	0

| Present State Bits | Inputs | | | | | PROM Addr | Next State | Outputs | | | | | | | |
	DO MULT	FLOW ERR	FNL ERR	PROD MSB	TEN			RND	PROD SHFT	PROD OUT	NO PN	EXCEP TION	IDLE CK	PROD CK	CLK PL
011	0	0	0	0	0	96	101	0	0	1	0	0	0	0	0
011	0	0	0	0	1	97	101	0	0	1	0	0	0	0	0
011	0	0	0	1	0	98	101	0	0	1	0	0	0	0	0
011	0	0	0	1	1	99	101	0	0	1	0	0	0	0	0
011	0	0	1	0	0	100	001	0	0	0	0	1	0	0	0
011	0	0	1	0	1	101	001	0	0	0	0	1	0	0	0
011	0	0	1	1	0	102	001	0	0	0	0	1	0	0	0
011	0	0	1	1	1	103	001	0	0	0	0	1	0	0	0
011	0	1	0	0	0	104	101	0	0	1	0	0	0	0	0
011	0	1	0	0	1	105	101	0	0	1	0	0	0	0	0
011	0	1	0	1	0	106	101	0	0	1	0	0	0	0	0
011	0	1	0	1	1	107	101	0	0	1	0	0	0	0	0
011	0	1	1	0	0	108	001	0	0	0	0	1	0	0	0
011	0	1	1	0	1	109	001	0	0	0	0	1	0	0	0
011	0	1	1	1	0	110	001	0	0	0	0	1	0	0	0
011	0	1	1	1	1	111	001	0	0	0	0	1	0	0	0
011	1	0	0	0	0	112	101	0	0	1	0	0	0	0	0
011	1	0	0	0	1	113	101	0	0	1	0	0	0	0	0
011	1	0	0	1	0	114	101	0	0	1	0	0	0	0	0
011	1	0	0	1	1	115	101	0	0	1	0	0	0	0	0
011	1	0	1	0	0	116	001	0	0	0	0	1	0	0	0
011	1	0	1	0	1	117	001	0	0	0	0	1	0	0	0
011	1	0	1	1	0	118	001	0	0	0	0	1	0	0	0
011	1	0	1	1	1	119	001	0	0	0	0	1	0	0	0
011	1	1	0	0	0	120	101	0	0	1	0	0	0	0	0
011	1	1	0	0	1	121	101	0	0	1	0	0	0	0	0
011	1	1	0	1	0	122	101	0	0	1	0	0	0	0	0
011	1	1	0	1	1	123	101	0	0	1	0	0	0	0	0
011	1	1	1	0	0	124	001	0	0	0	0	1	0	0	0
011	1	1	1	0	1	125	001	0	0	0	0	1	0	0	0
011	1	1	1	1	0	126	001	0	0	0	0	1	0	0	0
011	1	1	1	1	1	127	001	0	0	0	0	1	0	0	0
100	0	0	0	0	0	128	101	0	0	1	0	0	0	0	0
100	0	0	0	0	1	129	101	0	0	1	0	0	0	0	0
100	0	0	0	1	0	130	011	0	1	0	1	0	0	1	0
100	0	0	0	1	1	131	011	0	1	0	1	0	0	1	0
100	0	0	1	0	0	132	001	0	0	0	0	1	0	0	0
100	0	0	1	0	1	133	001	0	0	0	0	1	0	0	0
100	0	0	1	1	0	134	011	0	1	0	1	0	0	1	0
100	0	0	1	1	1	135	011	0	1	0	1	0	0	1	0
100	0	1	0	0	0	136	101	0	0	1	0	0	0	0	0
100	0	1	0	0	1	137	101	0	0	1	0	0	0	0	0
100	0	1	0	1	0	138	011	0	1	0	1	0	0	1	0
100	0	1	0	1	1	139	011	0	1	0	1	0	0	1	0
100	0	1	1	0	0	140	001	0	0	0	0	1	0	0	0
100	0	1	1	0	1	141	001	0	0	0	0	1	0	0	0
100	0	1	1	1	0	142	011	0	1	0	1	0	0	1	0
100	0	1	1	1	1	143	011	0	1	0	1	0	0	1	0

Present State Bits	Inputs					PROM Addr	Next State	Outputs							
	DO MULT	FLOW ERR	FNL ERR	PROD MSB	TEN			RND	PROD SHFT	PROD OUT	NO PN	EXCEP TION	IDLE CK	PROD CK	CLK PL
100	1	0	0	0	0	144	101	0	0	1	0	0	0	0	0
100	1	0	0	0	1	145	101	0	0	1	0	0	0	0	0
100	1	0	0	1	0	146	011	0	1	0	1	0	0	1	0
100	1	0	0	1	1	147	011	0	1	0	1	0	0	1	0
100	1	0	1	0	0	148	001	0	0	0	0	1	0	0	0
100	1	0	1	0	1	149	001	0	0	0	0	1	0	0	0
100	1	0	1	1	0	150	011	0	1	0	1	0	0	1	0
100	1	0	1	1	1	151	011	0	1	0	1	0	0	1	0
100	1	1	0	0	0	152	101	0	0	1	0	0	0	0	0
100	1	1	0	0	1	153	101	0	0	1	0	0	0	0	0
100	1	1	0	1	0	154	011	0	1	0	1	0	0	1	0
100	1	1	0	1	1	155	011	0	1	0	1	0	0	1	0
100	1	1	1	0	0	156	001	0	0	0	0	1	0	0	0
100	1	1	1	0	1	157	001	0	0	0	0	1	0	0	0
100	1	1	1	1	0	158	011	0	1	0	1	0	0	1	0
100	1	1	1	1	1	159	011	0	1	0	1	0	0	1	0
101	0	0	0	0	0	160	000	0	0	0	0	0	1	0	0
101	0	0	0	0	1	161	000	0	0	0	0	0	1	0	0
101	0	0	0	1	0	162	000	0	0	0	0	0	1	0	0
101	0	0	0	1	1	163	000	0	0	0	0	0	1	0	0
101	0	0	1	0	0	164	000	0	0	0	0	0	1	0	0
101	0	0	1	0	1	165	000	0	0	0	0	0	1	0	0
101	0	0	1	1	0	166	000	0	0	0	0	0	1	0	0
101	0	0	1	1	1	167	000	0	0	0	0	0	1	0	0
101	0	1	0	0	0	168	000	0	0	0	0	0	1	0	0
101	0	1	0	0	1	169	000	0	0	0	0	0	1	0	0
101	0	1	0	1	0	170	000	0	0	0	0	0	1	0	0
101	0	1	0	1	1	171	000	0	0	0	0	0	1	0	0
101	0	1	1	0	0	172	000	0	0	0	0	0	1	0	0
101	0	1	1	0	1	173	000	0	0	0	0	0	1	0	0
101	0	1	1	1	0	174	000	0	0	0	0	0	1	0	0
101	0	1	1	1	1	175	000	0	0	0	0	0	1	0	0
101	1	0	0	0	0	176	101	0	0	1	0	0	0	0	0
101	1	0	0	0	1	177	101	0	0	1	0	0	0	0	0
101	1	0	0	1	0	178	101	0	0	1	0	0	0	0	0
101	1	0	0	1	1	179	101	0	0	1	0	0	0	0	0
101	1	0	1	0	0	180	101	0	0	1	0	0	0	0	0
101	1	0	1	0	1	181	101	0	0	1	0	0	0	0	0
101	1	0	1	1	0	182	101	0	0	1	0	0	0	0	0
101	1	0	1	1	1	183	101	0	0	1	0	0	0	0	0
101	1	1	0	0	0	184	101	0	0	1	0	0	0	0	0
101	1	1	0	0	1	185	101	0	0	1	0	0	0	0	0
101	1	1	0	1	0	186	101	0	0	1	0	0	0	0	0
101	1	1	0	1	1	187	101	0	0	1	0	0	0	0	0
101	1	1	1	0	0	188	101	0	0	1	0	0	0	0	0
101	1	1	1	0	1	189	101	0	0	1	0	0	0	0	0
101	1	1	1	1	0	190	101	0	0	1	0	0	0	0	0
101	1	1	1	1	1	191	101	0	0	1	0	0	0	0	0

Present State Bits	Inputs DO MULT	FLOW ERR	FNL ERR	PROD MSB	TEN	PROM Addr	Next State	RND	PROD SHFT	PROD OUT	NO PN	EXCEP TION	IDLE CK	PROD CK	CLK PL
110	0	0	0	0	0	192	111	0	1	0	0	0	0	1	0
110	0	0	0	0	1	193	100	1	0	0	0	0	0	1	0
110	0	0	0	1	0	194	111	0	1	0	0	0	0	1	0
110	0	0	0	1	1	195	100	1	0	0	0	0	0	1	0
110	0	0	1	0	0	196	111	0	1	0	0	0	0	1	0
110	0	0	1	0	1	197	100	1	0	0	0	0	0	1	0
110	0	0	1	1	0	198	111	0	1	0	0	0	0	1	0
110	0	0	1	1	1	199	100	1	0	0	0	0	0	1	0
110	0	1	0	0	0	200	111	0	1	0	0	0	0	1	0
110	0	1	0	0	1	201	100	1	0	0	0	0	0	1	0
110	0	1	0	1	0	202	111	0	1	0	0	0	0	1	0
110	0	1	0	1	1	203	100	1	0	0	0	0	0	1	0
110	0	1	1	0	0	204	111	0	1	0	0	0	0	1	0
110	0	1	1	0	1	205	100	1	0	0	0	0	0	1	0
110	0	1	1	1	0	206	111	0	1	0	0	0	0	1	0
110	0	1	1	1	1	207	100	1	0	0	0	0	0	1	0
110	1	0	0	0	0	208	111	0	1	0	0	0	0	1	0
110	1	0	0	0	1	209	100	1	0	0	0	0	0	1	0
110	1	0	0	1	0	210	111	0	1	0	0	0	0	1	0
110	1	0	0	1	1	211	100	1	0	0	0	0	0	1	0
110	1	0	1	0	0	212	111	0	1	0	0	0	0	1	0
110	1	0	1	0	1	213	100	1	0	0	0	0	0	1	0
110	1	0	1	1	0	214	111	0	1	0	0	0	0	1	0
110	1	0	1	1	1	215	100	1	0	0	0	0	0	1	0
110	1	1	0	0	0	216	111	0	1	0	0	0	0	1	0
110	1	1	0	0	1	217	100	1	0	0	0	0	0	1	0
110	1	1	0	1	0	218	111	0	1	0	0	0	0	1	0
110	1	1	0	1	1	219	100	1	0	0	0	0	0	1	0
110	1	1	1	0	0	220	111	0	1	0	0	0	0	1	0
110	1	1	1	0	1	221	100	1	0	0	0	0	0	1	0
110	1	1	1	1	0	222	111	0	1	0	0	0	0	1	0
110	1	1	1	1	1	223	100	1	0	0	0	0	0	1	0
111	0	0	0	0	0	224	110	0	0	0	0	0	0	1	1
111	0	0	0	0	1	225	110	0	0	0	0	0	0	1	1
111	0	0	0	1	0	226	110	0	0	0	0	0	0	1	1
111	0	0	0	1	1	227	110	0	0	0	0	0	0	1	1
111	0	0	1	0	0	228	110	0	0	0	0	0	0	1	1
111	0	0	1	0	1	229	110	0	0	0	0	0	0	1	1
111	0	0	1	1	0	230	110	0	0	0	0	0	0	1	1
111	0	0	1	1	1	231	110	0	0	0	0	0	0	1	1
111	0	1	0	0	0	232	110	0	0	0	0	0	0	1	1
111	0	1	0	0	1	233	110	0	0	0	0	0	0	1	1
111	0	1	0	1	0	234	110	0	0	0	0	0	0	1	1
111	0	1	0	1	1	235	110	0	0	0	0	0	0	1	1
111	0	1	1	0	0	236	110	0	0	0	0	0	0	1	1
111	0	1	1	0	1	237	110	0	0	0	0	0	0	1	1
111	0	1	1	1	0	238	110	0	0	0	0	0	0	1	1
111	0	1	1	1	1	239	110	0	0	0	0	0	0	1	1

Present State Bits	Inputs					PROM Addr	Outputs								
	DO MULT	FLOW ERR	FNL ERR	PROD MSB	TEN		Next State	RND	PROD SHFT	PROD OUT	NO PN	EXCEP TION	IDLE CK	PROD CK	CLK PL
111	1	0	0	0	0	240	110	0	0	0	0	0	0	1	1
111	1	0	0	0	1	241	110	0	0	0	0	0	0	1	1
111	1	0	0	1	0	242	110	0	0	0	0	0	0	1	1
111	1	0	0	1	1	243	110	0	0	0	0	0	0	1	1
111	1	0	1	0	0	244	110	0	0	0	0	0	0	1	1
111	1	0	1	0	1	245	110	0	0	0	0	0	0	1	1
111	1	0	1	1	0	246	110	0	0	0	0	0	0	1	1
111	1	0	1	1	1	247	110	0	0	0	0	0	0	1	1
111	1	1	0	0	0	248	110	0	0	0	0	0	0	1	1
111	1	1	0	0	1	249	110	0	0	0	0	0	0	1	1
111	1	1	0	1	0	250	110	0	0	0	0	0	0	1	1
111	1	1	0	1	1	251	110	0	0	0	0	0	0	1	1
111	1	1	1	0	0	252	110	0	0	0	0	0	0	1	1
111	1	1	1	0	1	253	110	0	0	0	0	0	0	1	1
111	1	1	1	1	0	254	110	0	0	0	0	0	0	1	1
111	1	1	1	1	1	255	110	0	0	0	0	0	0	1	1

Index